Systems of Conservation L

Systems of conservation laws arise naturally in several area
understand them and their consequences (shock waves, finit
properly in mathematical terms requires, however, knowledge
This book sets up the foundations of the modern theory of co
the physical models and mathematical methods, leading to the
on this the author then takes the reader to the current state of kno
particular, he studies in detail viscous approximations, paying spec
profiles of shock waves. The maximum principle is considered fr
numerical schemes and also in terms of viscous approximation, wh
studied using the technique of compensated compactness. Small wave
geometrical optics methods. Finally, the initial–boundary problem is co
Throughout, the presentation is reasonably self-contained, with large num
and full discussion of all the ideas. This will make it ideal as a text for grad
the area of partial differential equations.

Denis Serre is Professor of Mathematics at the Ecole Normale Supérieure de Ly
a Member of the Institut Universitaire de France (1992–7).

Systems of Conservation Laws 1

Hyperbolicity, Entropies, Shock Waves

DENIS SERRE

Translated by

I. N. SNEDDON

CAMBRIDGE
UNIVERSITY PRESS

CAMBRIDGE UNIVERSITY PRESS
Cambridge, New York, Melbourne, Madrid, Cape Town, Singapore,
São Paulo, Delhi, Dubai, Tokyo, Mexico City

Cambridge University Press
The Edinburgh Building, Cambridge CB2 8RU, UK

Published in the United States of America by Cambridge University Press, New York

www.cambridge.org
Information on this title: www.cambridge.org/9780521582339

First published 1999

A catalogue record for this publication is available from the British Library

ISBN 978-0-521-58233-9 Hardback

To Paul and Fanny

Contents

Acknowledgments

This book would not have seen the light of day without a great deal of help. First of all that of the Institut Universitaire de France, by whom I was engaged, who assisted me by giving me the time and the freedom necessary to bring the first draft to a conclusion. Later my colleagues at the Ecole Normale Supérieure de Lyon gave similar support by accepting my release from normal duties for a considerable time so that I should be able to concentrate on this book. Finally and above all to my students, former students and friends, who have believed in using this work, who have supported me by discussing it often and have read it in detail. Their interest has been the most powerful of stimulants. I owe a considerable debt to Sylvie Benzoni, who has read the greater part of this book and whose severe criticism has constantly led me to improve the text.

I give heartfelt thanks also to Pascale Bergeret, Marguerite Gisclon, Florence Hubert, Christophe Cheverry, Hervé Gilquin, Arnaud Heibig, Peng Yue Jun, Julien Michel and Bruno Sévennec for their collaboration. Finally certain persons have taught me about topics which I did not properly know: Jean-Yves Chemin, Constantin Dafermos, Heinrich Freistühler, David Hoff, Sergiu Klainerman, Ling Hsiao, Tai-Ping Liu, Guy Métivier and Roberto Natalini.

Introduction

The conservation laws that are the subject of this work are those of physics or mechanics, when the state of the system considered is a field, that is a vector-valued function $(x, t) \mapsto u(x, t)$ of space variables $x = (x_1, \dots, x_d)$ and of the time t. The domain Ω covered by x is an open set of $\mathbb{R}^d$, with in general $1 \leq d \leq 3$. The scalar components $u_1, \dots, u_n$ of u are variables dependent on x and t: if Ω is bounded and in the absence of any exchange with the exterior,[1] the mean state of the system

$$\bar{u} := \frac{1}{|\Omega|} \int_\Omega u(x, t)\, \mathrm{d}x$$

is independent of the time and the system tends to a homogeneous equilibrium $u \equiv \bar{u}$ as the time increases. The fact that we speak of the mean indicates that the set $\mathscr{U}$ of admissible values of the field u is a convex set of $\mathbb{R}^n$.

A conservation law is a partial differential equation

$$\frac{\partial u_i}{\partial t} + \mathrm{div}_x \vec{q}_i = g_i,$$

where $g_i(x, t)$ represents the density (per unit volume) of the interaction with external fields. Among these fields, we can even find some which depend on u; for example the conservation of momentum of an electrically neutral continuum can be written

$$\frac{\partial \rho v_i}{\partial t} + \mathrm{div}_x(\rho v_i \vec{v} - \vec{T}^i) = \rho G_i,$$

where ρ is the mass density (ρ is one of the components of u), $\mathscr{T} = (\vec{T}^1, \dots, \vec{T}^d)$ is the strain tensor, $\vec{v}$ is the velocity and $\vec{G}$ the gravity field. Hence, in general we shall

[1] The boundary $\partial\Omega$ is thus impermeable and insulated, for example electrically, in short, there is no interaction with a field other than u.

have $g_i(x,t) = h_i(x,t,u(x,t))$ where h_i is a known function: here $g_i = u_1 G_{i-1}$ since ρ is u_1. We call $\vec{h}$ the *sources* of the system.

An equivalent formulation of a conservation law is given by an integral condition, which expresses the physical balance for the quantity represented by u_i in an arbitrary part ω of Ω:

$$\frac{\mathrm{d}}{\mathrm{d}t}\int_\omega u_i(x,t)\,\mathrm{d}x + \int_{\partial\omega} \vec{q}_i(x,t)\cdot\nu(x)\,\mathrm{d}x = \int_\omega g_i(x,t)\,\mathrm{d}x$$

where $\nu(x)$ denotes the outward unit normal at a point x on the boundary of ω. The vector field $\vec{q}_i$ is thus the *flux* of the variable u_i:

flux of mass if u_i is the mass density,
flux of energy if u_i is the energy density per unit volume,
electric current if u_i is the electric charge density, ...

The third formulation of a conservation law is also the most practical for finding the new equation when we have to effect a change of variables. We define a differential form α^i of degree d in $\Omega \times (0,T)$ by

$$\begin{aligned}\alpha^i := {} & u_i \mathrm{d}x_1 \wedge \mathrm{d}x_2 \wedge \cdots \wedge \mathrm{d}x_d - q_{i1}\mathrm{d}t \wedge \mathrm{d}x_2 \wedge \cdots \wedge \mathrm{d}x_d \\ & + \cdots (-)^d q_{id}\,\mathrm{d}t \wedge \mathrm{d}x_1 \wedge \cdots \wedge \mathrm{d}x_{d-1}.\end{aligned}$$

The conservation law is then written

$$\mathrm{d}\alpha^i = g_i\,\mathrm{d}t \wedge \mathrm{d}x_1 \wedge \cdots \wedge \mathrm{d}x_d.$$

This way of looking at the problem suggests that other conservation laws have a natural form $\mathrm{d}\alpha = \beta$ where α is a differential form of degree p, not necessarily equal to d. This is the case of Maxwell's electromagnetic equations or the Yang–Mills equations for which $p = 2$ and $d = 3$.

In this form, the conservation laws are intangible, in so far as the scales of time, length, velocity ... are compatible with a representation of the system by fields.[2] However, the description of the evolution of the state of the physical system is possible only if the system of equations

$$\partial_t u_i + \mathrm{div}_x \vec{q}_i = g_i, \quad 1 \le i \le n,$$

is closed under the *state laws*:

$$q_i := Q_i[u;\varepsilon].$$

These laws, in which ε denotes one or several dimensionless parameters,[3] describe

[2] Quantum effects are therefore excluded, but relativistic effects can, in general, be taken into account.
[3] Such as the inverse of a Reynolds number, a mean free path, a relaxation time.

in an empirical manner the behaviour of a continuum put into a given homogeneous state $u \in \mathcal{U}$. For example, a fixed mass of gas, in a prescribed volume and at an imposed temperature, exerts on the boundary a force whose density per unit surface area (the pressure) is constant and depends only on the thermodynamic parameters and on the nature of the gas; however, the complete ranges of time and of the space variables are not allowable and in certain cases, recourse must be had to a statistical description or to molecular dynamics. Care must always be taken, specially when an asymptotic analysis is being made, to ensure the validity of the model being used.

The description of the road traffic on a highway shows that the state law depends as much on human sciences as on physics: the average speed of vehicles is a function of the traffic density and reflects the average behaviour of human beings (the drivers) and depends on circumstances; again, there is nothing absolute about it as it varies according to material conditions (the reliability and security of the vehicles, the quality of circulation lanes), and the regulations in force and even the culture of a country.

The point common to all the models studied here is the fact that $f_i : u \mapsto Q_i[u, 0]$ is an ordinary local function: its value at (x, t) depends only on that of u at this point. By an abuse of notation, we may therefore write $f_i[u](x, t) = f_i(x, t, u(x, t))$ and, on this occasion, f_i denotes a function defined on $\mathcal{U}$ and with values in $\mathbb{R}$, which, in general, will be regular. In first approximation, the evolution of the system can be deduced from the knowledge of its initial state $u_0(\cdot) = (u_{01}, \ldots, u_{0n})$ given on Ω, in the solution of the system of non-linear partial differential equations

$$\partial_t u_i + \operatorname{div}_x f_i(u) = g_i(x, t, u), \quad 1 \leq i \leq n, \quad x \in \Omega, \quad t > 0, \qquad (0.1)$$

augmented by the appropriate boundary conditions. This is the *mixed problem* which when $\Omega = \mathbb{R}^d$ we rather call the *Cauchy problem.*

The apparent simplicity of these equations contrasts with the difficulty of the problems encountered when solving the Cauchy problem or the mixed problem, as much from the theoretical point of view as from that of numerical analysis. These two ways of considering the Cauchy problem are equally interesting and difficult. However, the present work is devoted only to the theoretical aspects, in particular because the numerical part is covered by a number of very good books. Let us cite at least those of Leveque [62], Godlewski and Raviart [34], Richtmyer and Morton [86], Sod [98], and Vichnevetsky and Bowles [109].

To illustrate the mathematical difficulties, let us say that there is not a satisfactory result concerning the existence of a solution of the Cauchy problem. For given regular initial data,[4] there exists a regular solution, but only during a finite time

[4] Let us say of class H^s, with $s > 1 + d/2$ with the result that u_0 is of class $\mathcal{C}^1$.

(Theorem 3.6.1) inversely proportional to $\nabla_x u_0$. Since, beyond a certain time, discontinuities in u must develop, this theorem is not satisfactory for applications. The results which concern weak solutions (those which have a chance of being defined for all time) are limited to the scalar case ($n = 1$) or to the one-dimensional case ($d = 1$)! Again in this latter case the restrictions themselves are severe: the global existence in time (Theorem 5.2.1) is known if the total variation of u_0 is sufficiently small (if $n = 2$, only if the product $\mathrm{TV}(u_0)\|u_0\|_\infty$ is supposed small). There is there a threshold effect, for a local result does not exist where the time of existence would depend on the scale of the data. This question is discussed by Temple and Young [105], who have obtained recently a result of this type for the system of gas dynamics.[5] For bounded initial data, but of arbitrary size, the situation is worse; only 2×2 systems (i.e. with $n = 2$ and $d = 1$) and related systems (see Chapter 12) have been tackled by the method of *compensated compactness*, under restrictive hypotheses and for results of relatively poor quality. Among these, the *Temple systems* gain from a suitable theory (see Chapter 13) in large part because they are a faithful generalisation of the scalar laws of conservation.

The appearance of discontinuities in finite time has led specialists in function spaces to pay particular attention to spaces such as L^∞ or BV (functions of bounded variation). It is in one or the other space that existence theorems have been obtained in one-dimensional space. The reason for their success is that these are algebras, which permits the treatment of the rather strong non-linearity of the equations. However, the work of Brenner [4], which is concerned with linear systems, shows that these spaces cannot be adapted to the multi-dimensional case. To the contrary, the spaces would have to be of Hilbert type, at least to be constructed on L^2. We are thus in the presence of a paradox which has up to the present not been resolved: *to find a function space which is an algebra, probably constructed on L^2 and which contains enough discontinuous functions.*

The study of discontinuous solutions, called *weak solutions*, makes use of the integral form of the conservation laws, in the equivalent form below:[6]

$$\iint_{\Omega\times\mathbb{R}^+} \left(u_i \frac{\partial \varphi}{\partial t} + f_i(u) \cdot \nabla_x \varphi \right) \mathrm{d}x\, \mathrm{d}t + \int_\Omega u_{i0} \varphi(\cdot, u)\, \mathrm{d}x = \iint_{\Omega\times\mathbb{R}^+} g_i \varphi\, \mathrm{d}x\, \mathrm{d}t$$

for every test function φ, of class $\mathscr{C}^\infty$ and with compact support in $\Omega \times \mathbb{R}^+$. We show easily the equivalence with the partial differential equation $\partial_i u_t + \mathrm{div}_x f_i(u) = g_i$ everywhere u is of class $\mathscr{C}^1$. On the other hand, when u is of class $\mathscr{C}^1$ on both sides of a hypersurface $\Sigma \in \mathbb{R}^{d+1}$, with the boundary values $u^+(x, t)$ on one side and $u^-(x, t)$ on the other, the integral formulation expresses a transmission condition,

[5] Their work is based on the particular structure of the system of gas dynamics and cannot be extended to systems of general conservation laws, by reason of an estimation due to Joly, Métivier and Rauch [47].

[6] The eventual boundary conditions have not been taken into account here, so as not to overburden the formulae.

called the *Rankine–Hugoniot condition*:

$$\nu_0[u_i] + \sum_{\alpha=1}^{d} \nu_\alpha[f_{i\alpha}(u)] = 0$$

where $(x, t) \mapsto (\nu_0, \ldots, \nu_d)$ is a normal vector field to Σ. This formula suggests that the role of the sources g_i in the propagation of discontinuities is negligible. This is the reason why these terms are omitted very frequently in this work.

A quick look at the systems of the form (0.1) suggests that they govern reversible phenomena, at least when $g \equiv 0$: if u is a solution so also is the function $\tilde{u}(x, t) := u(-x, -t)$. This is obvious if u is of class $\mathscr{C}^1$, it is also true for a weak solution. Nevertheless it is known that thermodynamics modelled by the equations of Euler which are the archetype of systems of conservation laws is the centre of irreversible processes. This paradox is bound to the lack of uniqueness of the solution of the Cauchy problem in the framework of weak solutions. The regular solutions are effectively reversible, but the discontinuous solutions are not. We attack there a central question of the theory: how to separate the wheat from the chaff, the solutions observed in nature (called 'physically admissible') from those that are only mathematical artefacts? There are two major types of reply to this question.

The first is descriptive and concerns only piecewise continuous solutions whose discontinuities occur along regular hypersurfaces of $\Omega \times \mathbb{R}^+$. These discontinuities are physically admissible if they obey a causality principle: the state of the system cannot contain more information than it has at the initial instant.[7] Mathematically, we consider the coupled system formed, on the one hand, partly of the conservation laws and partly of the hypersurface and, on the other hand, of the Rankine–Hugoniot condition, seen as an evolution equation for the location of the discontinuity. It is thus a free boundary problem, which can be transformed to a mixed problem in a fixed domain. We demand that this mixed problem be well-posed for increasing time. In dimension $d = 1$, an equivalent condition, at least if the amplitude of the jump in u is moderate, is the shock criterion of Lax, formed of four inequalities, described in §4.3. In a higher dimension, the characterisation of the admissible discontinuities, much more complex, is explained in Chapter 14. There are principally two kinds of 'good' discontinuities, according as the Lax inequalities are strict or two among them are equalities. Only the first type, called *shock waves*, are irreversible. The second type, reversible, bear the name *contact discontinuities*. Concerning thermodynamics, A. Majda [74, 73] has shown (see Chapter 14) that the shock waves are, in general, stable (that is, that the mixed problem introduced above is locally well-posed), while the contact discontinuities (the *vortex sheets*)

[7] or that the boundary conditions provide.

are strongly unstable.[8] This instability *à la Hadamard* is a stone in the garden of the mechanics of fluids; it renders Euler's equations unsuited to the prediction of flows and casts doubt on this model for thermodynamics.[9]

The second reply is of more general significance but manifests itself less practically in applications. To begin with is the criticism of the approximation made above. To simplify the matter, let us suppose ε to be a scalar. It is reasonable to replace $Q_i[u;\varepsilon]$ by $f_i(u)$ where the solution varies moderately, but it is debatable where it varies greatly. Now most often, the solution of the real problem (denoted by u^ε) is regular, of class $\mathscr{C}^1$, but varies rapidly in the neighbourhood of a discontinuity of u. Typically this neighbourhood has a width of the order of ε and the gradient of u^ε must be of the order of $1/\varepsilon$. In this narrow zone, what has been neglected is of the same order of magnitude as $f_i(u)$. We have in fact

$$Q_i[v;\varepsilon] - f_i(v) \sim \varepsilon B_i(v)\nabla_x v,$$

where $B(v)$ is a tensor with four indices. A description of the evolution more faithful than (0.1) is therefore

$$\partial_t u_i^\varepsilon + \operatorname{div}_x f_i(u^\varepsilon) = \varepsilon \operatorname{div}_x(B_i(u^\varepsilon)\nabla_x u^\varepsilon), \quad 1 \le i \le n. \tag{0.2}$$

The tensor B is such that the Cauchy problem for (0.2) is well-posed for $\varepsilon > 0$ and increasing time.[10] It represents, according to the case, the effect of a viscosity, that of thermal conduction, the Joule effect, ... In the model of road traffic, where the scalar u is the density of the vehicles, it represents the faculty of anticipation of the drivers as a function of the flow of traffic in the vicinity of their vehicle; it is this anticipation that causes irreversibility.

The system (0.2) is irreversible. This is the essential difference from (0.1), which is expressed quantitatively as follows. The undisturbed system is in general compatible, for the regular solutions, with a supplementary conservation law[11]

$$\partial_t E(u) + \operatorname{div}_x F(u) = \mathrm{d}E(u) \cdot g(x,t,u) \tag{0.3}$$

where $E\colon \mathscr{U} \to \mathbb{R}$ is strictly convex. This can always be reduced to the case in which E has positive values. The equation (0.3) then yields an *a priori* estimate of

[8] save in dimension $d = 1$ where the contact discontinuities are stable.

[9] It is not difficult to see that the vortex sheets necessarily appear, if $d > 2$, as a by-product of the interaction of multi-dimensional shocks. The case $d = 1$ is less clear.

[10] It would be wrong nevertheless to believe that the system (0.2) is parabolic, that is, that the operator $v \mapsto \operatorname{div}_x B(v)\nabla_x v$ is elliptic. Its symbol is generally positive but not positive definite.

[11] It is principally in this setting that this work is placed.

u in a Sobolev–Orlicz space via the differential equation[12]

$$\frac{\mathrm{d}}{\mathrm{d}t}\int_\Omega E(u(x,t))\,\mathrm{d}x = \int_\Omega \mathrm{d}E(u)\cdot g(x,t,u)\,\mathrm{d}x.$$

This allows us to control the value of the positive expression

$$\mathscr{E}(t) := \int_\Omega E(u(x,t))\,\mathrm{d}x.$$

For example, in the absence of sources, $\mathscr{E}(t)$ remains equal to $\mathscr{E}(0)$, which depends only on the initial condition. Of course, this calculation, in which we differentiate composite functions (for example $E \circ u$), does not have a rigorous basis for weak solutions. On the other hand, the solution of the perturbed problem is in general regular and satisfies the equation

$$\partial_t E(u^\varepsilon) + \mathrm{div}_x \vec{F}(u^\varepsilon) = \mathrm{d}E(u^\varepsilon)\cdot(\varepsilon\,\mathrm{div}_x(B(u^\varepsilon)\nabla_x u^\varepsilon) + g(x,t,u^\varepsilon)).$$

If $g \equiv 0$ and if the boundary values are favourable, we obtain at the best

$$\mathscr{E}'(t) = -\varepsilon \int_\Omega (\mathrm{D}^2 E(u^\varepsilon)\nabla_x u^\varepsilon \cdot B(u^\varepsilon)\nabla_x u^\varepsilon)\,\mathrm{d}x$$

which is negative for all the realistic examples. But, above all, the right-hand side does not tend to zero with ε, because we integrate an expression of the order of ε^{-1} (because of $\varepsilon(\nabla_x u^\varepsilon)^2$) over a zone whose measure is of the order of ε. The decay of $\mathscr{E}$, certainly preserved by passage to the limit when $\varepsilon \to 0^+$, thus will be strict in the presence of discontinuities. A criterion of the admissibility of solutions is thus

$$\iint_{\Omega\times\mathbb{R}^+} (E(u)\partial_t\varphi + F(u)\cdot\nabla_x\varphi)\,\mathrm{d}x\,\mathrm{d}t + \int_{\mathbb{R}} E(u_0)\varphi(\cdot,0)\,\mathrm{d}x \geq 0 \qquad (0.4)$$

for every *positive* test function $\varphi \in \mathscr{D}(\Omega\times\mathbb{R})$, with equality for a classical solution of (0.1). On the level of the discontinuities, (0.4) is translated as the jump condition[13]

$$\nu_0[E(u)] + \sum_{\alpha=1}^{d} \nu_\alpha[F_\alpha(u)] \leq 0. \qquad (0.5)$$

The equality in (0.5) is in general incompatible with the Rankine–Hugoniot condition except where this concerns the contact discontinuities.

[12] To simplify the exposition, no account has been taken of the boundary conditions. For example, the reader could assume that $\vec{F}\cdot\nu$ is null on the boundary.

[13] We remark that this condition is independent of the orientation of Σ.

A function E like that introduced above carries the name, in so far as it is a mathematical object, of *entropy*. By extension, we call a function E, not necessarily convex, in a conservation law compatible with (0.1), also an entropy. The vector field $\vec{F}$ is called the *entropy flux* associated with E. Again, this terminology is due to thermodynamics, as the form of the equations of motion in *lagrangian*[14] variables is a system of conservation laws compatible with a supplementary law in which E is equal to $-S$, the opposite of physical entropy of the fluid. This change of sign, which renders convex that which is concave and conversely, is a cultural difference between mathematicians and physicists. For the physicist, the entropy has a tendency to increase, while for the mathematician the opposite holds. In the eulerian representation, in which Ω is a domain in physical space, the difference is still more marked, as E corresponds to $-\rho S$. Despite this, the historical link between the physical theory and mathematics has led to the inequality (0.4) being called the *entropy condition*, when a system of conservation laws can be modelled on something other than the flow of a fluid. The weak solutions which satisfy (0.4) are called *entropy solutions*. By extension and in an improper manner, we again speak of entropy conditions with regard to the Lax shock condition, principally because in thermodynamics, the Lax inequalities express the fact that the entropy S, constant when it follows a particle, in fact grows when it crosses a shock wave.

The mechanism of dissipation, which makes $\mathscr{E}$ decrease and renders the evolution irreversible, is so central that it would not make sense to study theoretically (0.1), in isolation. This necessarily leads to the algebraic notion of a hyperbolic system in the linear case, but the understanding of the non-linear case calls for as much attention to be paid to the (partially) parabolic (0.2). This is why this book is not entitled *Hyperbolic systems of conservation laws*. Chapters 6, 7 and 15 are principally devoted to parabolic systems and these are involved in a significant way in Chapters 8 and 9.

Some references This volume owes a great deal to those which preceded it, in particular that of Majda [75]; this, at the same time short and profound, remains an essential reference and the energy which animates it gives birth to a sense of vocation. It is the only one to deal with nearly all the topics which deal with multi-dimensional or asymptotic problems. It is with this that we have tried to deal here, with more detail but less animation. Although dealing with many subjects, this work does not go on as long on classical problems as more specialised works. Thus, the reader who wishes to deepen his knowledge of the Riemann problem should read

[14] In these variables, a particle is represented by a fixed value of the variable x. This is therefore not a 'space variable' as strictly defined.

the text of Ling Hsiao and Tong Zhang [46]. The global methods, based on the Conley index, for studying the viscosity profiles, are found in Smoller [97]. A systematic study of the propagation and the interaction of non-linear waves is greatly developed by Whitham [112]; see also the monograph of Boillat [2]. For questions concerning the mechanics of fluids, with the description of multi-dimensional shocks, Courant and Friedrichs [11] should be consulted. Various types of singular perturbations (models of combustion, the incompressible limit) and the stability of multi-dimensional shocks are presented in Majda [75]. For mixed problems in a (partly) parabolic context, a good reference is Kreiss and Lorenz [55]. In the lecture notes by Hörmander [45] is found a simple presentation of the blow-up mechanisms for a general system (not necessarily rich) as well as the global (or nearly global) existence for a perturbation of the wave equation in dimension $d \geq 2$. The notes of Evans [20] give a view of the methods utilising weak convergence, which goes beyond mere compensated compactness. The decay of entropic solutions to N-waves is the subject of the memoir of Glimm and Lax [33]. Concerning the Cauchy problem and mixed problems for linear equations there are many references; let us, at least, cite Ivrii [48], Sakamoto [87] and again Kreiss and Lorenz [55]. The quasi-linear mixed problem in dimension $d = 1$, which includes the free boundary problems, is systematically studied in Li Ta-Tsien and Yu Wen-ci [65]. The geometrical aspects of the conservation laws, especially affine and convex, are the subject of the memoir of Sévennec [93].

The way the chapters of this book are ordered is merely an indication to the reader, since the chapters depend little on one another. The core of the theory is constituted by Chapters 2, 3, 4, 6 and 14. For a postgraduate course in which the aim is the solution of the Riemann problem for gas dynamics, Chapters 2 and 4 are indispensable but are not enough to give an advanced student a representative picture of the subject.

In spite of its length, this work does not pretend to be exhaustive. It leads to a blind alley on several questions, of which some are important. Uniqueness is the most important of these; the reason is that it is a matter of a subject which is much less advanced than that of existence (with, however, recent progress by A. Bressan), and on which we could not give a synthetic view. Likewise, this book does not tackle questions which touch on ‘pathology’: systems not strictly hyperbolic have other conditions for the admission of shock waves. At times, it has been mathematical rigour that has been neglected (with the hope that it is not too frequent for the taste of the reader): above all an attempt has been made to be the most descriptive possible, giving perhaps too many criteria and formulae, asymptotic analysis, and not enough proofs. Some new results will be found (few enough and none major in all cases) and lists of exercises which should satisfy those who believe in acquiring insight by the solution of examples. In spite of this range of descriptive material, there is not a word on the phenomenon of relaxation, nor on kinetic formulations, and not more on the description by N-waves of behaviour in large times, three important subjects of the theory. Perhaps, if the occasion arises, a future edition . . .

Lyon, November 1995

. . . cela peut durer pendant très longtemps,
si l'on ne fait pas d'omelette avant!

(Robert Desnos, Chantefables)

1
Some models

1.1 Gas dynamics in eulerian variables

Let us consider a homogeneous gas (all the molecules are identical with mass m) in a region Ω, whose coordinates $x = (x_1, \dots, x_d)$ are our 'independent' variables. From a macroscopic point of view, it is described by its mass density ρ, its momentum per unit volume $\vec{q}$ and its total energy per unit volume E. In a sub-domain ω containing at an instant N molecules[1] of velocities $\vec{v}^1, \dots, \vec{v}^N$ respectively, we have

$$\int_\omega \rho \, \mathrm{d}x = Nm, \qquad \int_\omega \vec{q} \, \mathrm{d}x = m \sum_{j=1}^{N} \vec{v}^j$$

from which it follows that $\vec{q} = \rho \vec{v}$, $\vec{v}$ being the mean velocity of the molecules.[2] Likewise, the total energy is the sum of the kinetic energy and of the rotational and vibrational energies of the molecules:

$$\int_\omega E \, \mathrm{d}x = \frac{1}{2} m \sum_{j=1}^{N} \|\vec{v}^2\| + \sum_{j=1}^{N} (e_V^j + e_R^j)$$

where e_V^j and e_R^j are positive. For a monatomic gas, such as He, the energy of rotation is null. The energy of vibration is a quantum phenomenon, of sufficiently weak intensity to be negligible at first glance. Applying the Cauchy–Schwarz inequality, we find that

$$\frac{1}{2} m \sum_{j=1}^{N} \|\vec{v}^j\|^2 \geq \frac{m}{2N} \left\| \sum_{j=1}^{N} \vec{v}^j \right\|^2$$

[1] N is a very large number, for example of the order of 10^{23} if the volume of ω is of the order of a unit, but the product mN is of the order of this volume.

[2] This can be suitably modified if there are several kinds of molecules of different masses.

which gives

$$\int_{\omega} E\,\mathrm{d}x \geq \frac{1}{2}\Big(\int_{\omega} \rho\,\mathrm{d}x\Big)^{-1} \Big\| \int_{\omega} \vec{q}\,\mathrm{d}x \Big\|^2 + \sum_{j=1}^{N} (e_{\mathrm{V}}^j + e_{\mathrm{R}}^j)$$
$$\geq \frac{1}{2}\Big(\int_{\omega} \rho\,\mathrm{d}x\Big)^{-1} \Big\| \int_{\omega} \rho\vec{v}\,\mathrm{d}x \Big\|^2.$$

This being true for every sub-domain, we can deduce that the quantity $E/\rho - \frac{1}{2}\|\vec{v}\|^2$ is positive. It is called the specific *internal energy* (that is per unit mass) and we denote it by e; we thus have

$$E = \frac{1}{2}\rho\|\vec{v}\|^2 + \rho e,$$

where the first term is (quite improperly) called the *kinetic energy* of the fluid. For the sequel it should be remembered that the internal energy can be decomposed into two terms $e_{\mathrm{k}} + e_{\mathrm{f}}$ where e_{k} is kinetic in origin and e_{f} is due to other degrees of freedom of the molecules.

The law of a perfect gas

A *perfect gas* obeys three hypotheses:

the vibration energy is null,

the velocities at a point (x, t) satisfy a gaussian distribution law

$$a \exp(-b\| \cdot - \vec{v}\|^2)$$

where a, b and $\vec{v}$ are functions of (x, t) (of course, $\vec{v}$ is the mean velocity introduced above),

the specific internal energy is made up among its different components *pro rata* with the degrees of freedom.

Comments (1) The gaussian distribution comes from the theorem of Laplace that considers the molecular velocities as identically distributed random variables when N tends to infinity. It is also the equilibrium distribution (when it is called 'maxwellian') in the Boltzmann equation, when it takes into account the perfectly elastic binary collisions.

(2) Several reasons characterise the gaussian as being the appropriate law. On the one hand, its set is stable by composition with a similitude O of $\mathbb{R}^d (\chi \mapsto \chi \circ O)$ and by multiplication by a scalar $(\chi \mapsto \lambda\chi)$. On the other, the components of the velocity are independent identically distributed random variables.

(3) The hypothesis of the equi-partition of energy is pretty well verified when there are a few degrees of freedom, for example for monatomic molecules (He), diatomic molecules (H_2, O_2, N_2) or rigid molecules ($H_2O, CO_2, C_2H_2, C_2H_4$). The more complex molecules are less rigid; they thus have more degrees of freedom, which are not equivalent from the energetic point of view.

(4) The equi-partition takes place also among the translational degrees of freedom. If the choice is made of an orthonormal frame of reference, each component $v_\alpha^j - v_\alpha$ of the relative velocity is responsible for the same fraction $e_{\mathrm{k}\alpha} = e_{\mathrm{k}}/d$ in the energy of kinetic origin.

Let β be the number of non-translational degrees of freedom. The hypothesis of equi-partition gives the following formula for each type of internal energy:

$$e_{\mathrm{k}1} = \cdots = e_{\mathrm{k}d} = \frac{1}{d} e_{\mathrm{k}}, \qquad e_{\mathrm{R}} = \frac{\beta}{d} e_{\mathrm{k}}$$

and thus $e = (d + \beta) e_{\mathrm{k}1}$.

The pressure p is the force exerted per unit area on a surface, by the gas situated on one side of it.[3] Take as surface the hyperplane $x_1 = 0$, the fluid being at rest ($\vec{v} \equiv 0$, a and b constants). Let A be a domain of unit area of this hyperplane. The force exerted on A by the gas situated to the left is proportional to the number M of particles hitting A per unit time, multiplied by the first component I_1 of the mean impulse of these.[4] On the one hand, M is proportional to the number N of particles multiplied by the mean absolute speed (the mean of $|v_1^\alpha|$) in the direction x_1. On the other hand, NI_1 is proportional to ρw_1^2, that is to $\rho e_{1\mathrm{k}}$. Nothing in this argument involves explicitly the dimension d and we therefore have $p = k\rho e_{1\mathrm{k}}$, where k is an absolute constant. A direct calculation in the one-dimensional case yields the result $k = 2$. Introducing the *adiabatic exponent*

$$\gamma = \frac{d + b + 2}{d + b}$$

there results the *law of perfect gases*

$$p = (\gamma - 1)\rho e.$$

The most current adiabatic exponents are $5/3$ and $7/5$ if $d = 3$, 2 and $5/3$ if $d = 2$ and 3 if $d = 1$. In applications air is considered to be a perfect gas for which $\gamma = 7/5$.

[3] In this argument, the surface in question is not a boundary, since it would introduce a reflexion and would eventually distort the gaussian distribution.

[4] This mean is not null as it is calculated solely from the set of molecules for which $v_1^j > 0$.

The Euler equations

The conservation laws of mass, of momentum and of energy can be written

$$\partial_t \rho + \mathrm{div}_x(\rho \vec{v}) = 0,$$

$$\partial_t(\rho v_i) + \mathrm{div}_x(\rho v_i \vec{v}) + \partial_i p = \sum_{j=1}^{d} \partial_j T_{ij}, \quad 1 \le i \le d,$$

$$\partial_t E + \mathrm{div}_x((E + p)\vec{v}) = \sum_{j=1}^{d} \partial_j (v_i T_{ij}) - \mathrm{div}_x \vec{q}$$

where $T - pI_d$ is the stress tensor and $\vec{q}$ the heat flux. In the last equation, two terms represent the power of the forces of stress. The conservation of the kinetic moment $\rho\vec{v} \wedge x$ implies that T is symmetric. We have seen that T is null for a fluid at rest and also when it is in uniform motion of translation. The simplest case is that in which T is a linear expression of the first derivatives $\nabla_x \vec{v}$, the coefficients being possibly functions of (ρ, e). The principle of frame indifference implies the existence of two functions α and β such that

$$T_{ij} = \alpha(\rho, e)\left(\frac{\partial v_i}{\partial x_j} + \frac{\partial v_j}{\partial x_i}\right) + \beta(\rho, e)(\mathrm{div}_x \vec{v})\delta_i^j \tag{1.1}$$

which clearly introduces second derivatives into the above equations. The tensor T represents the effects of viscosity and the linear correspondence is Newton's law. If α and β are null the conservation laws are called *Euler's equations*. In the contrary case they are called the *Navier–Stokes equations*.

Likewise, the heat flux is null if the temperature θ (defined later as a thermodynamic potential) is constant. The simplest law is that of Fourier, which can be written

$$\vec{q} = -k(\rho, e)\nabla_x \theta,$$

with $k \ge 0$.

For a regular flow, a linear combination of the equations yields the reduced system

$$\partial_t \rho + \mathrm{div}(\rho \vec{v}) = 0,$$

$$\partial_t v_i + \vec{v} \cdot \nabla_x v_i + \rho^{-1} \partial_i p = \rho^{-1}\,\mathrm{div}(T_i.),$$

$$\partial_t e + \vec{v} \cdot \nabla_x e + \rho^{-1} p \,\mathrm{div}\, \vec{v} = \rho^{-1}\left(\sum_{i,j} T_{ij}\partial_j v_i - \mathrm{div}\,\vec{q}\right).$$

Let us linearise this system in a constant solution, in a reference frame in which the velocity is null:

$$\partial_t R + \rho \,\mathrm{div}\, \vec{V} = 0,$$

$$\partial_t V_i + \rho^{-1}(p_\rho \partial_i R + p_e \partial_i \chi) = \rho^{-1}(\alpha \Delta V_i + (\alpha + \beta)\partial_i \,\mathrm{div}\, \vec{V}),$$

$$\partial_t \chi + \rho^{-1} p \,\mathrm{div}\, \vec{V} = \rho^{-1} k \Delta(\theta_\rho R + \theta_e \chi).$$

The last equation can be transformed to

$$\partial_t(\theta_\rho R + \theta_e \chi) + \lambda \operatorname{div} \vec{V} = \frac{k\theta_e}{\rho}\Delta(\theta_\rho R + \theta_e \chi).$$

A necessary condition for the Cauchy problem for this linear system to be well-posed is the (weak) ellipticity of the operator

$$(R, \vec{V}, \xi) \mapsto (0, \alpha\Delta\vec{V} + (\alpha+\beta)\nabla \operatorname{div} \vec{V},\ k\theta_e\Delta\xi)$$

which results in the inequalities

$$k\theta_e \geq 0, \qquad \alpha \geq 0, \qquad 2\alpha + \beta \geq 0. \tag{1.2}$$

The entropy

In the absence of second order terms, the flow satisfies

$$p(\partial_t \rho + \vec{v} \cdot \nabla\rho) = \rho^2(\partial_t e + \vec{v} \cdot \nabla e)$$

which suggests the introduction of a function $S(\rho, e)$, without critical point, such that

$$p\frac{\partial S}{\partial e} + \rho^2\frac{\partial S}{\partial \rho} = 0.$$

Such a function is defined up to composition on the left by a numerical function: if $h\colon \mathbb{R} \to \mathbb{R}$ and if S works, then $h \circ S$ does too, provided that h' does not vanish. Such a function satisfies the equation

$$(\partial_t + \vec{v} \cdot \nabla)S = 0,$$

as long as the flow is regular, this signifies that S is constant along the trajectories[5] of the particles. On taking account of the viscosity and of the thermal conductivity, it becomes

$$\rho(\partial_t + \vec{v} \cdot \nabla)S = S_e \sum_{i,j}(T_{ij}\partial_j v_i) + \operatorname{div}(k\nabla\theta),$$

that is to say

$$\partial_t(\rho S) + \operatorname{div}(\rho S\vec{v}) = S_e\left(\frac{1}{2}\alpha\sum_{i,j}(\partial_i v_j + \partial_j v_i)^2 + \beta(\operatorname{div}\vec{v})^2\right) + S_e \operatorname{div}(k\nabla\theta).$$

Free to change S to $-S$, we can suppose that S_e is strictly positive. The name *specific entropy* is given to S. The effect of the viscosity is to increase the integral

[5] We refer to the mean trajectory.

of ρS. The second law of thermodynamics states that the thermal diffusion behaves in the same sense, that is that

$$\int_\omega S_e \operatorname{div}(k\nabla\theta)\,\mathrm{d}x \ge 0$$

if there is no exchange of heat across $\partial\omega$ (Neumann condition $\partial\theta/\partial\nu = 0$). Otherwise, this integral is compensated by these exchanges. In other terms, after integration by parts, we must have

$$\int_\omega k\nabla\theta \cdot \nabla S_e\,\mathrm{d}x \le 0,$$

without restriction on ω. Thus $\nabla\theta\cdot\nabla S_e$ must be negative at every point and naturally for every configuration. It is then deducted that θ is a decreasing function of S_e. Free to compose θ on the left with an increasing function,[6] there is no loss of generality if we assume that $\theta = 1/S_e$, which gives the thermodynamic relation

$$\theta\,\mathrm{d}S = \mathrm{d}e + p\mathrm{d}\left(\frac{1}{\rho}\right), \qquad \theta \ge 0,$$

in which $1/\theta$ appears as an integrating factor of the differential form $\mathrm{d}e + p\mathrm{d}(1/\rho)$. For a perfect gas are chosen as usual $\theta = e$ and $S = \log e - (\gamma - 1)\log\rho$.

Barotropic models

A model is *barotropic* if the pressure is, because of an approximation, a function of the density only. There are three possible reasons: the flow is isentropic or it is isothermal, or again it is the shallow water approximation.

For a regular flow without either viscosity or conduction of heat (that makes up many of the less realistic hypotheses), we have $(\partial_t + \vec{v}\cdot\nabla)S = 0$: S is constant along the trajectories. If, in addition, it is constant at the initial instant, we have $S = \text{const}$. As $S_e > 0$, we can invert the function $S(\cdot\,,\rho)$: we have $e = \mathscr{E}(S,\rho)$, with the result that also p is a function of (S,ρ). In the present context, p must be a function of ρ alone and similarly this is true of all the coefficients of the system, for example α and β. The conservation of mass and that of momentum thus form a closed system of partial differential equations (here again we have taken account of the newtonian viscosity[7]):

$$\partial_t\rho + \operatorname{div}(\rho\vec{v}) = 0,$$
$$\partial_t(\rho v_i) + \operatorname{div}(\rho v_i\vec{v}) + \partial_i p(\rho) = \operatorname{div}(\alpha(\nabla v_i + \partial_i\vec{v})) + \partial_i(\beta\operatorname{div}\vec{v})\cdot$$

[6] This does not affect Fourier's law, as k is changed with the result that the product $k\nabla\theta$ is not.
[7] One more odd choice!

The equation of the conservation of energy becomes a redundant equation.[8] We shall use it as the 'entropy' conservation law of the inviscid model. We call this the *isentropic model*:

$$\partial_t \rho + \operatorname{div}(\rho \vec{v}) = 0,$$
$$\partial_t(\rho v_i) + \operatorname{div}(\rho v_i \vec{v}) + \partial_i p(\rho) = 0.$$

Its mathematical entropy is the mechanical energy $\frac{1}{2}\rho(\|\vec{v}\|^2 + e(\rho))$, associated with the 'entropy flux' $\rho(\frac{1}{2}\|\vec{v}\|^2 + e(\rho))\vec{v} + p(\rho)\vec{v}$. For a perfect gas, the hypothesis $S = \text{const.}$, states that $e^{\gamma-1} = c\rho$ and furnishes the *state law* $p = \kappa\rho^\gamma$. This, then, is called a *polytropic gas*.

The *isothermal* model is reasonable when the coefficient of thermal diffusion is large relative to the scales of the time and space variables. For favourable boundary conditions, the entropic balance gives

$$\frac{\mathrm{d}}{\mathrm{d}t}\int_\Omega \rho S \,\mathrm{d}x \geq -\int_\Omega k\nabla\theta \cdot \nabla S_e \,\mathrm{d}x = \int_\Omega k\frac{\|\nabla\theta\|^2}{\theta^2}\,\mathrm{d}x.$$

According to the conservation laws, we can add to ρS an affine function of the variables $(\rho, \rho\vec{v}, E)$ in the preceding inequality. Meanwhile, experience shows that the mapping $(\rho, \rho\vec{v}, E) \mapsto \rho S$ is concave.[9] We can thus choose an affine function η_0 with the result that $\eta := \rho S + \eta_0$ is negative. If the domain Ω is the whole space $\mathbb{R}^d$, the fluid being at rest at infinity, we can also take η to be null at infinity. Finally

$$\int_\Omega k\frac{\|\nabla\theta\|^2}{\theta^2}\,\mathrm{d}x \leq -\int_\Omega \eta_{t=0}\,\mathrm{d}x\cdot$$

The right-hand side is a datum of the problem, supposed finite. If k is large, we see that it is all right to approach θ by a constant; that it is a constant and not a function of time is not clear but is currently assumed. Again, the pressure and the viscosity become functions of ρ only, and the conservation of mass and that of momentum form a closed system: the mechanical energy is taken as the mathematical entropy of the system. For a perfect gas, $e = \theta$ is constant, with the result that the state law is linear: $p = \kappa\rho$.

The isothermal approximation is reasonable enough in certain régimes, because, for a gas, for instance, the thermal effects are always more significant than the viscous effects. A general criterion regarding these approximations is however that the shocks of the barotropic models are not the same as those of the Euler equations: the Rankine–Hugoniot condition is different.

[8] Or rather incompatible, if we have included the newtonian viscosity.

[9] In fact, this concavity is the condition for the Cauchy problem of the linearised Euler equations to be well-posed. It no longer holds if we model a fluid with several phases.

The third barotropic model describes the flow in a shallow basin, that is, in one whose horizontal dimensions are great with respect to its depth. The domain Ω is the horizontal projection of the basin: we thus have $d = 1$ or $d = 2$. The fluid is *incompressible* with density ρ_0. We do not take the vertical displacements into account. The variables treated are the horizontal velocity (averaged over the height) $\vec{v}(x,t)$ and the height of the fluid $h(x,t)$. The pressure is considered to be the integral of the hydrostatic pressure $\rho_0 g z$ where z is the vertical coordinate. We therefore have $p = \rho_o g h^2/2$. The conservation of mass and that of momentum give the system

$$\partial_t(\rho_0 h) + \operatorname{div}(\rho_0 h \vec{v}) = 0,$$

$$\partial_t(\rho_0 h v_i) + \operatorname{div}(\rho_0 v_i \vec{v}) + \frac{1}{2} g \partial_i(\rho_0 h^2) = 0, \quad 1 \leq i \leq d.$$

Comments Dividing by ρ_0, we recover the isentropic model of a perfect gas for which $\gamma = 2$.

We have not taken into account the effects of viscosity and this is an error: they are responsible for a boundary layer on the base of the basin which implies a resistance to the motion. That resistance makes itself manifest in the model by a source term in the second equation of the form $-f(h, |\vec{v}|)v_i$, with $f > 0$.

One way of obtaining these equations from the Euler equations is to integrate the latter with respect to z (but not x). We then make the hypothesis that certain means of products are the products of means, that is that the vertical variations in ρ and $\vec{v}$ are weak.

The relativistic models of a gas, though much more complicated than those which have preceded, are also those of systems of conservation laws. We shall not give a detailed presentation here. By way of an example, we shall consider the simplest among those systems: a barotropic fluid, isentropic, one-dimensional and in special relativity; the conversation of mass and that of momentum give

$$\partial_t\left(\frac{p+\rho c^2}{c^2}\frac{v^2}{c^2-v^2} + \rho\right) + \partial_x\left((p+\rho c^2)\frac{v}{c^2-v^2}\right) = 0,$$

$$\partial_t\left((p+\rho c^2)\frac{v}{c^2-v^2}\right) + \partial_x\left((p+\rho c^2)\frac{v^2}{c^2-v^2} + p\right) = 0.$$

For more general models the reader should consult Taub [102].

1.2 Gas dynamics in lagrangian variables

Writing the equations of gas dynamics in lagrangian coordinates is very complicated if $d \geq 2$; in addition it furnishes a system which does not come into the spirit of this book. This is why we limit ourselves to the one-dimensional case ($d = 1$). We shall make a change of variables $(x, t) \mapsto (y, t)$ which depends on the solution. The conservation law of mass

$$\rho_t + (\rho v)_x = 0$$

is the only one which makes no appeal to any approximation. It expresses that the differential form $\alpha := \rho\,dx - \rho v\,dt$ is closed and therefore exact.[10] We thus introduce a function $(x, t) \mapsto y$, defined to within a constant by $\alpha = dy$. We have $dx = v\,dt + \tau\,dy$, where we have denoted by $\tau = \rho^{-1}$ the *specific volume* (which is rather a specific length here).

Being given another conservation law $\partial_t u_i + \partial_x q_i = 0$, which can be written $d(q_i\,dt - u_i\,dx) = 0$, we have that

$$d((q_i - u_i v)\,dt - u_i \tau\,dy) = 0,$$

that is

$$\partial_t(u_i\tau) + \partial_y(q_i - u_i v) = 0.$$

The system, written in the variables (y, t), is thus formed of conservation laws. Let us look at for example the momentum $u_2 = \rho v$. In the absence of viscosity, we have $q_2 = \rho v^2 + p(\rho, e)$. From this comes

$$\partial_t v + \partial_y P(\tau, e) = 0,$$

where $P(\tau, e) := p(\tau^{-1}, e)$. Similarly, for the energy, $u_3 = \frac{1}{2}\rho v^2 + \rho e$ and $q_3 = (u_3 + p)v$:

$$\partial_t\left(\frac{1}{2}v^2 + e\right) + \partial_y(P(\tau, e)v) = 0.$$

The conservation of mass gives nothing new since it was already used to construct the change of variables. With $u_1 = \rho$ and $q_1 = \rho v$, we only obtain the trivial equation $1_t + 0_y = 0$. To complete the system of equations for the unknowns (τ, v, e) we have to involve a trivial conservation law. For example with $u_4 \equiv 1$ and $q_4 \equiv 0$, we obtain

$$\partial_t \tau = \partial_y v.$$

We note that in lagrangian variables the perfect gas law is written $P = (\gamma - 1)e/\tau$.

If we take into account the thermal and viscous effects, then $q_2 = \rho v^2 + p(\rho, e) - \nu(\rho, e)v_x$. As $\tau v_x = v_y$ we obtain

$$\partial_t v + \partial_y P(\tau, e) = \partial_y\left(\frac{\nu}{\tau}\partial_y v\right).$$

[10] These assertions are correct even $(\rho, \rho v)$ are no better than locally integrable.

Similarly, $q_3 = (u_3 + p)v - \nu v v_x - k\theta_x$ gives

$$\partial_t\left(\frac{1}{2}v^2 + e\right) + \partial_y(Pv) = \partial_y\left(\frac{\nu}{\tau}(v\partial_y v)\right) + \partial_y\left(\frac{k}{\tau}\partial_y\theta\right).$$

Criticism of the change of variables

Although this change of variable is perfectly justified, even if (e, v) is bounded without more regularity as well as v^{-1} (see D. Wagner [110]), it raises a major difficulty if the vacuum is somewhere part of the space. In this case, the jacobian ρ of $(x, t) \mapsto (y, t)$ vanishes and it is no longer a change of variable. The specific volume then reduces to a Dirac mass, with norm equal to the length of the interval of the vacuum. It becomes critical to give sense to the equations (it is nothing other than the conservation law of a mathematical difficulty). The equations in eulerian coordinates are also ill-posed in the vacuum: the velocity cannot be defined and the fluxes q_2 and q_3 are singular. Indeed, returning to the variables $u = (\rho, \rho v, E)$, we have $q_2 = u_2^2/u_1 + p$, which makes no sense for $\rho = 0$.

1.3 The equation of road traffic

Let us consider a highway (a unique sense of circulation will be sufficient for our purpose), in which we take no account of entries or exits. We represent the vehicle traffic as the motion of a one-dimensional continuous medium, which is reasonable if the physical domain which we consider is very great in length in comparison with the length of the cars. In normal conditions, we have a conservation law of 'mass'

$$\partial_t \rho + \partial_x q = 0,$$

where $q = \rho v$ is the flux, or flow, and v is the mean velocity. Unlike the case of a fluid there is no conservation law of momentum or of energy. The drivers choose their velocities according to the traffic conditions. It results in a relation $v = V(\rho)$ where V is the speed limit if ρ is small. The function $\rho \mapsto V$ is decreasing and vanishes for a saturation value $\rho_{\rm m}$, for which neighbouring vehicles are bumper-to-bumper. The space of the states is therefore $\mathcal{U} = [0, q_{\rm m}]$.

This model is a typical example of a scalar conservation law. The state law $q(\rho) = \rho V(\rho)$ has the form indicated in Fig. 1.1. We notice that each possible value of the flow corresponds to two possible densities, of different velocities, with the exception of the maximal flow.

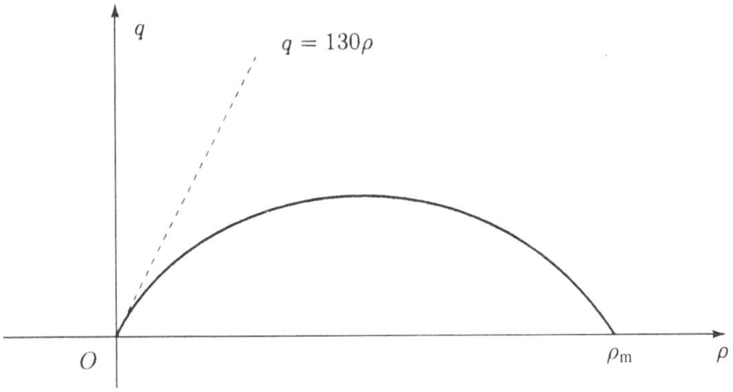

Fig. 1.1: Road traffic: flux vs density (in France).

A more precise model is obtained by taking the drivers' anticipation into account. If they observe an upstream increase in the density (respectively a diminution), they show a tendency to brake (respectively to accelerate) slightly. In other terms, $v - V(\rho)$ is of the opposite sign to that of ρ_x. The simplest state law which takes account of this phenomenon is $v = V(\rho) - \varepsilon\rho_x$, with $0 < \varepsilon \ll 1$, which leads to the weakly parabolic equation

$$\rho_t + q(\rho)_x = \varepsilon(\rho\rho_x)_x.$$

1.4 Electromagnetism

Electromagnetism is a typically three-dimensional phenomenon ($d = 3$), which brings vector fields into play: the electric intensity E, the electric induction D, the magnetic intensity H, the magnetic induction B, the electric current j and the heat flux q. Denoting by e the internal energy per unit volume, the conservation laws are

Faraday's law

$$\partial_t B + \text{curl } E = 0,$$

with which is associated the compatibility condition $\text{div } B = 0$ (absence of magnetic charge),

Ampère's law

$$\partial_t D - \text{curl } H + j = 0,$$

conservation of energy

$$\partial_t E + \text{div}(E \wedge H + q) = 0.$$

Maxwell's equations

In the first instance let us neglect the current and the heat flux (which is correct for example in the vacuum). Combining the three laws, we obtain

$$\partial_t e = H \cdot \partial_t B + E \cdot \partial_t D.$$

If the system formed by the laws of Faraday and Ampère is closed by the state laws

$$H = \mathcal{H}(B, D), \qquad E = \mathcal{E}(B, D),$$

from the conservation of energy it is then deduced that

$$\mathcal{H}(B, D) \cdot \mathrm{d}B + \mathcal{E}(B, D) \cdot \mathrm{d}D$$

is an exact differential. Following Coleman and Dill [9], we can then postulate the existence of a function $W: \mathbb{R}^3 \times \mathbb{R}^3 \to \mathbb{R}$ such that

$$H_j = \frac{\partial W}{\partial B_j}, \qquad E_j = \frac{\partial W}{\partial D_j}, \quad j = 1, 2, 3.$$

We have $e = W(B, D)$; the conservation laws are called *Maxwell's equations*:

$$\partial_t B + \operatorname{curl} \frac{\partial W}{\partial D} = 0, \qquad \partial_t D - \operatorname{curl} \frac{\partial W}{\partial B} = 0.$$

These lead to *Poynting's formula*

$$\partial_t W(B, D) + \operatorname{div}(E \wedge H) = 0,$$

which shows that W is an entropy of the system, generally convex. Some other entropies of the system, not convex, are the components of $B \wedge D$.

Now, taking into account the charge and the heat, the complete model is the following:

$$\partial_t B + \operatorname{curl} \frac{\partial W}{\partial D} = 0, \qquad \partial_t D - \operatorname{curl} \frac{\partial W}{\partial B} = -j,$$
$$\partial_t (W(B, D) + \varepsilon_0) + \operatorname{div}(E \wedge H + q) = 0,$$

where ε_0 is the purely calorific part of the internal energy.[11] For a regular solution we have

$$\partial_t \varepsilon_0 + \operatorname{div} q = E \cdot j,$$

where the right-hand side represents the work done by the electromagnetic force

[11] We have made the hypothesis that the underlying material is fixed in the reference frame. For a material in accelerated motion, see for example the following section.

(the Joule effect). We notice that transfer between the two forms of energy is possible. In the vacuum, the current is zero and there is neither temperature, nor heat flux; next, following Feynman [21] (Chapter 12.7 of the first part of vol. II), the Maxwell equations are linear in a large range of the variables. The energy W is thus a quadratic form:

$$W(B, D) = \frac{1}{2}\left(\frac{1}{\mu_0}\|B\|^2 + \frac{1}{\varepsilon_0}\|D\|^2\right).$$

The constants of electric and magnetic permittivity have the values (in S.I. units) $\varepsilon_0 = (36\pi \cdot 10^9)^{-1}$ and $\mu_0 = 4\pi \cdot 10^{-7}$. Their product is c^{-2}, the inverse of the square of the velocity of light.

In material medium, conducting and isotropic, the state law has the same form but with constants $\varepsilon > 0$ and $\mu > 0$ of greater value. The number $(\varepsilon\mu)^{-1/2}$ is again equal to the velocity of propagation of plane waves in the medium. In media which are poor conductors (dielectrics) the state law is no longer linear. The isotropy manifests itself by the condition

$$W(RB, RD) = W(B,D), \qquad \forall R \in O_3(\mathbb{R}).$$

This implies the existence of a function w of three variables, such that

$$W(B, D) = w(\|B\|, \|D\|, B \cdot D).$$

Finally, paramagnetic bodies present phenomena of memory (with *hysteresis*), which do not come into the body of systems with conservation laws.

Plane waves

Henceforth, let us neglect the thermodynamic effects as well as the electric current. For a plane wave which is propagating in the x_1-direction we have $\partial_2 = \partial_3 = 0$, with the result that $\partial_t B_1 = \partial_t D_1 = 0$. There remain four equations, in which we write $x = x_1$, the unique space variable:

$$\partial_t B_2 - \partial_x \frac{\partial W}{\partial D_3} = 0, \qquad \partial_t B_3 + \partial_x \frac{\partial W}{\partial D_2} = 0,$$

$$\partial_t D_2 + \partial_x \frac{\partial W}{\partial B_3} = 0, \qquad \partial_t D_3 - \partial_x \frac{\partial W}{\partial B_2} = 0.$$

Let us look at the simple case in which W is a function of $\rho := (\|B\|^2 + \|D\|^2)^{1/2}$ only. Introducing the functions $y := B_2 + D_3 + \mathrm{i}(B_3 - D_2)$, $z := B_2 - D_3 + \mathrm{i}(B_3 + D_2)$, we have $y_t - (\varphi(\rho)y)_x = 0$, $z_t + (\varphi(\rho)z)_x = 0$. The polar coordinates (r, s, α, β), defined by $y = r \exp \mathrm{i}\alpha$ and $z = s \exp \mathrm{i}\beta$, enable us to simplify the system into

$$\alpha_t - \varphi(\rho)\alpha_x = 0, \qquad r_t - (\varphi(\rho)r)_x = 0,$$
$$\beta_t + \varphi(\rho)\beta_x = 0, \qquad s_t + (\varphi(\rho)s)_x = 0,$$

with the connection $2\rho^2 = r^2 + s^2$.

1.5 Magneto-hydrodynamics

Magneto-hydrodynamics (abbreviated as M.H.D.) studies the motion of a fluid in the presence of an electromagnetic field. As it is a moving medium, the field acts on the acceleration of the particles, while the motion of the charges contributes to the evolution of the field. This coupling is negligible in a great number of situations but comes into action in a Tokamak, a furnace with induction, or in the interior of a star.

The fluid is described by its density, its specific internal energy, its pressure, and its velocity. If no account is taken of the diffusion processes, we write the conservation laws of mass, of momentum, of energy and Faraday's law as follows:

$$\rho_t + \operatorname{div}(\rho v) = 0,$$
$$(\rho v_i)_t + \operatorname{div}(\rho v_i v) + \frac{\partial}{\partial x_i}\left(p + \frac{1}{2}\|B\|^2\right) - \operatorname{div}(B_i \cdot B) = 0, \quad 1 \leq i \leq 3,$$
$$\left(\rho\left(\frac{1}{2}\|v\|^2 + \varepsilon\right) + \frac{1}{2}\|B\|^2\right)_t + \operatorname{div}\left(\rho\left(\frac{1}{2}\|v\|^2 + \varepsilon\right) + pv + E \wedge B\right) = 0,$$
$$B_t + \operatorname{curl} E = 0.$$

We see from these equations that the magnetic field exerts a force on the fluid particles and contributes to the internal energy of the system. The fact that the electric field does not is the result of an approximation, the same as we made in disregarding Ampère's law.

There are two state laws: on the one hand $p = p(\rho, e)$, which always has the form $P = (\gamma - 1)\rho e$ for a perfect gas; on the other hand, $E = B \wedge v$. This expresses a local equilibrium: the acceleration of the particles taken individually is of the form $f + (E + v \wedge B)/m$ where m is the mass of a particle of unit charge and f is the force due to the binary interactions. As $m \ll 1$ and since the velocity of the fluid remains moderate,[12] $E + v \wedge B$ is very small.

[12] Under this hypothesis, the fluid is seen as a dielectric.

For a sharper description, we take account of the processes of diffusion: the viscosity, Fourier's law certainly, even Ohm's law:[13]

$$E = B \wedge v + \eta j + \chi(j \wedge B).$$

Finally we take Ampère's law into account, but we neglect in it the derivative $\partial_t E$ considering that E varies slowly in time:

$$j = \text{curl } B.$$

Each of the phenomena which we come to take into account is studied by adding one or several of the second order terms in the laws of conservation. Whether the factors such as η, χ, k, α and β can be considered as small or not depends on the scale of the problems studied.

Plane waves in M.H.D.

Again, we consider the solutions for which $\partial_2 = \partial_3 = 0$ and $\beta := B_1$ is constant. This behaviour is established when the initial condition satisfies it. In the sequel we write

$$z := v_1, \qquad w := (v_2, v_3), \qquad b := (B_2, B_3), \qquad x := x_1.$$

In Faraday's law $B_t + \text{curl } E = 0$, the component in the direction of x_1 and the compatibility condition $\text{div } B = 0$ are trivial. There remain seven equations in place of eight, which is correct since B_1 is no longer an unknown:

$$\partial_t \rho + \partial_x(\rho z) = 0,$$

$$\partial_t(\rho z) + \partial_x\left(\rho z^2 + p(\rho, e) + \frac{1}{2}\|b\|^2\right) = 0,$$

$$\partial_t(\rho w) + \partial_x(\rho z w - \rho b) = 0,$$

$$\partial_t\left(\rho\left(\frac{1}{2}z^2 + \frac{1}{2}\|w\|^2 + e\right) + \frac{1}{2}\|b\|^2\right) + \partial_x\left(\rho z\left(\frac{1}{2}z^2 + \frac{1}{2}\|w\|^2 + e\right)\right.$$
$$\left. + (p + \|b\|^2)z - \beta b \cdot w\right) = 0,$$

$$\partial_t b + \partial_x(zb - \beta w) = 0.$$

The system is simpler in lagrangian coordinates (y, t), defined by $dy = \rho(dx - z\,dt)$ – see §1.2. Denoting by $\tau = 1/\rho$ the specific volume, these equations are

[13] Which replaces the hypothesis $E = B \wedge v$.

transformed to

$$\tau_t = z_y,$$
$$z_t + \left(p(1/\tau, e) + \frac{1}{2}\|b\|^2\right)_y = 0,$$
$$w_t - \beta b_y = 0,$$
$$\left(\frac{1}{2}z^2 + \frac{1}{2}\|w\|^2 + e + \frac{1}{2}\|b\|^2\right)_t + \left((p + \frac{1}{2}\tau\|b\|^2)z - \beta b \cdot w\right)_y = 0,$$
$$(\tau b)_t - \beta w_y = 0.$$

A combination of these equations gives, for a regular solution, $e_t + pz_y = 0$ or again $e_t + p\tau_t = 0$, that is to say

$$S(\tau, e)_t = 0,$$

S being the thermodynamic entropy ($\theta\, \mathrm{d}S = \mathrm{d}e + p\, \mathrm{d}\tau$). The analogous calculation in eulerian variables yields the transport equation

$$(\partial_t + z\partial_x)S = 0,$$

which shows that ρS is an entropy, in the mathematical sense, of the model.

A simplified model of waves

Let us consider the system of plane waves of M.H.D. in eulerian variables to fix the ideas, with $\beta \neq 0$. It admits in general seven distinct velocities of propagation $\lambda_1 < \lambda_2 < \cdots < \lambda_7$ among which $\lambda_4 = z$, $\lambda_2 = z - \beta\rho^{-1/2}$, and $\lambda_6 = z + \beta\rho^{-1/2}$ (λ_2 and λ_6 are the speeds of the *Alfven waves*). The four remaining speeds are the roots of the quartic equation

$$((\lambda - z)^2 - c^2)((\lambda - z)^2 - \beta^2/\rho) = (\lambda - z)^2\|b\|^2/\rho,$$

$c = c(\rho, e)$ being the speed of sound in the absence of an electromagnetic field. However, when b vanishes, we have $\lambda_3 = \lambda_2$ and $\lambda_5 = \lambda_6$. This coincidence of two speeds and the non-linearity of the equations induce a resonance. For waves of small amplitude, this phenomenon can be described by an asymptotic development.

First of all, a choice of a galilean frame of reference allows the assumption that the base state u_0, constant, satisfies $w_0 = 0$ (we already have $b_0 = 0$) and $z_0\sqrt{\rho_0} = \beta_0$. We thus have $\lambda_2(u_0) = \lambda_3(u_0) = 0$: the resonance occurs along curves (in the physical plane) with small velocities. If $u - u_0$ is of the size $\varepsilon \ll 1$ this velocity is also of the order of ε, which leads to the change of the time variable

$s := \varepsilon t$, so $\partial_t = \varepsilon \partial_s$. The other hypotheses are

on the one hand $\rho = \rho_0 + \varepsilon \rho_1 + \cdots$, $z = z_0 + \varepsilon z_1 + \cdots$, $e = e_0 + \varepsilon e_1 + \cdots$, on the other hand $w = \sqrt{\varepsilon} \cdot (w^1(s, x) + \varepsilon w^2(s, t, x) + \cdots)$, $b = \sqrt{\varepsilon} \cdot (b^1(s, x) + \varepsilon b^2(s, t, x) + \cdots)$. We note that, although $\sqrt{\varepsilon}$ is great compared with ε, these hypotheses ensure that λ_2 and λ_3 are of the order of ε.

The examination of the terms of order ε in the conservation laws shows that ρ_1, z_1, e_1 and w^1 are explicit functions of b^1. Finally, the terms of order $\varepsilon^{3/2}$ in Faraday's law, averaged with respect to the slow variable t to eliminate b^2, furnish a system which governs the evolution of $U := b^1$:

$$\partial_t U + \sigma \partial_x (\|U\|^2 U) = 0, \tag{1.3}$$

where σ is a constant which depends only on (ρ_0, e_0). In this book, we shall copiously use the system (1.3) to illustrate the various theories, but we shall also make appeal to a slightly more general one:

$$\partial_t U + \partial_x (\varphi(\|U\|) U) = 0$$

where $\varphi \colon \mathbb{R}^+ \to \mathbb{R}$ is a given smooth function.

1.6 Hyperelastic materials

We shall consider a deformable solid body, which occupies, at rest, a reference configuration which is an open set $\Omega \subset \mathbb{R}^d$. We describe its motion by a mapping $(x, t) \mapsto (y, t)$, $\Omega \to \mathbb{R}^d$, where y is the position at the instant t of the particle which was situated at rest at x in the reference configuration. We define the velocity $v \colon \Omega \to \mathbb{R}^d$ and the deformation tensor $u \colon \Omega \to M_d(\mathbb{R})$ by

$$v = \frac{\partial y}{\partial t}, \qquad u_{\alpha j} = \frac{\partial y_\alpha}{\partial x_j}.$$

In the first instance we write the compatibility conditions

$$\partial_t u_{\alpha j} = \partial_j v_\alpha, \qquad \partial_k u_{\alpha j} = \partial_j u_{\alpha k}, \quad 1 \le \alpha, j, k \le d.$$

A material is said to be *hyperelastic* if it admits an internal energy density of the form $W(u)$ and if the forces due to the deformation derive from this energy (principle of *virtual work*):

$$f = (f_1, \ldots, f_d), \qquad f_\alpha = -\frac{\delta \mathscr{E}}{\delta y_\alpha}.$$

Here $\delta / \delta y$ denotes the variational derivative of $\mathscr{E}[y] := \int W(\nabla y)\, \mathrm{d}x$:

$$f_\alpha = \sum_{j=1} \partial_j \frac{\partial W}{\partial u_{\alpha j}}.$$

The fundamental law of dynamics is written

$$\partial_t v_\alpha = f_\alpha + g_\alpha,$$

where g represents the other forces, due to gravity or to an electromagnetic field (but here we do not consider any coupling). Finally, $U := (u, v)$ obeys a system of conservation laws of first order (for which $n = d(d+1)$)

$$\partial_t u_{\alpha j} = \partial_j v_\alpha, \quad 1 \le \alpha, j \le d,$$

$$\partial_t v_\alpha = \sum_{k=1}^{d} \partial_k \frac{\partial W}{\partial u_{\alpha k}} + g_\alpha, \quad 1 \le \alpha \le d.$$

These equations can be linear, when W is a quadratic polynomial, but this type of behaviour is not realistic. In fact, the energy is defined only for $u \in \mathrm{GL}_d(\mathbb{R})$ with $\det(u) > 0$ (the material does not change orientation), and must tend to infinity when the material is compressed to a single point:

$$\lim_{u \to 0_n} W(u) = +\infty.$$

The models of elasticity are thus fundamentally non-linear. Other restrictions on the form of W are due to the principle of *frame indifference*:

$$W(u) = W(Ru), \quad R \in \mathrm{SO}_d(\mathbb{R}), \tag{1.4}$$

and, if the material is isotropic,

$$W(u) = W(uR), \quad R \in \mathrm{SO}_d(\mathbb{R}). \tag{1.5}$$

From (1.4), there exists a function $w: S^+ \to \mathbb{R}$, on the cone S^+ of positive definite symmetric matrices such that

$$W(u) = w(u^{\mathrm{T}} u).$$

If, in addition, (1.5) holds, then the function $S \mapsto w(S)$ depends only on the eigenvalues of S.

We find an entropy of the system in writing the conservation of energy:

$$\partial_t \left(\frac{1}{2} \|v\|^2 + W(u) \right) = \sum_{\alpha, j=1}^{d} \partial_j \left(v_\alpha \frac{\partial W}{\partial u_{\alpha j}} \right).$$

The total mechanical energy $(v, u) \mapsto \frac{1}{2}\|v\|^2 + W(u)$ is not always convex.[14] However, it is in the 'directions compatible' with the constraint $\partial_k u_{\alpha j} = \partial_j u_{\alpha k}$. In other

[14] There are obstructions due to the invariances mentioned above and to the fact that W tends to infinity at 0 and at infinity. See [8] Theorem 4.8-1 for a discussion.

words, W is convex on each straight line $\bar{u} + \mathbb{R}z$ where z is of rank one (the *Legendre–Hadamard condition*). This reduced concept of convexity is appropriate for problems with constraints. In particular, the mechanical energy furnishes an *a priori* estimate. A constitutive law currently used is that of St Venant and Kirchhoff:

$$w(S) = \frac{1}{2}\lambda(\operatorname{Tr} E)^2 + \mu \operatorname{Tr}(E^2), \quad E := \frac{1}{2}(S - I_n).$$

On the other hand, other entropies do not have this convexity property; for all $k \leq d$

$$\partial_t(v \cdot u_k) = \partial_k\left(\frac{1}{2}\|v\|^2 - W(u)\right) + \sum_{j=1}^{d} \partial_j\left(u_{\alpha k}\frac{\partial W}{\partial u_{\alpha j}}\right).$$

Strings and membranes

More generally, we can consider a material for which $x \in \Omega$ (with $\Omega \subset \mathbb{R}^d$) but with $(y, t) \in \mathbb{R}^p$ with $p \geq d$. The case $p = d$ is that described above. When $p = 3$ and $d = 2$ it is a mater of a membrane or a shell, while $p \geq 2$ and $d = 1$ corresponds to a string. For a membrane or a string, the equations are the same as in the preceding paragraph, but the Greek suffixes go from 1 to p instead of from 1 to d. There are then $n = p(d+1)$ unknowns and just as many equations of evolution.

Let us look at the case of string: u is a vector and $W(u) = \varphi(\|u\|)$, because of frame indifference, φ being a state law. We have

$$\frac{\partial W}{\partial u_\alpha} = \frac{1}{r}\varphi'(r)u_\alpha, \qquad r := \|u\|,$$

with the result that $\mathrm{d}W$ is the product of φ' (called the *tension* of the string) by the unit tangent vector to it: $r^{-1}u$. There are four or six equations:

$$u_t = v_x, \qquad v_t = (r^{-1}\varphi'(r)u)_x + g.$$

1.7 Singular limits of dispersive equations

The systems of conservation laws which are presented here proceed from completely integrable dispersive partial differential equations. We take as an example the Korteweg–de Vries (KdV) equation

$$u_t + 6uu_x = u_{xxx}, \tag{1.6}$$

but there are others, of which the best known is the cubic non-linear Schrödinger equation.

Certain solutions of (1.6) are progressive periodic waves: they have the form $u = u(x - ct)$ with $u''' = 6uu' - cu'$, with the result that

$$\frac{1}{2}u'^2 = u^3 - \frac{1}{2}cu^2 - au - b,$$

where a and b are constants of integration. The triplet (a, b, c) defines a unique periodic solution (to within a translation) when the polynomial equation $P(X) := X^3 - \frac{1}{2}cX^2 - aX - b = 0$ has real roots: $u_1 < u_2 < u_3$. We then have $\min u(x) = u_1$ and $\max u(x) = u_2$.

What are of interest here are such periodic solutions of the KdV equation, which are, in first approximation, modulated by the slow variables $(s, y) := (\varepsilon t, \varepsilon x)$ with $0 < \varepsilon \ll 1$.

$$u^\varepsilon(x, t) = u_0(a(s, y), b(s, y), c(s, y); x - c(s, y)t) + \varepsilon u_1(s, y, x, t) + O(\varepsilon^2).$$

We require that u_1 and u_0 be smooth functions and that u_1 be almost periodic with respect to (x, t).

The choice of the parameters (a, b, c) is not the most practical from the point of view of calculations. We proceed to construct another set, with the aid of the expressions

$$i_1 := u, \quad i_2 := \frac{1}{2}u^2, \qquad i_3 := \frac{1}{2}u_x^2 + u^3.$$

These are invariants of the KdV equation in the sense that sufficiently smooth solutions[15] satisfy

$$\partial_t i_k + \partial_x j_k = 0, \quad 1 \le k \le 3,$$

with

$$j_1 = 3u^2 - u_{xx}, \qquad j_2 = 2u^3 + \frac{1}{2}u_x^2 - uu_{xx},$$

$$j_3 = \frac{1}{2}u_{xx}^2 + 6uu_x^2 - u_x u_{xxx} - 3u^2 u_{xx} + \frac{9}{2}u^4.$$

Let $(a_1, a_2, a_3) \in \mathbb{R}^3$ be a triplet such that there exists a function $w \in H^1(S^1)$, $S^1 = \mathbb{R}/\mathbb{Z}$, with

$$\int_0^1 w\, d\xi = a_1, \quad \int_0^1 \frac{1}{2}w^2\, d\xi = a_2, \quad \int_0^1 w^3\, d\xi < a_3.$$

Then the set $X(a_1, a_2, a_3)$ of the couples $(v, Y) \in H^1(S^1) \times (0, +\infty)$ such that

$$\int_0^1 v\, d\xi = a_1, \qquad \int_0^1 \frac{1}{2}v^2\, d\xi = a_2, \qquad \int_0^1 \left(v^3 + \frac{1}{2Y^2}v'^2\right) d\xi = a_3$$

[15] There is no interest in the question of smoothness here; let us say that it is does not cause trouble.

is not empty. It corresponds (via $(v, Y) \mapsto u(\cdot/Y)$) to the periodic functions of $H^1(\mathbb{R})$, the period not being fixed *a priori*, with the prescribed means

$$\langle u\rangle = a_1, \qquad \frac{1}{2}\langle u^2\rangle = a_2, \qquad \left\langle u^3 + \frac{1}{2}u'^2\right\rangle = a_3.$$

It can be shown without difficulty that the mapping

$$X(a_1, a_2, a_3) \to \mathbb{R}, \qquad (v, Y) \mapsto Y,$$

attains its lower bound (strictly positive), which is denoted by $S(a_1, a_2, a_3)$. An optimal pair (v, S) defines, via $u(x) := v(x/S)$, a progressive periodic wave of the KdV equation. In general, (v, S) is unique to within a translation, with the result that if a differential polynomial P is given, then the mean $\langle P(u, u_x, \ldots, \partial_x^m u)\rangle$ is perfectly determined and depends only on (a_1, a_2, a_3). Those which we shall need are the functions

$$J_k(\vec{a}) = \langle j_k(u)\rangle.$$

For example,

$$J_1(\vec{a}) = \langle 3u^2 - u_{xx}\rangle = \langle 3u^2\rangle = 6a_2.$$

The two other functions have much less explicit expressions, which involve elliptic functions.

Let us denote by $U(\vec{a}, x, t)$ 'the' periodic solution such that $\langle i_k\rangle = a_k$, U being of class $\mathscr{C}^\infty$ with respect to each of its five variables. The modulated solutions which we consider are written

$$u^\varepsilon(x, t) = U(\vec{a}(\varepsilon x, \varepsilon t); x, t) + \varepsilon u_1(\varepsilon x, \varepsilon t; x, t) + O(\varepsilon^2).$$

Our purpose is to determine the evolution of $\vec{a}$ as a function of (y, s). We write for that the conservation laws

$$\partial_t i_k[u^\varepsilon] + \partial_x j_k[u^\varepsilon] = 0, \quad 1 \le k \le 3.$$

In these, the terms of order ε^0 are absent because $(x, t) \mapsto U$ is an exact solution of the KdV equation. There remain

$$\partial_s i_k[U] + \partial_y j_k[U] + \partial_t \cdots + \partial_x \cdots = O(\varepsilon),$$

where the imprecise expressions are smooth and almost periodic in (x, t). We eliminate their derivatives in x or t by taking the mean in Bohr's sense (with respect to (x, t)) of this equality:

$$\partial_s\langle i_k[U]\rangle + \partial_y\langle j_k[U]\rangle = O(\varepsilon).$$

As the left-hand side does not depend on ε, there only remains

$$\partial_s \vec{a}_k + \partial_y J_k(\vec{a}) = 0, \quad 1 \leq k \leq 3, \tag{1.7}$$

which makes up a closed system of three conservation laws.

Remarks We do not have to use the solutions of (1.7) before the formation of shocks. In fact, if $\vec{a}$ is discontinuous along a curve, the asymptotics cannot be justified and the periodic solutions have to be replaced by more complicated, almost periodic solutions. The equations of modulation are then made up of $2p + 1$ conservation laws in place of three (see [61]).

The validity of the asymptotics is closely linked to the hyperbolicity of (1.7), which allows it to have local smooth solutions. This property has been studied by Levermore [63].

The invariants i_1, i_2, i_3 are only the first of a denumerable list $(i_k)_{k\geq 1}$, where i_k is a polynomial in $(u, u_x, \ldots, \partial_x^{k-2} u)$. The expressions $I_k(\vec{a}) := \langle i_k(U, \ldots, U^{(k-1)})\rangle$ are thus entropies of the system (1.7): $\partial_s I_k + \partial_y J_k = 0$. Other entropies exist, in particular

$$\partial_s S(a) - \partial_y(cS) = 0,$$

where $c = c(\vec{a})$ is the speed of the progressive wave U (see the book by Whitham [112]).

1.8 Electrophoresis

Electrophoresis is a procedure of separating ions in an aqueous solution, by means of an electromagnetic field. We refer the reader to the article by Fife and Geng [22] for more general models than that presented here.

The medium is one-dimensional ($d = 1$). The ions represent a negligible fraction of the total mass, with the result that we can suppose the solution to be at rest. Each kind of ion (there are $n + 1$) has density $u_i(x, t) \geq 0$ for $0 \leq i \leq n$. The unknown of the problem is $U := (u_0, \ldots, u_n)$. The flux of mass of the ion of the ith kind is

$$f_i := \mu_i z_i E u_i - d_i \partial_x u_i$$

where $\mu_i > 0$ is the mobility, z_i the charge and $d_i \geq 0$ the diffusivity (these numbers are constants). The electric current is thus

$$J = -z \cdot f = -\sum_{i=1}^{n} z_i f_i.$$

The electric field is given by Ampère's law $\beta \partial_x E = -z \cdot u$. When $0 < \beta \ll 1$, we are led to make the hypothesis of electric neutrality:

$$z \cdot u = \left(\sum_{i=1}^{n} z_i u_i \right) \equiv 0. \tag{1.8}$$

The conservation law of the ith type of ion is

$$\partial_t u_i + \partial_x f_i = 0.$$

We therefore deduce from (1.8) that $\partial_x (z \cdot f) = 0$, or $\partial_x J = 0$. In this context the current J is imposed by the experimentalist; it will, in general, be constant. It is a datum of the problem, which allows the expression of E as a function of U via

$$E \sum_{j=1}^{n} \mu_j z_j^2 u_j = \sum_{j=1}^{n} z_j d_j \partial_x u_j - J.$$

Finally the vector $(u_0, \dots, u_n)$ obeys the system of conservation laws

$$\partial_t u_i - \partial_x \frac{\mu_i z_i J u_i}{\sum_{j=0}^{n} \mu_j z_j^2 u_j} = \partial_x \left(\sum_{j=0}^{n} b_{ij} \partial_x u_j \right), \quad 1 \le i \le n, \tag{1.9}$$

where

$$b_{ij} = d_i \delta_i^j - \frac{\mu_i z_i z_j d_j u_i}{\sum_{k=0}^{n} \mu_k z_k^2 u_k}.$$

We notice that the above system is not completely parabolic since $z^{\mathrm{T}} B = 0$; this comes from the electric neutrality, which renders the unknowns dependent on each other. We obtain a system conforming more with the general body of this book in eliminating one of the unknowns, for example u_0, and writing the conservation laws for $u := (u_1, \dots, u_n)$.

Let us look at the example where $z_0 = 1$ and $z_i = -1$ otherwise. Then $u_0 = \sum_{i \ge 1}^{n} u_i$. If we neglect the diffusion of the ions ($d_i = 0$), the system becomes

$$\partial_t v_i + \partial_x \frac{\alpha_i v_i}{\sum_{k=1}^{n} v_k} = 0, \quad 1 \le k \le n,$$

where $\alpha_i = \mu_i J > 0$ and $v_k := (\mu_k + \mu_0) u_k$. This system has very strong properties, which render the study of the Cauchy problem easy (see Chapter 13). When diffusion is taken into account, the right-hand side has to be replaced by

$$\partial_x \left(\sum_{j=1}^{n} \beta_{ij} \partial_x u_j \right),$$

with $\beta_{ij} = b_{ij} + b_{i0}$. In the equi-diffusive case in which $d_i = D\mu_i$, with D a constant, we have

$$\beta_{ij} = D\mu_i\left(\delta_i^j + \frac{\mu_0 - \mu_j}{S}u_i\right), \quad S := \sum_{k=1}^{n}(\mu_0 + \mu_k)u_k.$$

It is easily shown (this is a variant of Gerschgörin's theorem) that the eigenvalues of β all lie in the right half-plane $\Re z > 0$, since $D > 0$ and $u_i > 0$ for all i. The system is then parabolic. It is one of the rare natural examples where the diffusion matrix is invertible.

Let us mention another procedure for the separation of the constituents of a mixture, which makes use of gravity and a temperature gradient in a column, which furnishes equations very close to those we have established here. This procedure, called *chromatography*, separates the solvents according to their molar masses.

2

Scalar equations in dimension $d = 1$

In this chapter we consider a scalar unknown function $u(x, t)$. The equation governing it is a conservation law, completed by an initial condition:

$$\left.\begin{aligned} &u_t + f(u)_x = 0, && x \in \mathbb{R}, t > 0, \\ &u(x, 0) = u_0(x), && x \in \mathbb{R}. \end{aligned}\right\} \tag{2.1}$$

2.1 Classical solutions of the Cauchy problem

A classical solution of the Cauchy problem is a solution of class $\mathscr{C}^1$ for $t > 0$, continuous for $t \geq 0$, which satisfies (2.1) pointwise. When u_0 is also of class $\mathscr{C}^1$, a classical solution is of class $\mathscr{C}^1$ for $t \geq 0$. To avoid the related phenomena of propagation with infinite speed, we suppose in addition that u_0 is bounded on $\mathbb{R}$.

The linear case

First of all let us examine the case in which f is given by the formula $f(u) = cu$, c being a constant. Then

$$\frac{\mathrm{d}}{\mathrm{d}t}(u(x + ct, t)) = (u_t + cu_x)(x + ct, t) = 0.$$

Thus, $t \mapsto u(x + ct, t)$ is a constant, with value $u_0(x)$. Replacing x by $x - ct$ we obtain

$$u(x, t) = u_0(x - ct)$$

for the unique solution of (2.1). For all initial data, there therefore exists one and only one solution which has the same regularity.

Non-linear case. The method of characteristics

We now abandon the linear hypothesis. The flux f is a given function of class $\mathscr{C}^\infty$. We write $c(u) = f'(u)$.

Let $u \in \mathscr{C}^1$ be a solution of the Cauchy problem. We define the *characteristic curves*, or simply the *characteristics*, in the band $\mathbb{R} \times [0, T]$ as the integral curves $t \mapsto (X(t), t)$ of the differential equation

$$\frac{\mathrm{d}X}{\mathrm{d}t} = c(u(X, t)).$$

In the linear case, c is constant with the result that the characteristics are *a priori* known straight lines. This is no longer true in the general case, where they depend on the solution itself. Let us calculate

$$\frac{\mathrm{d}}{\mathrm{d}t}u(X(t), t) = \frac{\mathrm{d}X}{\mathrm{d}t}u_x(X, t) + u_t(X, t) = (u_t + c(u)u_x)(X, t),$$

the last equality being the conservation law. Thus, u is constant along each characteristic, taking the value $u_0(y)$ where $(y, 0)$ is the base of the latter. It follows that the slope of this curve has the constant value $c(u_0(y))$. This is thus a straight line:

$$X(t) = y + tc(u_0(y)).$$

The method of characteristics, which considers the solution of (2.1) by leading to an algebraic equation, is therefore the following.

Being given $(x, t) \in \mathbb{R} \times [0, T]$, find y, a solution of the equation

$$y + tc \circ u_0(y) = x.$$

Then put $u(x, t) = u_0(y)$.

Let $F^t: \mathbb{R} \to \mathbb{R}$ be the function defined by $F^t(y) = y + tc \circ u_0(y)$. If u_0 is continuous, so also is $F^t(y)$. Since $F^t(\pm\infty) = \pm\infty$, the mean value theorem ensures the existence of a value y such that $F^t(y) = x$. But this non-linear equation can have several solutions, thus preventing the construction of a classical solution. We make that precise in the statement of the proposition below.

Proposition 2.1.1 *Let $u_0 \in \mathscr{C}^1(\mathbb{R})$ be, together with its derivative, bounded. We define $T^* = +\infty$ if $c \circ u_0$ is increasing,*

$$T^* = -\left(\inf \frac{\mathrm{d}}{\mathrm{d}x} c \circ u_0\right)^{-1}$$

otherwise. Then (2.1) possesses one and only one solution of class $\mathscr{C}^1$ in the band $\mathbb{R} \times [0, T^*)$ and does not possess one in any greater band $\mathbb{R} \times [0, T]$.

Proof Let $p = c \circ u_0$. We have $F^{t'} = 1 + tp' \geq 1 - t/T^* > 0$, for $t \in [0, T^*)$. The solution of $F^t(y) = x$ is then unique since F^t is strictly increasing. In addition, the implicit function theorem ensures that $(x, t) \mapsto y(x, t)$ is of class $\mathscr{C}^1$. Let us then verify that $u(x, t)$ defined by $u = u_0(y(x, t))$ is a solution of (1.1). First of all we have $y(x, 0) = x$ so that $u(x, 0) = u_0(x)$. Then, in differentiating the equation $F^t(y) = x$, we obtain $F^{t'}(y)y_t = -p(y)$ and $F^{t'}(y)y_x = 1$. Hence

$$F^{t'}(y(x,t))(u_t + c(u)u_x) = u_0'(y)F^{t'}(y)(y_t + c(u)y_x)$$
$$= u_0'(y)(-p(y) + c(u_0(y))) = 0.$$

■

Blow-up in finite time

When classical solutions are provided by the method of characteristics there is no other that can be constructed, at least for $0 \leq t < T^*$. Let us now show that that solution cannot be prolonged beyond that.

Let $T > 0$ be such that there exists a regular solution on $\mathbb{R} \times [0, T]$. We differentiate the quantity $v = c'(u)u_x$ along the characteristics. The following formula is obtained by differentiating the conservation law with respect to x.

$$\frac{\mathrm{d}}{\mathrm{d}t} v(X, t) = -c'(u)^2 u_x^2 = -v^2.$$

This is an equation of Ricatti type. If $T^* < \infty$, there exist values of y for which $p'(y) = c'(u_0(y))u_0'(y) < 0$. For these, the function $v(X(t), t)$ blows up at the time $-1/p'(y)$ because its initial value is $p'(y)$. More precisely, for $t < \max(T, T^*)$

$$\inf_{x \in \mathbb{R}} v(x, t) = (t + (\inf p')^{-1})^{-1}.$$

Thus, we have $T \leq -(\inf p')^{-1}$ which proves the proposition.

Remark The phenomenon of blow-up which we have just described is of non-linear origin since it does not occur in the linear case. Notice that the hypothesis $T^* < \infty$ supposes that c is not constant. The exercises 2.1 and 2.2 also describe the effects of the non-linearity.

2.2 Weak solutions, non-uniqueness

Classical solutions are not sufficient to resolve (1.1), so we turn to weak solutions, that is say solutions in the sense of distributions. This choice is consistent with the underlying physics of this type of problem. In particular, the most interesting solutions being piecewise continuous, we find indirectly from distributions the transmission conditions which are introduced, for example, into fluid mechanics.

To give a meaning to the conservation law, it is enough that u and $f(u)$ be distributions. Since f, in general, is not linear, we must suppose that u is a measurable function so that $f(u)$ is defined pointwise. We then shall say that u is a *weak solution* of the equation $u_t + f(u)_x = 0$ in an open set ω of $\mathbb{R}^2$ if $u \in L^1_{\text{loc}}(\omega)$, $f(u) \in L^1_{\text{loc}}(\omega)$ and if for every test function $\varphi \in \mathscr{D}(\omega)$, we have

$$\iint_\omega \left(u\frac{\partial \varphi}{\partial t} + f(u)\frac{\partial \varphi}{\partial x} \right) \mathrm{d}x\, \mathrm{d}t = 0.$$

We shall say that u is a *weak solution* of the Cauchy problem (2.1) in the band $Q = \mathbb{R} \times [0, T]$ if $u \in L^1_{\text{loc}}(Q)$, $f(u) \in L^1_{\text{loc}}(Q)$, and if for all test functions $\varphi \in \mathscr{D}(Q)$.

$$\iint_Q \left(u\frac{\partial \varphi}{\partial t} + f(u)\frac{\partial \varphi}{\partial x} \right) \mathrm{d}x\, \mathrm{d}t + \int_{\mathbb{R}} u_0(x)\varphi(x, 0)\, \mathrm{d}x = 0.$$

In the account given below, we consider the simple case in which $u \in L^\infty(Q)$, which ensures that $u \in L^1_{\text{loc}}(Q)$, $f(u) \in L^1_{\text{loc}}(Q)$, and which is consistent with the maximum principle which we shall establish. This choice is nevertheless not a natural one once we consider systems of conservation laws since the maximum principle is then an exception. In addition, the quantities which have a physical meaning are those that are involved in Green's formula, that is $\int u\, \mathrm{d}x$ (mass in a domain at a given instant) and $\int f(u)\, \mathrm{d}t$ (flux of mass across a boundary during a given time). We thus see that a *natural* space for u is $\mathscr{C}((0, T); L^1_{\text{loc}}(\mathbb{R}))$.

The reader should be able easily to verify that the notion of a weak solution extends that of a classical solution: every classical solution of (1.1) is also a weak solution.

The Rankine–Hugoniot condition

Let us consider a pair (u, q) of functions, piecewise continuous in the domain ω, whose line of discontinuity lies along a regular curve Γ, which separates ω into two connected components $\omega_\pm$. We assume that (u, q) is of class $\mathscr{C}^1$ in ω_- and in ω_+. Finally, we denote by $u_+(x, t)$ the limit of $u(y, s)$ when (y, s) tends to $(x, t) \in \Gamma$ and stays in ω_+. In the same way we define $q_+(x, t)$ then $u_-(x, t)$ and $q_-(x, t)$ along Γ, and we write $[h](x, t) = h_+(x, t) - h_-(x, t)$, the *jump across* Γ of each piecewise continuous function h.

We now wish to translate into simple terms the equation $u_t + q_x = 0$.

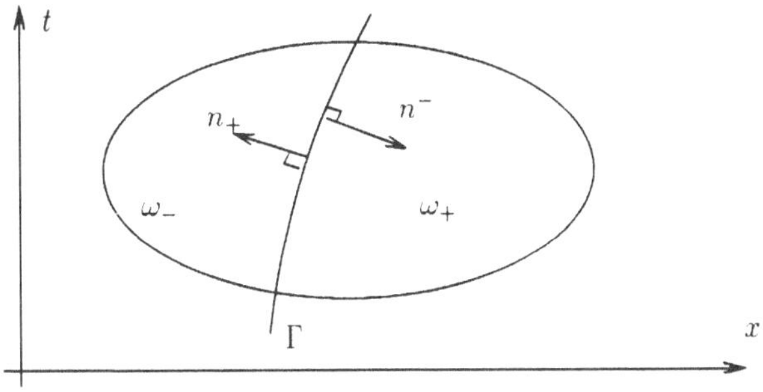

Fig. 2.1: Curves of discontinuity, unit normals.

Lemma 2.2.1 *Under the above hypothese, the pair (u, q) satisfy the equation in the distributional sense in ω if and only if*

(1) *On the one hand, u and q satisfy the equation pointwise in ω_+ and ω_-,*
(2) *On the other hand, the jump condition $[u]n_t + [q]n_x = 0$ is satisfied along Γ, where n is a unit normal vector to Γ in (x, t).*

Proof Let us begin with the necessary condition. Let (u, q) be a solution of the equation in ω. We have

$$\iint_\omega \left(u\frac{\partial \varphi}{\partial t} + q\frac{\partial \varphi}{\partial x} \right) \mathrm{d}x\, \mathrm{d}t = 0.$$

First of all choosing test functions whose support is in ω_-, we see that (u, q) is a weak solution in ω_-. In the same way we have the result for ω_+. With a general test function, we calculate the integral with the aid of Green's formula:

$$\begin{aligned} 0 &= \left(\iint_{\omega_-} + \iint_{\omega_+} \right) \left(u\frac{\partial \varphi}{\partial t} + q\frac{\partial \varphi}{\partial x} \right) \mathrm{d}x\, \mathrm{d}t \\ &= \int_{\partial\omega_-} (u_- n_t^- + q_- n_x^-)\varphi\, \mathrm{d}s - \int_{\omega_-} \varphi(u_t + q_x)\, \mathrm{d}x\, \mathrm{d}t \\ &\quad + \int_{\partial\omega_+} (u_+ n_t^+ + q_+ n_x^+)\varphi\, \mathrm{d}s - \int_{\omega_+} \varphi(u_t + q_x)\, \mathrm{d}x\, \mathrm{d}t. \end{aligned}$$

In the above formula, $\partial\omega_\pm$ denote the boundaries of the domains $\omega_\pm$, and $n_\pm$ are their unit normal vectors, pointing outwards (see Fig. 2.1). The preceding argument shows that the integrals over $\omega_\pm$ are null. On the other hand, the border of ω_- is made up of part of the boundary of ω on which φ is zero and also of Γ. The remaining part is therefore

$$\int_\Gamma (u_- n_t^- + q_- n_x^-)\varphi\, \mathrm{d}s + \int_\Gamma (u_+ n_t^+ + q_+ n_x^+)\varphi\, \mathrm{d}s = 0.$$

However, $n_- = -n_+$ along Γ, with the result that

$$\int_\Gamma ([u]n_t + [q]n_x)\varphi \, \mathrm{d}s = 0.$$

This equality being true for every smooth function (say, of class $\mathscr{C}^1$) in ω it is true when we replace φ by any smooth function, defined and with compact support on Γ. From this we deduce the jump condition (2).

Conversely, the same calculation, taken in the reverse sense, shows that these conditions are sufficient. ■

When $q = f(u)$, the jump condition is called the *Rankine–Hugoniot condition*. Let M be a Lipschitz constant of f in the larger interval $[a, b]$ in which u takes its values. We have $|[f(u)]| \leq M|[u]|$ with the result that $|n_t| \leq M|n_x|$. The curve of discontinuity can thus be parametrised by the variable t in the form

$$\Gamma = \{(X(t), t) : t \in]t_1, t_2[\}.$$

The Rankine–Hugoniot condition can then be written in the definitive form

$$[f(u)] = \frac{\mathrm{d}X}{\mathrm{d}t}[u].$$

We see again that the speed c plays a rôle like that in the method of characteristics, since provided that $[u]$ is not null, the mean value theorem is given by

$$\frac{\mathrm{d}X}{\mathrm{d}t} = c(\bar{u}(t))$$

where $\bar{u}(t)$ is a number between $u_-(X(t), t)$ and $u_+(X(t), t)$. In particular, when the amplitude $[u]$ of the discontinuity is weak, its speed approaches that of the neighbouring characteristics.

Examples (1) The simplest discontinuous solutions are of the form

$$u = \begin{cases} u_-, & x < \sigma t, \\ u_+, & x > \sigma t, \end{cases}$$

where $\sigma = (f(u_+) - f(u_-))/(u_+ - u_-)$. Indeed, u satisfies the equation trivially outside of the straight line $x = \sigma t$.

(2) For Burgers' equation $f(u) = \frac{1}{2}u^2$, the speed of propagation of the discontinuities is $\mathrm{d}X/\mathrm{d}t = \frac{1}{2}(u_+ + u_-)$.

(3) For the model of road traffic, this speed is a new concept, distinct from the speed of the vehicles. The discontinuities arise from the sudden variations in the density of the traffic, for example between traffic moving below a point and blocked

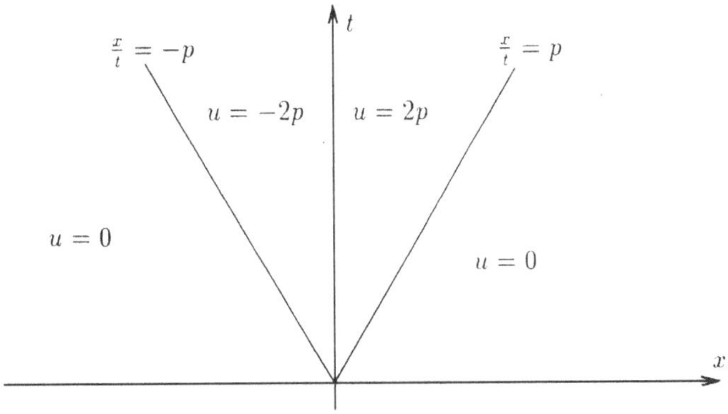

Fig. 2.2: Non-trival solution of a trivial Cauchy problem.

above. This point actually moves since it is nothing but $X(t)$ except if the flow of vehicles is the same below as above, the growth of the speed compensating exactly the diminution in the density.

Non-uniqueness for the Cauchy problem

To construct a Cauchy problem which admits more than one weak solution, we clearly must choose a non-linear flux. We choose the simplest example, the Burgers equation. If $u \equiv 0$, we have a trivial classical solution $u \equiv 0$. Here is another solution, using the example treated above (see Fig. 2.2):

$$u(x,t) = \begin{cases} 0, & x < -pt, \\ -2p, & -pt < x < 0, \\ 2p, & 0 < x < pt, \\ 0, & pt < x. \end{cases}$$

This example gives, in fact, an infinity of solutions of the same Cauchy problem, parametrised by the positive real number p. We can, for the moment, conclude that between the framework of the classical solution for which the existence is missing, and that of weak solutions whose uniqueness is not guaranteed, it is necessary to find an intermediate theory for which the Cauchy problem is well-posed in the sense of Hadamard, that is to say satisfies the following three conditions.

(1) For each given initial datum in a function space Y, the Cauchy problem admits a solution in a function space Z which is contained in $L^\infty_{\text{loc}}(\mathbb{R}; Y)$.
(2) That solution is unique.
(3) The mapping $Y \to Z$ which with a given initial datum associates a solution is continuous.

2.3 Entropy solutions, the Kružkov existence theorem

Approximate solutions; entropy inequalities

The examination of various models has suggested that the conservation laws are only a simplification of a more complex reality and, for example, should better be written

$$u_t + f(u)_x = \varepsilon u_{xx}. \tag{2.2}$$

Here the positive number ε is a diffusion coefficient, $\varepsilon \ll 1$. The Cauchy problem for (2.2) can be shown to have one and only one classical solution u^ε which satisfies the maximum principle. Let us assume also that the sequence $\{u^\varepsilon\}$ converges almost everywhere to a function u when $\varepsilon \to 0$ (this is proved under sufficiently general hypotheses [86], see also Theorem 5.4.1).

Lemma 2.3.1 *Suppose that $u_0 \in \mathscr{C}_b(\mathbb{R})$. If $u^\varepsilon(x,t) \to u(x,t)$ almost everywhere in $Q = \mathbb{R} \times (0,T]$ then u is a weak solution of (2.1).*

Proof As has been noted, u^ε is a classical solution, with the result that all the integrations by parts are admissible. In fact, $u^\varepsilon \in \mathscr{C}^\infty(Q) \cap \mathscr{C}(\bar{Q})$ and for every test function φ with support in $\mathbb{R} \times (-\infty, T)$

$$\begin{aligned} 0 &= \iint_Q \varphi\left(\varepsilon u^\varepsilon_{xx} - u^\varepsilon_t - f(u^\varepsilon)_x\right) \mathrm{d}x\, \mathrm{d}t \\ &= \iint_Q (u^\varepsilon(\varepsilon\varphi_{xx} + \varphi_t) + f(u^\varepsilon)\varphi_x)\, \mathrm{d}x\, \mathrm{d}t + \int_{\mathbb{R}} u_0(x)\varphi(x,0)\, \mathrm{d}x. \end{aligned}$$

From the maximum principle, and since $u_0 \in \mathscr{C}_b(\mathbb{R})$, the family $\{u^\varepsilon\}$ is bounded by a constant. The theorem of dominated convergence thus ensures that, when $\varepsilon \to 0$,

$$\iint_Q u^\varepsilon(\varepsilon\varphi_{xx} + \varphi_t)\, \mathrm{d}x\, \mathrm{d}t \to \iint_Q u\varphi_t\, \mathrm{d}x\, \mathrm{d}t$$

and

$$\iint_Q f(u^\varepsilon)\varphi_x\, \mathrm{d}x\, \mathrm{d}t \to \iint_Q f(u)\varphi_x\, \mathrm{d}x\, \mathrm{d}t.$$

Finally, we obtain the desired result

$$\iint_Q (u\varphi_t + f(u)\varphi_x)\, \mathrm{d}x\, \mathrm{d}t + \int_{\mathbb{R}} u_0(x)\varphi(x,0)\, \mathrm{d}x = 0.$$

■

The following calculus leads to a radically new procedure whose importance is such that it dominates the whole theory of hyperbolic systems of conservation laws: the entropy inequality, which is one of the means of recognising, from among the weak solutions, the solutions of physical origin. For a scalar equation, this criterion has the advantage, beyond taking account of a residual diffusion, of resolving an essential mathematical problem, the uniqueness of the Cauchy problem, while preserving the existence.

The concept of entropy, or rather the notion of the entropy–entropy-flux pair, refers to the pair of regular functions (E, F) defined on the space of the states u, for which every classical solution of $u_t + f(u)_x = 0$ also satisfies $E(u)_t + F(u)_x = 0$. For the moment, let us say that in the scalar case, every regular function E is an entropy of which the corresponding flux is given to within a constant by $F' = f'E'$. If E is convex, then u^ε satisfies

$$\begin{aligned} E(u^\varepsilon)_t + F(u^\varepsilon)_x &= E'(u^\varepsilon)\big(u^\varepsilon_t + f(u^\varepsilon)_x\big) = \varepsilon E'(u^\varepsilon)u^\varepsilon_{xx} \\ &= \varepsilon E(u^\varepsilon)_{xx} - \varepsilon E''(u^\varepsilon)\big(u^\varepsilon_x\big)^2 \le \varepsilon E(u^\varepsilon)_{xx}. \end{aligned}$$

Integrating this inequality, multiplied by a test function with positive values, over Q :

$$\begin{aligned} 0 &\le \iint_Q \varphi(\varepsilon E(u^\varepsilon)_{xx} - E(u^\varepsilon)_t - F(u^\varepsilon)_x)\,\mathrm{d}x\,\mathrm{d}t \\ &= \iint_Q (E(u^\varepsilon)(\varepsilon\varphi_{xx} + \varphi_t) + F(u^\varepsilon)\varphi_x)\,\mathrm{d}x\,\mathrm{d}t + \int_{\mathbb{R}} E(u_0(x))\varphi(x,0)\,\mathrm{d}x \\ &\to \iint_Q (E(u)\varphi_t + F(u)\varphi_x)\,\mathrm{d}x\,\mathrm{d}t + \int_{\mathbb{R}} E(u_0(x))\varphi(x,0)\,\mathrm{d}x, \end{aligned}$$

with the preceding hypotheses.

Proposition 2.3.2 *Under the hypotheses of Lemma 2.3.1, the solution u of (2.1) also satisfies the following inequalities, for all pairs (entropies $= E$, flux $= F$) with E continuous and convex:*

$$\iint_Q (E(u)\varphi_t + F(u)\varphi_x)\,\mathrm{d}x\,\mathrm{d}t + \int_{\mathbb{R}} E(u_0(x))\varphi(x,0)\,\mathrm{d}x \ge 0,$$

for all $\varphi \in \mathscr{D}(\mathbb{R} \times (-\infty, T))$, and $\varphi \ge 0$.

Proof It only remains to pass from convex entropies of class $\mathscr{C}^2$ to continuous convex entropies, and then defining what an entropy flux is in this context.

Let E be a convex function. Then E is locally a uniform limit of convex functions of class $\mathscr{C}^\infty$, for example functions $E_n = E * \rho_n$, convolution products by $\rho_n(s) = n\rho(ns)$ where $\rho \in \mathscr{D}(\mathbb{R})$. The convexity is assured by $\rho \ge 0$. Let F_n be the associated flux, fixed by

$$F_n(s) = \int_0^s f'(y)E_n'(y)\,\mathrm{d}y.$$

we have

$$F_n(s) = f'(s)E_n(s) - f'(0)E_n(0) - \int_0^s f''(y)E_n(y)\,\mathrm{d}y,$$

which shows that F_n converges locally uniformly to the continuous function

$$F(s) = f'(s)E(s) - f'(0)E(0) - \int_0^s f''(y)E(y)\,\mathrm{d}y.$$

The entropy inequality, stated in the proposition, is already true for the pairs (E_n, F_n). A new passage to the limit when $n \to \infty$ shows that it is still true for (E, F). ■

Definition 2.3.3 We say that a weak solution of (2.1) is an *entropy* (or *admissible*) solution if it satisfies the entropy inequalities for every convex continuous entropy E of flux F:

$$\iint_Q (E(u)\varphi_t + F(u)\varphi_x)\,\mathrm{d}x\,\mathrm{d}t + \int_{\mathbb{R}} E(u_0(x))\varphi(x, 0)\,\mathrm{d}x \geq 0 \tag{2.3}$$

for all $\varphi \in \mathscr{D}^+(\mathbb{R} \times (-\infty, T))$.

We note that taking into account that test functions may not take the value zero on $\mathbb{R} \times \{0\}$ is essential in the entropy inequalities. If, as in certain existing articles or books, we suppress the initial integral in (2.3) in restricting the entropy inequalities to test functions $\varphi \in \mathscr{D}^+(\mathbb{R} \times (0, T))$, we lose the property of uniqueness (see Theorem 2.3.5) in letting abnormal solutions continue to exist. A significant example is Exercise 2.12. As in the definition of a weak solution of the Cauchy problem, (2.3) expresses at the same time an initial condition, in the form of an inequality, and a partial differential relation in the open set $\mathbb{R} \times (0, T)$. In using the formalism of distributions and that of the traces of certain functional spaces, these two conditions can be written

$$\mathrm{tr}_{t=0} E(u) \leq E(u_0) = E(\mathrm{tr}_{t=0} u), \tag{2.4}$$

$$E(u)\varphi_t + F(u)\varphi_x \leq 0. \tag{2.5}$$

Let $k \in \mathbb{R}$, the function $u \mapsto |u - k|$ is convex and continuous, its flux being equal (to within a constant) to $(f(u) - f(k))\,\mathrm{sgn}(u - k)$, where

$$\mathrm{sgn}(s) = \begin{cases} 1, & s > 0, \\ 0, & s = 0, \\ -1, & s < 0. \end{cases}$$

An entropic solution thus satisfies the inequalities

$$\iint_Q (\varphi_t |u-k| + \varphi_x (f(u) - f(k)) \operatorname{sgn}(u-k)) \, dx \, dt \\ + \int_{\mathbb{R}} |u_0(x) - k| \varphi(x, 0) \, dx \geq 0. \tag{2.6}$$

Conversely, let us suppose that $u \in L^\infty$ satisfies (2.6). Let φ belong to $\mathscr{D}^+(\mathbb{R} \times (-\infty, T))$. The function u_0 and the solution u are assumed to take values in a bounded interval (a, b). For $k = a$, we have $|u - k| = u - a$ and $|u_0 - k| = u_0 - a$, with the result that

$$\iint_Q (\varphi_t u + \varphi_x f(u)) \, dx \, dt + \int_{\mathbb{R}} u_0(x) \varphi(x, 0) \, dx \\ \geq a \left(\iint_Q \varphi_t \, dx \, dt + \int_{\mathbb{R}} \varphi(x, 0) \, dx \right) + f(a) \iint_Q \varphi_x \, dx \, dt = 0$$

on integrating the second term by parts. Similarly, taking $k = b$, $|u - k| = b - u$ and $|u_0 - k| = b - u_0$, we obtain the inequality opposite to the preceding one. Thus for $\varphi \in \mathscr{D}^+(\mathbb{R} \times (-\infty, T))$

$$\iint_Q (\varphi_t u + \varphi_x f(u)) \, dx \, dt + \int_{\mathbb{R}} u_0(x) \varphi(x, 0) \, dx = 0.$$

By linearity, this is again true without sign condition on φ: u is a weak solution of (2.1).

Let us now show that u is indeed an entropic solution. Being given a convex continuous entropy E, of flux F, there exists for all $\alpha > 0$ an entropy E_α of flux F_α which satisfies

$E(s) \leq E_\alpha(s) \leq E(s) + \alpha$ for $s \in [a, b]$,
E_α is convex, piecewise affine: $E_\alpha(s) = b_0 + b_1 s + \Sigma_j a_j |s - k_j|$, with $a_j > 0$.

For that it is enough to interpolate E linearly on a sufficiently fine grid. Certainly, $F_\alpha(s) = b_1 f(s) + \Sigma_j a_j (f(s) - f(k_j)) \operatorname{sgn}(s - k_j)$. In making use of (2.6) and (2.1), we see that (2.3) is valid for E_α. As E_α and F_α converge uniformly to E and F on $[a, b]$ we can pass to the limit in the integrals, with the result that (2.3) is valid for E. Finally, we have

Proposition 2.3.4 *A bounded measurable function u on $\mathbb{R} \times (0, T)$ is an entropy solution of (2.1) if and only if it satisfies (2.6) for all $k \in \mathbb{R}$ and all $\varphi \in \mathscr{D}^+(\mathbb{R}^+ \times [0, T))$.*

Irreversibility

The definition 2.3.3 introduces the concept of irreversibility in the solution of (2.1). Previously, a weak solution u was reversible in the sense that the function v defined by $v(x,t) = u(-x, s-t)$ was also a weak solution in the band $\mathbb{R} \times (0, s)$ for the given initial function $v_0(x) =: u(-x, s)$.

Exercise Prove this result rigorously, that is to say with test functions.

On the other hand, the entropy inequalities change when we pass from u to v (verify this likewise), with the result that an entropy solution of (2.1) is not reversible, that is to say that v is an entropy solution only if the entropy inequalities are indeed equalities

$$\iint_Q (E(u)\varphi_t + F(u)\varphi_x)\,\mathrm{d}x\,\mathrm{d}t + \int_{\mathbb{R}} E(u_0(x))\varphi(x,0)\,\mathrm{d}x = 0,$$
$$\forall \varphi \in \mathscr{D}^+(\mathbb{R} \times (-\infty, T)).$$

We presume that the reversible solutions are in fact a little more regular than the others at least if the flux f is sufficiently non-linear, say if f'' is not identically zero. A weak form of this statement is found in §2.4.

However, in the linear case, we can show that there is an equivalence between the notions of weak solution and of entropy solution. A good method is to make use of the theorem of existence and uniqueness, 2.3.5 below. This states that if u_0 is a bounded, measurable function, then the entropy solution exists (it is in fact unique, but that does not play a part in this argument). This is a weak solution, but we show easily by duality (this is nothing but an application of the Hahn–Banach theorem) that the weak solution of (2.1) is unique in the linear case. Thus every weak solution is an entropy solution. In particular, since it is reversible, it satisfies entirely the entropy *equalities*. By linearity, these equalities hold without a convexity condition on the entropy and with no sign condition on the test function.

Existence and uniqueness for the Cauchy problem

The Cauchy problem is well-posed in the class of entropy solutions for the scalar conservation laws. Although Kružkov's result is valid for an equation with variable coefficients and in several space dimensions, we begin by enunciating a version which corresponds to the framework which we have adopted until now.

Theorem 2.3.5 (Kružkov [79]) *For every bounded measurable function u_0 on $\mathbb{R}$, there exists one and only one entropy solution of (2.1) in $L^\infty(Q) \cap \mathscr{C}([0, T);$*

$L^1_{loc}(\mathbb{R}))$. *It satisfies the maximum principle*

$$\|u\|_{L^\infty(Q)} = \|u_0\|_{L^\infty(\mathbb{R})}. \tag{2.7}$$

The theorem is, in fact, more complete than that, but so as not to overload the statement of the theorem, we have preferred to summarise below the principal properties of the solution.

Proposition 2.3.6 *Let u_0 and v_0 be two bounded measurable functions and u and v the entropy solutions associated with them. Let* $M = \sup\{|f'(s)|; s \in [\inf(u_0, v_0), \sup(u_0, v_0)]\}$. *Then:*

(P1) *For all $t > 0$ and every interval $[a, b]$, we have*

$$\int_a^b |v(x,t) - u(x,t)|\,dx \le \int_{a-Mt}^{b+Mt} |v_0(x) - u_0(x,t)|\,dx.$$

(P2) *In particular, if u_0 and v_0 coincide on $\{x: |x - x_0| < d\}$, then u and v coincide on the triangle $\{(x,t): |x - x_0| + Mt < d\}$.*

(P3) *If $u_0 - v_0 \in L^1(\mathbb{R})$, then $u(t) - v(t) \in L^1(\mathbb{R})$ (writing $u(t) := u(\cdot, t)$) and*

$$\|v(t) - u(t)\|_{L^1(\mathbb{R})} \le \|v_0 - u_0\|_{L^1(\mathbb{R})},$$

$$\int_{\mathbb{R}} (v(x,t) - u(x,t))\,dx = \int_{\mathbb{R}} (v_0(x) - u_0(x))\,dx.$$

(P4) *If $u_0 \in L^1(\mathbb{R})$, then $u(t) \in L^1(\mathbb{R})$, for all $t > 0$, and*

$$\|u(t)\|_{L^1(\mathbb{R})} \le \|u_0\|_{L^1(\mathbb{R})}, \qquad \int_{\mathbb{R}} u(x,t)\,dx = \int_{\mathbb{R}} u_0(x)\,dx.$$

(P5) *If $u_0(x) \le v_0(x)$ for almost all $x \in \mathbb{R}$, then also $u(x,t) \le v(x,t)$.*

(P6) *If $u_0 \in \mathrm{BV}(\mathbb{R})$, the space of functions of total bounded variation, then $u(t) \in \mathrm{BV}(\mathbb{R})$ and*

$$TV(u(t)) \le TV(u_0).$$

Comments (1) Theorem 2.3.5 allows us to construct an operator $S(t)$ which with a given initial value u_0 associates at the instant $t > 0$ the entropy solution $u(t)$. The family $(S(t))_{t \ge 0}$ is a semi-group because

the conservation law does not involve the time explicitly (if u is a solution, then $u^s := u(\cdot, \cdot + s)$ is also a solution, for the given initial value $u(s)$).

we verify easily that if v is the entropy solution for the initial value $u(s)$, then the function $\bar{u}$ defined by

$$\bar{u}(x,t) = \begin{cases} u(x,t), & t < s, \\ v(x, t-s), & t > s \end{cases}$$

is also an entropy solution of (2.1), with the result that $S(t+s)u_0 = S(t)S(s)u_0$, thanks to uniqueness.

(2) The property (P3) expresses the fact that $t \mapsto S(t)$ is a contraction semigroup in $L^1(\mathbb{R}) \cap L^\infty(\mathbb{R})$, with respect to the L^1-norm. However, the above results are much more general since $S(t)$ is defined on $L^\infty(\mathbb{R})$. We do not know of a non-decoupled system of at least two conservation laws which possess such a contraction property, even for a metric structure. It is suspected that it does not exist, which renders difficult the question of uniqueness for systems (cf. [104]).

(3) The property (P1) obviously contains the uniqueness property of the entropy solution, but it contains a more precise fact, plain from (P2): the value of the entropy solution u at a point (x,t) depends only on the restriction of u_0 to an interval $[x - Mt, x + Mt]$. A perturbation (sufficiently small) with compact support disjoint from this interval does not modify the value $u(x,t)$. We call the *domain of dependence* of (x,t) the smallest compact set K such that, for every bounded function a with compact support disjoint from K, the solution of the Cauchy problem with initial condition $v_0 =: u_0 + \varepsilon a$ coincides with u at (x,t) for ε sufficiently small. The domain of dependence of (x,t) is thus included in $[x - Mt, x + Mt]$, but this is not necessarily an interval once shocks are developed, for the latter create shadow zones. More generally we can define the domain of dependence of (x,t) at the instant $t_0 < t$ by considering the Cauchy problem with a given initial condition at the time t_0. A symmetrical notion is that of the *domain of influence* of a point (x_0, t_0), made up of points (x,t) with $t > t_0$ of which (x_0, t_0) belongs to the domain of dependence.

(4) The property (P4) is a trivial consequence of (P3), which leads also to (P5) by the following calculation:

$$\begin{aligned} \|v(t) - u(t)\|_{L^1(\mathbb{R})} \le \|v_0(t) - u_0(t)\|_{L^1(\mathbb{R})} &= \int_{\mathbb{R}} (v_0(x) - u_0(x))\,\mathrm{d}x \\ &= \int_{\mathbb{R}} (v(x,t) - u(x,t))\,\mathrm{d}x, \end{aligned}$$

that is to say $v - u \ge 0$.

(5) Similarly, (P3) implies (P6). If $u_0 \in \mathrm{BV}(\mathbb{R})$, then $u_0(\cdot + h) - u_0$ is integrable for all $h \in \mathbb{R}$ and

$$\mathrm{TV}(u_0) = \lim_{h \to 0} h^{-1} \int_{\mathbb{R}} |u_0(x+h) - u_0(x)|\,\mathrm{d}x.$$

Thus, $u(\cdot + h, t) - u(\cdot, t)$, is integrable, and

$$\begin{aligned}\mathrm{TV}(u(t)) &= \lim_{h\to 0} h^{-1} \int_{\mathbb{R}} |u(x+h,t) - u(x,t)|\,\mathrm{d}x \\ &\le \lim_{h\to 0} h^{-1} \int_{\mathbb{R}} |u_0(x+h) - u_0(x)|\,\mathrm{d}x \\ &= \mathrm{TV}(u_0).\end{aligned}$$

(6) The monotonic property (P5) implies that to a monotonic given initial function there corresponds a solution having the same monotonic property with respect to x.

(7) Kružkov's theorem is in reality of much more general power. It applies to scalar equations with $d \ge 2$ space variables when the fluxes depend explicitly on the space and time variables and in the presence of a source term:

$$\partial_t u + \sum_{1\le\alpha\le d} \partial_{x_\alpha} f^\alpha(x,t,u) = g(x,t,u), \quad x \in \mathbb{R}^d \tag{2.8}$$

The sole hypothesis, except the regularity of f and g is that $g - \mathrm{div}_x f$ is uniformly bounded with respect to $x \in \mathbb{R}^d$. There exists one and only one entropy solution $u \in L^\infty(Q) \cap \mathscr{C}([0,T); L^1(\mathbb{R}^d))$ of the Cauchy problem for (2.8), that is to say satisfying for every positive test function φ

$$\begin{aligned}0 \le &\iint_{(0,T)\times\mathbb{R}^d} (|u-k|\varphi_t + \nabla_x\varphi\cdot(f(x,t,u) - f(x,t,k))\,\mathrm{sgn}(u-k))\,\mathrm{d}x\,\mathrm{d}t \\ &+ \int_{\mathbb{R}^d} |u_0 - k|\varphi(x,0)\,\mathrm{d}x \\ &+ \iint_{(0,T)\times\mathbb{R}^d} (g(x,t,u) - (\mathrm{div}_x f)(x,t,k))\varphi\,\mathrm{sgn}(u-k)\,\mathrm{d}x\,\mathrm{d}t.\end{aligned}$$

The properties (P1) to (P6) remain true in this context provided that $g \equiv 0$ and that f depends only on u, with the following adaptation for (P1). If ω is a bounded open set of $\mathbb{R}^d$

$$\int_\omega |v(x,t) - u(x,t)|\,\mathrm{d}x \le \int_{\omega+B(0;Mt)} |v_0(x) - u_0(x)|\,\mathrm{d}x.$$

In addition, f being vector-valued, $|f'|$ denotes the norm of f' in the definition of M. The statements remain valid whatever the norm chosen.

Application: admissible discontinuities

Piecewise smooth entropy solutions

Since the entropy inequalities are trivial for classical solutions, it is important to understand them for less smooth solutions. The simplest, best understood and most useful case is that of piecewise smooth solutions. This case is placed in the same class of solutions as §2.2. We have seen that the curve of discontinuity is parametrised by the time: $t \mapsto X(t)$. We fix naturally ω_- and ω_+ by $x < X(t)$ and $x > X(t)$ respectively. If $\varphi \in \mathscr{D}^+(\omega)$ and if E is a convex entropy of class $\mathscr{C}^1$ of flux F,

$$\begin{aligned}
0 &\leq \iint_\omega (\varphi_t E(u) + \varphi_x F(u))\,\mathrm{d}x\,\mathrm{d}t \\
&= \left(\iint_{\omega_-} + \iint_{\omega_+}\right)(\varphi_t E(u) + \varphi_x F(u))\,\mathrm{d}x\,\mathrm{d}t \\
&= -\left(\iint_{\omega_-} + \iint_{\omega_+}\right)(E(u)_t + F(u)_x)\varphi\,\mathrm{d}x\,\mathrm{d}t + \int_\Gamma ([E(u)]n_t^- + [F(u)]n_x^-)\varphi\,\mathrm{d}s \\
&= \int_\Gamma ([E(u)]n_t^- + [F(u)]n_x^-)\varphi\,\mathrm{d}s.
\end{aligned}$$

As φ is an arbitrary positive function, we deduce that along Γ

$$[E(u)]n_t^- + [F(u)]n_x^- \geq 0,$$

that is

$$[F(u)] \leq \frac{\mathrm{d}X}{\mathrm{d}t}[E(u)]. \tag{2.9}$$

By continuity, we deduce also the following inequality:

$$[(f(u) - f(k))\,\mathrm{sgn}(u-k)] \leq \frac{\mathrm{d}X}{\mathrm{d}t}[|u-k|], \quad k \in \mathbb{R}. \tag{2.10}$$

It is easy to make the reverse argument. First of all, (2.10) and the Rankine–Hugoniot condition imply (2.9) (indeed, we see that (2.10) leads to the Rankine–Hugoniot condition). Then we show that a piecewise smooth function, which is a classical solution outside of Γ and which satisfies (2.9) along Γ, is an entropy solution. Finally we obtain the following result.

Proposition 2.3.7 *Let u be a piecewise $\mathscr{C}^1$ function in Q, whose discontinuities are carried by the union Γ of Lipschitz curves, pairwise disjoint. Then u is an entropy solution of (2.1) if and only if u is a classical solution outside of Γ and satisfies (2.10) on Γ.*

Oleĭnik's condition

Now let us analyse the condition (2.10) in detail. We can fix a point of Γ and suppose that $u_+ \neq u_-$. Let I be the interval with extremities u_- and u_+. In choosing k outside of I, we obtain successively two inequalities which, together, express the Rankine–Hugoniot condition. We thus have

$$\frac{\mathrm{d}X}{\mathrm{d}t} = \frac{[f(u)]}{[u]}.$$

Finally we take $k \in I$, that is to say $k = au_- + (1-a)u_+$, $a \in [0, 1]$. Then

$$\left(f(u_+) + f(u_-) - 2f(k) - \frac{[f(u)]}{[u]}(u_+ + u_- - 2k) \right) \operatorname{sgn}(u_+ - u_-) \leq 0.$$

But $u_+ + u_- - 2k = (2a-1)(u_+ - u_-)$, with the result that

$$\begin{aligned} &(af(u_-) + (1-a)f(u_+) - f(au_- + (1-a)u_+)) \\ &\quad \times \operatorname{sgn}(u_+ - u_-) \leq 0, \quad a \in [0, 1]. \end{aligned} \tag{2.11}$$

There are therefore two cases, according to the sign of $u_+ - u_-$.

Case $u_- < u_+$ then a discontinuity is admissible if and only if the graph of f, restricted to the interval $[u_-, u_+]$, is situated above its chord.

Case $u_- > u_+$ then a discontinuity is admissible if and only if the graph of f, restricted to the interval $[u_+, u_-]$, is situated below its chord.

Examples 2.3.8 If f is convex, its graph is always below its chord; a discontinuity is thus admissible if and only if $u_+ < u_-$.

If f is concave, its graph is always above its chord; a discontinuity is thus admissible if and only if $u_- < u_+$.

In the general case, an admissible discontinuity is reversible if and only if f is affine between u_- and u_+.

Shocks

An important consequence of (2.11) is *Lax's inequality* (also called *Lax's shock condition*). On dividing (2.11) by $a|u_+ - u_-|$ (respectively $(1-a)|u_+ - u_-|$), then on making a tend to 0 (respectively to 1), we obtain

$$f'(u_+) \leq \frac{\mathrm{d}X}{\mathrm{d}t} \leq f'(u_-). \tag{2.12}$$

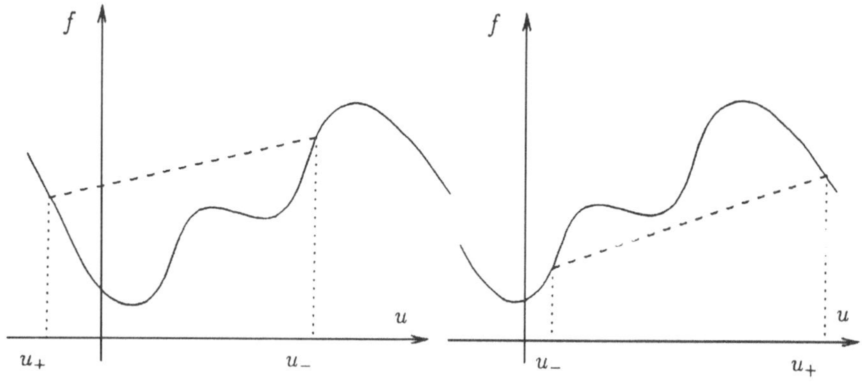

Fig. 2.3: Admissible discontinuities.

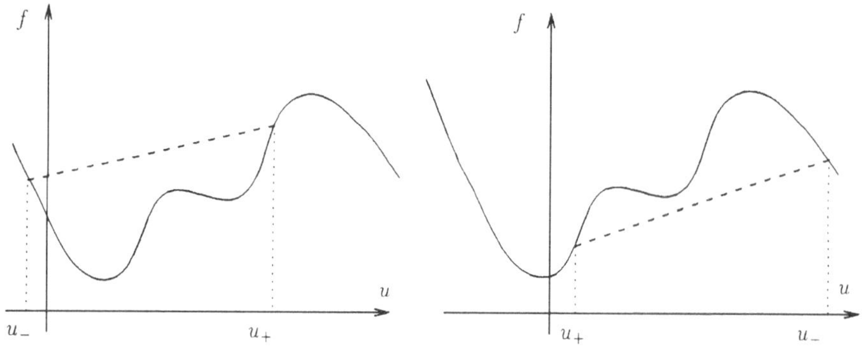

Fig. 2.4: Inadmissible discontinuities.

These inequalities express that the characteristics, which are straight lines in ω_+ or ω_-, defined, if need be, up to Γ, usually cannot emerge from Γ. A borderline case is the one of a tangentially emerging characteristic. In most cases, these inequalities are strict, for example if f is strictly convex or concave. In this favourable case, the characteristics can only penetrate into Γ coming from the past and not leaving towards the future. More precisely, from a point of Γ two characteristics and only two can be drawn, both directed towards the past. In addition, not being able to encounter another discontinuity by going back in time, they end up at a point $(x_0, 0)$. This reasoning is still correct if f'' is of constant sign, a characteristic then being able to coincide with Γ. Finally:

Proposition 2.3.9 *If f is concave or convex, the entropy solution of (2.1) can be calculated by the method of characteristics.*

We shall see below that in this case, there exists an explicit formula, due to Lax ([59]), giving the unique entropy solution of (2.1). For a more general flux, there does not always exist such a formula, in particular the method of characteristics does not work because these do not necessarily return to the initial point. In fact, there exist admissible discontinuities for which for example $f'(u_+) = \mathrm{d}X/\mathrm{d}t$ while $f''(u_+) \neq 0$. We can then construct a Cauchy problem and its entropy solution of which a characteristic emerges from Γ towards the future (see Exercise 2.9).

Definition 2.3.10 We shall say that a discontinuity $(u_-, u_+, \sigma = \mathrm{d}X/\mathrm{d}t)$ is a *shock* if the inequalities of (2.12) are strict. We shall say that it is a *semi-characteristic shock* (to the left or to the right according to the side on which the characteristics are tangents) if one is strict and the other is an equality. Finally we shall say it is a *characteristic shock* if the two are equalities without f being affine from u_- to u_+. If f is affine between u_- and u_+, we shall speak of a *contact discontinuity*.

One of the principal properties of shocks is to induce a little more smoothness, because it propagates along the characteristics. For example if u is of class $\mathscr{C}^1$ to the left of Γ, with a continuous limit u_- on this side, if Γ is of class $\mathscr{C}^1$ and if $f'(u_-) \neq \mathrm{d}X/\mathrm{d}t$, then this continuous extension is of class $\mathscr{C}^1$ up to Γ. We shall see in an exercise that this cannot be true in a semi-characteristic shock to the left.

2.4 The Riemann problem

Self-similar solutions. Rarefactions

The *Riemann problem* is the Cauchy problem in the particular case of a given initial condition of the form

$$u_0(x) = \begin{cases} u_{\mathrm{L}}, & x < 0, \\ u_{\mathrm{R}}, & x > 0. \end{cases}$$

The rôle of the Riemann problem is to furnish all the solutions of the Cauchy problem which are invariant under the group of homotheties $(x, t) \mapsto (ax, at)$, a group which leaves invariant all the conservation laws of the first order. More precisely, let $a > 0$. If u_0 is as above, and if u is the corresponding entropy solution, then $u^a := u(ax, at)$ is the entropic solution for the given $u_0(ax)$, here $u_0(x)$. From Theorem 2.3.5, this solution is unique: $u^a \equiv u$. Choosing $a = t^{-1}$, we obtain $u(x, t) = u(x/t, 1)$. We must therefore seek the solution of the Riemann problem among the self-similar solutions of the form $u(x, t) = v(x/t)$ with $v \in L^\infty$. The system (2.1) then reduces, in the sense of distributions, to

$$\left.\begin{aligned} f(v)_\xi &= \xi v_\xi, \\ v(-\infty) &= u_{\mathrm{L}}, \\ v(+\infty) &= u_{\mathrm{R}}. \end{aligned}\right\} \tag{2.13}$$

Since we seek the entropic solution, we have also

$$[(f(v) - f(k))\mathrm{sgn}(v - k)]_\xi \leq -\xi |v - k|_\xi .$$

The sense to give to the conditions at infinity for v is trivial, for because of property (P1) of propagation with finite speed, we see that $u \equiv u_L$ for $x < Mt$. Here, $M = \sup\{|f'(s)|: s \in I(u_L, u_R)\}$ where $I(u_L, u_R)$ is the interval with ends u_L and u_R. In fact we therefore have $v(\xi) = u_L$ for $\xi < -M$; similarly $v(\xi) = u_R$ for $\xi > M$.

As to the general Cauchy problem, we must first of all consider the smooth solutions of (2.13). But we must take care that these lead to solutions of (2.1) that are singular at the origin.

If $v \in \mathscr{C}(\mathbb{R})$ is piecewise of class $\mathscr{C}^1$, we develop (2.13): $(c(v)-\xi)(\mathrm{d}v/\mathrm{d}\xi) = 0$. This equality is trivial in certain zones, for example for $|x| > M$ since v is constant there. Where $v' \neq 0$, we have $c(v(\xi)) = \xi$, which leads to the following definition.

Definition 2.4.1 A rarefaction wave is a self-similar solution $u(x, t) = v(x/t)$ of class $\mathscr{C}^1$ in a wedge $at < x < bt$, which thus satisfies $c(u(x, t)) = x/t$.

In a rarefaction wave, v is injective, thus monotonic, and takes its values in an interval in which c is monotonic, that is to say where f is either convex or concave. For example, if f is convex on $\mathbb{R}$, the rarefaction waves are increasing.

Another type of self-similar wave has already been encountered in the preceding, it consists of shocks. These are defined by $v(\xi) = v_-$ for $\xi < \sigma$, $v(\xi) = v_+$ for $\xi > \sigma$, v_-, v_+, and σ being linked by the Rankine–Hugoniot condition and satisfying Oleĭnik's condition.

Thus we have three kinds of self-similar solutions: the rarefaction waves, the shocks and the constants. The solution of the Riemann problem does not make use of any other.

The solution of the Riemann problem

In the Riemann problem, the initial datum is monotonic, with the result that the solution u is also monotonic with respect to the space variable. The same is true for $\xi \mapsto v(\xi) = u(\xi, 1)$, which thus has, at the most, a denumerable number of discontinuities. These will be the shocks. The construction of the solution is the following. For $u_L = u_R$, the solution is constant. If not, we denote by χ the characteristic function of the interval $I(u_L, u_R)$ with values 0 and ∞.

Case $u_L < u_R$ Let $g = \sup\{h \text{ convex}: h \leq f + \chi\}$. Then $d := g'$, defined on $[u_L, u_R]$, is increasing and we put

$$d(v(\xi)) = \xi, \qquad \xi \in [d(u_L), d(u_R)].$$

This formula defines v in a unique manner except on the set of critical values of d which is denumerable. For $\xi < d(u_L)$, we put $v = u_L$, while for $\xi > d(u_R)$ we put $v = u_R$.

Case $u_L > u_R$ Here, g is the smallest of the concave functions which bound $f - \chi$ above. Its derivative d, defined on $[u_R, u_L]$ is decreasing and therefore allows us to define v on $[d(u_L), d(u_R)]$ by $d(v(\xi)) = \xi$.

Let us show that this construction solves (2.13) as well as satisfying the entropy inequalities. By symmetry it is enough to consider the case $u_L < u_R$. For $\xi = x/t \in [d(u_L), d(u_R)]$, we have $f(v) = g(v)$ almost everywhere since f and g coincide except on the critical values of d. For every $s \in [u_L, u_R]$, we thus have

$$f(s) \geq g(s) \geq g(v) + \xi(s - v) = f(v) + \xi(s - v), \tag{2.14}$$

the second inequality being due to the convexity of g.

Choosing $s = v(\xi + a)$, $a \neq 0$ in (2.14), then dividing by $|a|$ and letting a tend to zero through positive and negative values respectively, we obtain (2.13). The entropy inequality is thus satisfied for $k \leq u_L$, and for $k \geq u_R$ $u_L \leq v \leq u_R$.

If $u_L < k < u_R$, the monotonicity of v ensures the existence of a real number ξ_0 such that $v(\xi) \leq k$ for $\xi < \xi_0$ and $v(\xi) \geq k$ for $\xi > \xi_0$. Let us define $w := (f(v) - f(k))\mathrm{sgn}(v - k) - \xi|v - k|$. Owing to (2.13), we have $w_\xi = -|v - k|$ on the open set $\mathbb{R} - \{\xi_0\}$. To deduce the entropy inequality $w_\xi + |v - k| \leq 0$, it is enough to establish Oleĭnik's inequality $[w] \leq 0$ at ξ_0, which comes from (2.14) when we choose $\xi = \xi_0 + a$, $s = k$ with $a > 0$ and we make a tend to zero.

2.5 The case of f convex. The Lax formula

If the flux is strictly convex, Lax's inequality for the discontinuities is strict ($f'(u_+) < \sigma < f'(u_-)$) and it reduces to $u_- > u_+$. As they cannot be tangent to a shock curve, the characteristics are not able to emerge. For a piecewise smooth solution, we see that the characteristic passing through a point (x_0, t_0) is a straight line which exists at least in the past on the time interval $[0, t_0]$. In fact, it is limited only by its encounter with a shock, which can only be produced in the future, owing to Lax's inequality. If (x_0, t_0) is on a shock curve, there are two such characteristics, one to each side of the shock; at one such point, the calculation of the characteristics furnishes two results, which correspond to the two degrees of freedom, the value of u and the shock speed. Otherwise, there is only a single characteristics.

The Hamilton–Jacobi equation

Let u be the entropy solution of (2.1). This conservation law is the compatibility condition which ensures the existence of a function v satisfying $v_x = u$, $v_t = -f(u)$.

Since u is measurable and bounded, v is Lipschitz and satisfies the above equations almost everywhere. We thus have $v_t + f(v_x) = 0$ almost everywhere; this is a particular case of the Hamilton–Jacobi equation.

Since f is convex, we have for all $s \in \mathbb{R}$.

$$v_t \leq f'(s)(s - v_x) - f(s),$$

which has first integral

$$v(y + f'(s)t, t) \leq v_0(y) + t(sf'(s) - f(s)), \quad s, y \in \mathbb{R}, \quad t > 0. \quad (2.15)$$

As $s \to f'(s)$ is strictly increasing, we can make the change of variables $(y, s, t) \mapsto (y, x, t)$ with $x = y + tf'(s)$. We derive $s = b((x - y)/t)$ where b is the derivative of the convex conjugate function $g = f^*$ of f for $g' \circ f' = \mathrm{id}$. We then have $sf'(s) - f(s) = g((x - y)/t)$. On minimising the right-hand side of (2.15) with respect to y keeping (x, t) fixed we therefore obtain the inequality

$$v(x, t) \leq V(x, t) =: \inf_{y \in \mathbb{R}} (v_0(y) + tg((x - y)/t)). \quad (2.16)$$

Let $(x, t) \in \mathbb{R} \times \mathbb{R}^+$ and let us choose a characteristic $\tau \mapsto X(\tau)$ which ends on (x, t). We have seen that it is defined on $[0, t]$; we write $z = X(0)$. The calculation below, made along the characteristic, shows that the inequality (2.16) is optimal. A rigorous justification of this calculation for every entropy solution will be found in [60].

$$\frac{\mathrm{d}}{\mathrm{d}\tau} v(X, \tau) = v_t + c(u)v_x = c(u)u - f(u) = g(c(u)).$$

We have also $\mathrm{d}X/d\tau = c(u)$, which is constant. Finally,

$$v(x, t) = v_0(z) + tg(c(u)) = v_0(z) + tg((x - z)/t).$$

Theorem 2.5.1 (Lax [59]) *If f is strictly convex and $u_0 \in L^\infty(\mathbb{R})$, then the entropy solution of (2.1) is given by $u = v_x$ where*

$$v(x, t) = \sup_{y \in \mathbb{R}} \left(v_0(y) + tf^* \left(\frac{x - y}{t} \right) \right),$$

v_0 being a primitive of u_0.

A dual formula to Lax's

Lax's formula is rendered possible because in the case in which f is convex, we have seen that the characteristics all issue from the axis $t = 0$. For a general flux f, this is no longer possible because characteristics can originate in a semi-characteristic shock. However, it will be seen in Exercise 2.9 that this cannot happen if the initial datum is monotonic. We thus have in this case an explicit formula, which is the dual of that of Lax in the sense that one is of the form $v(x,t) = \inf_{y\in\mathbb{R}} \sup_{s\in\mathbb{R}}(\ldots)$ while the other is written $v(x,t) = \sup_{s\in\mathbb{R}} \inf_{y\in\mathbb{R}}(\ldots)$.

Proposition 2.5.2 (Kunik [57]) *Let u_0 be an increasing given initial function, the flux f being an arbitrary smooth function. Then $u = v_x$ where v is given by the formula*

$$v(x,t) = \sup_{s\in\mathbb{R}}(sx - tf(s) - v_0^*(s)).$$

Here, v_0 is a primitive (convex by hypothesis) of u_0 and v_0^* is its convex conjugate function.

For a discussion of this formula, the reader should see [92].

2.6 Proof of Theorem 2.3.5: existence

The approach by semi-groups

An original proof of Kružkov's theorem, due to Crandall [14], is based on the particular properties of the scalar case, notably the fact that the solution of the Cauchy problem furnishes a semi-group of contractions in $L^1 \cap L^\infty(\mathbb{R})$, which is described by the properties (P3) and (2.7). In the one-dimensional case, which is ours, we can again take advantage of the order structure on $\mathbb{R}$ to have the shortest of the proofs of existence of an entropy solution of (2.1).

We begin by considering a simplified case, that in which, as well as u_0 being merely integrable on $\mathbb{R}$, the flux f is uniformly monotonic ($\inf\{f'(s): s \in \mathbb{R}\} > 0$) and satisfies $f'' = 0$ outside a compact set. We define an unbounded operator A in the following manner.

Domain of A: $D(A) = W^{1,1}(\mathbb{R}) = \{v \in L^1(\mathbb{R}),\ v_x \in L^1(\mathbb{R})\}$.
Graph of A: $Av = f(v)_x, \quad \forall v \in D(A)$.

As f is Lipschitz, $v \mapsto f(v)$ is a continuous operator of $L^1(\mathbb{R})$ into itself, with the result that the graph of A is closed (we say that A is *closed*). The construction of an entropy solution of (2.1) is made by approaching the conservation law by a difference equation. We seek a solution $u_\varepsilon \in \mathscr{C}(\mathbb{R}^+; L^1(\mathbb{R}))$ of the following equation:

$$\left.\begin{aligned} &\frac{u_\varepsilon(t) - u_\varepsilon(t-\varepsilon)}{\varepsilon} + Au_\varepsilon(t) = 0, \quad && t \geq 0, \\ &u_\varepsilon(t) = u_0, && t < 0. \end{aligned}\right\} \tag{2.17}$$

We then make use of an abstract theorem from the theory of semi-groups [13].

Theorem 2.6.1 *Let X be a Banach space and A a closed operator with domain $D(A)$ dense in X, accretive and such that* $(\mathrm{id}_X + \lambda A)(D(A)) = X$ *for all* $\lambda > 0$. *Then, for all ε and every $u_0 \in X$, the problem (2.17) possesses a unique solution $u_\varepsilon \in \mathscr{C}(\mathbb{R}^+; X)$. In addition, $u_\varepsilon(t)$ converges in X, uniformly on every compact set of $\mathbb{R}^+$ to a function $t \mapsto S(t)u_0$. The family $(S(t))_{t\geq 0}$ is a contraction semi-group in X, that is to say,*

(1) $S(t)S(s) = S(t+s), \quad \forall t, s \geq 0,$
(2) $\|S(t)v - S(t)w\| \leq \|v - w\|, \quad t \geq 0, \quad v, w \in X,$
(3) $S(0) = \mathrm{id}_X,$
(4) $(t, v) \mapsto S(t)v$ *is continuous on* $\mathbb{R}^+ \times X$.

The accretivity of which mention is made in the hypotheses of the theorem is the following property.

Definition 2.6.2 The operator A is said to be *accretive* if for all $v, w \in D(A)$ and every $\lambda \geq 0$, we have

$$\|v - w\| \leq \|v + \lambda Av - w - \lambda Aw\|.$$

We shall make use of the Banach space $X = L^1(\mathbb{R})$.

Accretivity of A

The following lemma is classical, but we give a proof for the convenience of the reader. In all that follows sgn s vanishes if $s = 0$.

Lemma 2.6.3 *Let $z \in W^{1,1}(\mathbb{R})$. Then $z_x \operatorname{sgn} z = |z|_x$.*

Proof Let j be a Lipschitz function and $(j_n)_{n\geq 0}$ a sequence of Lipschitz functions of class $\mathscr{C}^1$, such that j_n converge uniformly to j and j_n' converge pointwise to j' while staying uniformly bounded. For the lemma, we choose $j = \operatorname{sgn}$ and $j_n(s) = \sqrt{(s^2 + n^{-1})}$. As $W^{1,1} \subset \mathscr{C}(\mathbb{R})$, we have $j_n(z) \in W^{1,1}(\mathbb{R})$ and $j_n(z)_x = j_n'(z)z_x$.

Let us pass to the limit in this equality. On the one hand $j_n(z) \to j(z)$ uniformly, so $j_n(z)_x \to j(z)_x$ in $\mathscr{D}'(\mathbb{R})$. On the other hand, $j_n'(z)z_x \to j'(z)z_x$ pointwise, staying bounded in $L^1(\mathbb{R})$. We can therefore apply the theorem of dominated convergence. The convergence is valid in $L^1(\mathbb{R})$, so in the sense of distributions. ■

Now let $v, w \in D(A)$. The lemma and the growth of f give $|f(v) - f(w)|_x = (f(v) - f(w))_x \operatorname{sgn}(v - w)$. Hence

$$\begin{aligned}\|v + \lambda Av - w - \lambda Aw\| &\geq \int_{\mathbb{R}} (v + \lambda f(v)_x - w - \lambda f(w)_x) \operatorname{sgn}(v - w)\,\mathrm{d}x \\ &= \int_{\mathbb{R}} (|v - w| + \lambda |f(v) - f(w)|_x)\,\mathrm{d}x \\ &= \int_{\mathbb{R}} |v - w|\,\mathrm{d}x,\end{aligned}$$

since $\int_{\mathbb{R}} z_x\,\mathrm{d}x = 0$ for all $z \in W^{1,1}(\mathbb{R})$. Finally, A is accretive.

The range of $\mathrm{id}_X + \lambda A$

Let $\lambda > 0$. The equation $(\mathrm{id}_X + \lambda A)v = h$, for $h \in X = L^1(\mathbb{R})$ and $v \in D(A)$, leads to the ordinary differential equation

$$\frac{\mathrm{d}v}{\mathrm{d}x} = g(x, v) =: \frac{h(x) - v}{\lambda f'(v)}. \tag{2.18}$$

The function g being uniformly Lipschitz with respect to v and uniformly integrable with respect to x, the Cauchy problem for (2.18) possesses one and only one solution defined on $\mathbb{R}$. We denote by v_n the solution of (2.18) on $(-n, \infty)$ which satisfies the initial condition $v_n(-n) = 0$. Every solution of (2.18) satisfies $\mathrm{d}|v|/\mathrm{d}x \leq \Omega|h| - \omega|v|$ where $0 < \omega < 1/(\lambda f'(z)) < \Omega$ for all z. We deduce that, for $x > 0$,

$$|v(x)| \leq \mathrm{e}^{-\omega x}|v(0)| + \Omega \int_0^x \mathrm{e}^{\omega(y-x)}|h(y)|\,\mathrm{d}y$$

and hence that $\lim_{x\to+\infty} v(x) = 0$ (by applying the theorem of dominated convergence to the integral). Making use of the same differential inequality, we derive now

$$\omega \int_{-n}^{+\infty} |v_n(x)|\,\mathrm{d}x \leq \Omega\|h\|_1$$

which shows that the sequence $(v_n)_{n\in\mathbb{N}}$, continued by 0 for $x < -n$, is bounded in $L^1(\mathbb{R})$.

Similarly, we have $\mathrm{d}|v|/\mathrm{d}x \geq -\omega|h| - \Omega|v|$ which on integrating and taking account of the preceding estimate gives $|v_n(x)| \leq (\omega + \Omega^2/\omega)\|h\|_1$. This sequence is thus also bounded in $L^\infty(\mathbb{R})$. Making use of the differential equation, we see that the sequence is uniformly equi-continuous on every compact set. It thus admits a cluster point when $n \to \infty$, for the topology of uniform convergence on every compact set. Denote this limit by v, which is continuous. We can pass to the limit in the integrated form of the differential equation, which shows that v is a solution of (2.18). In addition, by Fatou's lemma, v is integrable on $\mathbb{R}$ with $\omega \int_{\mathbb{R}} |v(x)|\,\mathrm{d}x \le \Omega \|h\|_1$. We thus have $v \in X$. Since in addition $v_x = g(\cdot, v) \in X$, indeed we have $v \in D(A)$ and $v + \lambda A v = h$. Thus, $\mathrm{id}_X + \lambda A$ is surjective. As A is accretive, $\mathrm{id}_X + \lambda A$ is equally injective.

Passage to the limit

We can thus apply Theorem 2.6.1. The limiting solution u_ε therefore exists, is unique and converges uniformly on every compact set $[0, T]$ to a function $t \mapsto u(t)$ with values in $L^1(\mathbb{R})$: $u \in \mathscr{C}(\mathbb{R}^+; L^1(\mathbb{R}))$. Let $k \in \mathbb{R}$; then applying Lemma 2.6.3 to $z = f(u_\varepsilon(t)) - f(k)$, we have

$$\begin{aligned} |u_\varepsilon(t) - k| + \varepsilon |f(u_\varepsilon(t)) - f(k)|_x &= (u_\varepsilon(t-\varepsilon) - k)\,\mathrm{sgn}(u_\varepsilon(t) - k) \\ &\le |u_\varepsilon(t-\varepsilon) - k|. \end{aligned}$$

For $\varphi \in \mathscr{D}^+(\mathbb{R}^2)$, we therefore have

$$\begin{aligned} 0 &\le \iint_{\mathbb{R}\times\mathbb{R}^+} \varphi(t) \left(\frac{|u_\varepsilon(t-\varepsilon) - k| - |u_\varepsilon(t) - k|}{\varepsilon} - |f(u_\varepsilon(t)) - f(k)|_x \right) \mathrm{d}x\,\mathrm{d}t \\ &= \iint_{\mathbb{R}\times\mathbb{R}^+} \left(|u_\varepsilon(t) - k| \frac{\varphi(t+\varepsilon) - \varphi(t)}{\varepsilon} + |f(u_\varepsilon(t)) - f(k)| \varphi_x \right) \mathrm{d}x\,\mathrm{d}t \\ &\quad + \frac{1}{\varepsilon} \int_0^\varepsilon \mathrm{d}t \int_{\mathbb{R}} \varphi(x,t) |u_0(x) - k|\,\mathrm{d}x \\ &\overset{\varepsilon\to 0}{\to} \iint_{\mathbb{R}\times\mathbb{R}^+} (|u - k|\varphi_t + |f(u) - f(k)|\varphi_x)\,\mathrm{d}x\,\mathrm{d}t + \int_{\mathbb{R}} \varphi(x,0)|u_0(x) - k|\,\mathrm{d}x, \end{aligned}$$

because of the uniform convergence of $\varphi(t+\varepsilon) - \varphi(t)/\varepsilon$ to φ_t and of $\varepsilon^{-1} \int_0^\varepsilon \varphi(x,t)\,\mathrm{d}t$ to $\varphi(x, 0)$. Finally, the existence of at least an entropy solution is demonstrated in the case $u_0 \in L^1(\mathbb{R})$ when f'' is null outside a compact set and $\inf f' > 0$.

Let us notice immediately an essential property of this solution – the maximum principle.

Proposition 2.6.4 *Let $u_0 \in L^1 \cap L^\infty(\mathbb{R})$. The solution, limit of u_ε, of (2.1) satisfies $\|u\|_\infty \le \|u_0\|_\infty$.*

Proof As a result of Theorem 2.6.1 it is enough to show that if $h \in L^1 \cap L^\infty$ then the solution of $v + \varepsilon A v = h$ satisfies $\|v\|_\infty \le \|h\|_\infty$. But $v \in W^{1,1}(\mathbb{R})$ is continuous and tends to zero at infinity, so attains its bounds. There, v_x is null and $v = h$. ■

The general case

If $u_0 \in L^1(\mathbb{R})$ and if f'' is null outside a compact set, then $\inf f' > -\infty$ and we are led to the preceding case by choosing $r > -\inf f'$. The function $f_r(s) =: f(s)+rs$ satisfies the hypotheses of the preceding paragraphs. The Cauchy problem for f_r and u_0 therefore possesses an entropy solution u_r, which furnishes an entropy solution of (2.1) by $u(x,t) =: u_r(x+rt,t)$. (Exercise: prove in detail this assertion.)

If $u_0 \in L^1 \cap L^\infty(\mathbb{R})$ and f is an arbitrary function of class $\mathscr{C}^2$, let us put $M = \|u_0\|_\infty$. We choose a function g of class $\mathscr{C}^2$ such that $g(s) = f(s)$ for $|s| < M$ and $g''(s) = 0$ for $|s| > M+1$. The solution already constructed of the Cauchy problem for u_0 and g is valid since $\|u_0\| \leq M$, with the result that $f(u) \equiv g(u)$. To be precise, that shows that u is a weak solution of (2.1) and also that u satisfies the entropy inequalities for $|k| \leq M$. But for $|k| > M$, these inequalities are trivial since $\|u_0\| \leq M$ and u is a weak solution.

It remains for us to prove the existence of a solution under the hypotheses of Theorem 2.3.5, that is when, $u_0 \in L^\infty(\mathbb{R})$. For that, we make use of the properties of uniqueness and of propagation with finite speed which will be proved in the following section.

Let M be as defined above. For $T > 0$ and $y \in \mathbb{R}$, we denote by $u^{y,T}$ the solution given by the method of semi-groups when we choose the prescribed initial condition to be $u_0\chi[y-MT, y+MT] \in L^1(\mathbb{R})$ (here χ takes values 0 and 1). Its restriction to the triangle $\Delta_{y,T} =: \{(x,t): |x-y| + Mt < MT\}$ is denoted by $v^{y,T}$. From (P2) $v^{y,T}$ and $v^{z,S}$ and coincide on the intersection of $\Delta_{y,T}$ and $\Delta_{z,S}$. We therefore construct a function on $\mathbb{R} \times \mathbb{R}^+$ by putting $u(x,t) = u^{y,T}(x,t)$ if $(x,t) \in \Delta_{y,T}$, which is an entropy solution of (2.1). In fact, the localization by a test function φ makes the variational formulation for u equivalent to that for $u^{y,T}$ by choosing (y,T) such that $\operatorname{supp}\varphi \cap (\mathbb{R} \times \mathbb{R}^+)$ is contained in $\Delta_{y,T}$.

2.7 Proof of Theorem 2.3.5: uniqueness

The essential idea of the proof of uniqueness is the inequality

$$|u-v|_t + ((f(u)-f(v))\operatorname{sgn}(u-v))_x \leq 0,$$

which is satisfied by two entropic solutions of (2.1). The entropy inequalities are deduced in the following sub-section. On integrating over a domain of the form $|x-x_0| + Mt < b$, where M bounds the speed of propagation of the waves above, we deduce the property (P1) of Proposition 2.3.6.

An inequality for two entropy solutions

Proposition 2.7.1 *Let u and v be two entropy solutions of (2.1) of which the initial values are respectively u_0 and v_0. For all $\varphi \in \mathscr{D}^+(Q)$, where $Q = \mathbb{R} \times [0, T)$, we have*

$$\iint_Q (|u - v|\varphi_t + (f(u) - f(v))\,\mathrm{sgn}(u - v)\varphi_x)\,\mathrm{d}x\,\mathrm{d}t \tag{2.19}$$
$$+ \int_{\mathbb{R}} |u_0(x) - v_0(x)|\,\varphi(x, 0)\,\mathrm{d}x \geq 0.$$

Proof Let $\Phi \in \mathscr{D}^+(Q \times Q)$. We apply (2.6) with the solution u, with the constant $k = v(y, s)$ and with the test function $\Phi(\cdot, \cdot, y, s)$, then we integrate the inequality obtained with respect to (y, s) over Q. We make the same calculation replacing u by v and conversely, with the test function $\Phi(x, t, \cdot, \cdot)$. The sum of the two inequalities obtained is:

$$\begin{aligned} 0 \leq & \iiiint_{Q\times Q} |u(x, t) - v(y, s)|(\Phi_t + \Phi_s)(x, t, y, s)\,\mathrm{d}t\,\mathrm{d}s\,\mathrm{d}x\,\mathrm{d}y \\ & + \iiiint_{Q\times Q} \mathrm{sgn}(u(x, t) - v(y, s))(f(u(x, t)) \\ & - f(v(y, s)))(\Phi_x + \Phi_y)\,\mathrm{d}t\,\mathrm{d}s\,\mathrm{d}x\,\mathrm{d}y \\ & + \iint_{\mathbb{R}\times Q} |u_0(x) - v(y, s)|\Phi(x, 0, y, s)\,\mathrm{d}x\,\mathrm{d}y\,\mathrm{d}s \\ & + \iint_{Q\times\mathbb{R}} |u(x, t) - v_0(y)|\Phi(x, t, y, 0)\,\mathrm{d}x\,\mathrm{d}t\,\mathrm{d}y. \end{aligned} \tag{2.20}$$

Let $\varphi \in \mathscr{D}^+(Q)$. We apply (2.20) to the function

$$\Phi_\varepsilon = \varphi(x, t)\chi_\varepsilon(x - y, t - s) \tag{2.21}$$

where $\chi_\varepsilon(x, t) = \varepsilon^{-2}\chi(x/\varepsilon, t/\varepsilon)$ is a positive approximation to the Dirac mass at the origin: $\chi \in \mathscr{D}^+(\mathbb{R}^2)$, $\int_{\mathbb{R}^2} \chi\,\mathrm{d}x\,\mathrm{d}t = 1$. In fact we shall choose χ to be of the form $\theta(x)\eta(t)$, the support of η being in $[-2, -1]$.

We now apply a technical lemma.

Lemma 2.7.2 *Let F be a locally Lipschitz function on $\mathbb{R}^2$. Then for Φ_ε of the form (2.21):*

(1) *When $\varepsilon \to 0$, the integral*

$$\iiiint_{Q\times Q} F(u(x, t), v(y, s))\Phi_\varepsilon(x, t, y, s)\,\mathrm{d}x\,\mathrm{d}t\,\mathrm{d}y\,\mathrm{d}s \tag{2.22}$$

converges to the integral

$$\iint_Q F(u(x,t), v(x,t))\varphi(x,t)\,dx\,dt. \tag{2.23}$$

(2) *When $\varepsilon \to 0$, the integral*

$$\iiint_{\mathbb{R}\times Q} F(u_0(x), v(y,s))\Phi_\varepsilon(x,0,y,s)\,dx\,dy\,ds \tag{2.24}$$

tends to the integral

$$\int_{\mathbb{R}} F(u_0(x), v_0(x))\Phi(x,0)\,dx. \tag{2.25}$$

In fact if Φ is of the form (2.21), then $(\Phi_t+\Phi_s)(x,t,y,s) = \varphi_t(x,t)\chi_\varepsilon(x-y,t-s)$ and $(\Phi_x+\Phi_y)(x,t,y,s) = \varphi_x(x,t)\chi_\varepsilon(x-y,t-s)$ are again of form (2.21). As the functions coming into play with u and v in the integrals are locally Lipschitz on $\mathbb{R}^2$ we can pass to the limit in the first three terms of (2.20) making use of the lemma. Also, the last integral is zero because of the factor $\eta(t/\varepsilon) = 0$. The proposition is clearly proved. ∎

Let us pass to the proof of the lemma which is only a result from measure theory.

Proof (1) First of all, $\int_Q \chi_\varepsilon(x-y,t-s)\,dy\,ds = 1$ because of the condition on the support of η. Thus (2.23) has the value

$$\iiiint_{Q\times Q} F(u(x,t), v(x,t))\Phi_\varepsilon(x,t,y,s)\,dx\,dt\,dy\,ds.$$

The functions u and v are bounded and F is locally Lipschitz with the result that

$$|F(u(x,t), v(x,t)) - F(u(x,t), v(y,s))| \le C|v(x,t) - v(y,s)|.$$

It is enough therefore to show that

$$\begin{aligned} I_\varepsilon(v) =: &\iiiint_{Q\times Q} |v(x,t) - v(y,s)|\Phi_\varepsilon(x,t,y,s)\,dx\,dt\,dy\,ds \\ = &\iiiint_{Q\times\mathbb{R}^2} |v(x,t) - v(x+\varepsilon y, t+\varepsilon s)|\varphi(x,t)\chi(y,s)\,dx\,dt\,dy\,ds \end{aligned}$$

converges to zero when $\varepsilon \to 0$, the second equality occurring for ε small enough. Let $U \Subset Q$ be a compact neighbourhood of the support of φ. For ε sufficiently small, the integral I_ε is borne only by U and we have the upper bound

$$I_\varepsilon(v) \le 2\|\varphi\|_\infty \|v\|_{L^1(U)}.$$

Similarly, the theorem of dominated convergence shows that if v is continuous $I_\varepsilon \to 0$ as $\varepsilon \to 0$. To conclude the proof in the general case we thus choose $\delta > 0$ arbitrarily small and w continuous and bounded such that $\|v - w\| \le \delta$. Since $I_\varepsilon(v) \le I_\varepsilon(v - w) + I_\varepsilon(w)$, we have that $\limsup_{\varepsilon\to 0} I_\varepsilon(v) \le \delta$, which implies the stated result.

(2) Similarly $\int_Q \chi_\varepsilon(x - y, -s)\,\mathrm{d}y\,\mathrm{d}s = 1$ and we are led back to the convergence to 0 of the integral

$$\iiint_{\mathbb{R}\times Q} |v_0(x) - v(y, s)|\varphi(x, 0)\chi_\varepsilon(x - y, -s)\,\mathrm{d}x\,\mathrm{d}y\,\mathrm{d}s.$$

The same method as above shows that

$$\iiint_{\mathbb{R}\times Q} |v(x, s) - v(y, s)|\varphi(x, 0)\chi_\varepsilon(x - y, -s)\,\mathrm{d}x\,\mathrm{d}y\,\mathrm{d}s.$$

tends to zero with ε. It therefore remains to show the convergence to zero of

$$J_\varepsilon(v) =: \iint_{\mathbb{R}\times Q} |v_0(x) - v(x, s)|\varphi(x, 0)\chi_\varepsilon(x - y, -s)\,\mathrm{d}x\,\mathrm{d}y\,\mathrm{d}s.$$

From the choice (2.21), the integral with respect to y is harmless:

$$J_\varepsilon(v) =: \iint_Q |v_0(x) - v(x, s)|\varphi(x, 0)\varepsilon^{-1}\eta(-s/\varepsilon)\,\mathrm{d}x\,\mathrm{d}s.$$

Next, since v is continuous on $(0, T)$ with values in $L^1_{\mathrm{loc}}(\mathbb{R})$, we may replace v_0 by $v(\cdot, 0)$ and we conclude by noting that

$$\begin{aligned} J_\varepsilon(v) &\le \|\varphi\|_\infty \int_1^2 \|v(0) - v(s)\|_{L^1(V)}\eta(-s/\varepsilon)\,\mathrm{d}s/\varepsilon \\ &= \|\varphi\|_\infty \int_0^\infty \|v(0) - v(\varepsilon s)\|_{L^1(V)}\eta(-s)\,\mathrm{d}s \end{aligned}$$

where V is an interval containing the support of $\varphi(\cdot, 0)$. ∎

Integration of the inequality (2.19)

Let (a, b) be an open interval of $\mathbb{R}$ and $s > 0$ a real number. We begin by constructing a trapezium B of Q whose horizontal sections B_t are such that

(1) $B_s = (a, b)$,
(2) if $t < \tau$, then B_t contains the domain of dependence of B_τ (see Fig. 2.5).

For that, we denote by I the smallest interval containing the essential values of u and v. Since these solutions are supposed bounded, the same is true of I. Next we choose the number M which bounds above the absolute value of the velocity

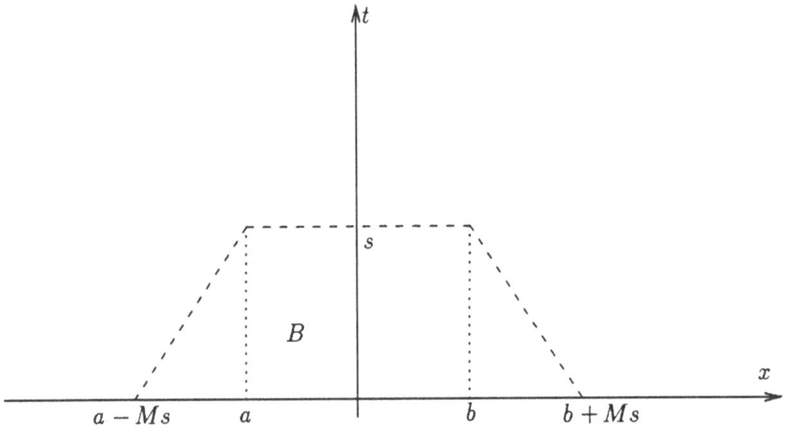

Fig. 2.5: The domain of integration.

of wave propagation: $M =: \sup_{r\in I}|f'(r)|$. Finally we define $B = \{(x,t) \in Q: a - M(s-t) < x < b + M(s-t)\}$.

Now we turn to the proof of the property (P1). By a simple translation we can always suppose that $a = -b$. Let $d > b$ and choose an even function $\theta \in \mathscr{D}^+(\mathbb{R})$ decreasing on $\mathbb{R}^+$ and such that $\theta(y) = 1$ for $|y| < b$ and $\theta(y) = 0$ for $|y| > d$. We also choose a function $\chi \in \mathscr{D}^+((-\infty, T))$ such that $\chi(0) = 1$ and $\chi(t) = 0$ for $t > s'$, where $s < s' < s + b/M$. We apply (2.19) with the test function $\varphi(x,t) =: \chi(t)\theta(|x| + Mt)$. Writing $F(u,v) = (f(u) - f(v))\operatorname{sgn}(u-v)$, we have

$$|u-v|\varphi_t + F(u,v)\varphi_x = \chi'(t)\theta(|x|+Mt)|u-v| \\ + \chi(t)\theta'(|x|+Mt)(M|u-v| + F(u,v)\operatorname{sgn} x).$$

Now $|F(u,v)| \le M|u-v|$, so the last bracket is positive. As $\chi \ge 0$ and $\theta' \le 0$, we have

$$|u-v|\varphi_t + F(u,v)\varphi_x \le \chi'(t)\theta(|x|+Mt)|u-v|.$$

Substituting this result in (2.19), we obtain

$$\iint_Q (|u-v|\chi'(t)\theta(|x|+Mt))\,\mathrm{d}x\,\mathrm{d}t + \int_{\mathbb{R}} |u_0 - v_0|\theta(|x|)\,\mathrm{d}x \ge 0. \quad (2.26)$$

Now, let us make d tend to b. The theorem of dominated convergence allows us to pass to the limit in the two integrals which renders the formula correct when $\theta(|x| + Mt)$ is simply the characteristic function of the set B. Let us write then $h(t) =: \|u(t) - v(t)\|_{L^1(B_t)}$. Since $t \mapsto u(t)$ is continuous with values in $L^1_{\mathrm{loc}}(\mathbb{R})$, h

is continuous and we deduce from (2.26) the inequality

$$\int_0^s h(t)\chi'(t)\,\mathrm{d}t + h(0) \geq 0, \tag{2.27}$$

for every function $\chi \in \mathscr{D}^+((-\infty, s'))$ such that $\chi(0) = 1$. This property implies classically that h is decreasing on $[0, s']$. We thus have

$$\int_{-b}^{b} |u(x,s) - v(x,s)|\,\mathrm{d}x = h(s) \leq h(0) = \int_{-b-Ms}^{b+Ms} |u_0(x) - v_0(x)|\,\mathrm{d}x,$$

which is exactly the inequality sought.

End of the proof of Proposition 2.3.6

It remains only to show that the property (P3) is valid. We thus assume that $u_0 - v_0 \in L^1(\mathbb{R})$. Then the property (P1) implies that for every bounded interval I,

$$\int_I |u(x,t) - v(x,t)|\,\mathrm{d}x \leq \|u_0 - v_0\|_{L^1(\mathbb{R})}.$$

By means of Fatou's lemma we then deduce that $u(t) - v(t) \in L^1(\mathbb{R})$ and that

$$\|u(t) - v(t)\|_1 \leq \|u_0 - v_0\|_1.$$

Next, we express that u and v are weak solutions of (2.1). Choosing a test function of the form $\varphi_\varepsilon(x,t) = \chi(t)\theta(\varepsilon x)$ where $\chi \in \mathscr{D}((-\infty, T))$ and $\theta(x) \in \mathscr{D}(\mathbb{R})$ with $\theta \equiv 1$ in a neighbourhood of the origin, we have

$$\iint_Q ((u - v)\varphi_{\varepsilon t} + (f(u) - f(v))\varphi_{\varepsilon x})\,\mathrm{d}x\,\mathrm{d}t \tag{2.28}$$
$$+ \int_{\mathbb{R}} (u_0 - v_0)(x)\varphi_\varepsilon(x, 0)\,\mathrm{d}x = 0,$$

that is to say,

$$\iint_Q (u - v)\chi'(t)\theta(\varepsilon x)\,\mathrm{d}x\,\mathrm{d}t + \varepsilon \iint_Q (f(u) - f(v))\chi(t)\theta'(\varepsilon x)\,\mathrm{d}x\,\mathrm{d}t \tag{2.29}$$

$$+\chi(0)\int_{\mathbb{R}} (u_0 - v_0)\theta(\varepsilon x)\,\mathrm{d}x = 0. \tag{2.30}$$

Each factor $u(t) - v(t)$, $u_0 - v_0$ and $f(u(t)) - f(v(t))$ is integrable on $\mathbb{R}$ uniformly with respect to t, the last because we have $|f(u) - f(v)| \leq M|u - v|$. When $\varepsilon \to 0$, the theorem of dominated convergence allows us to pass to the limit in each of the three integrals, the second tending to zero. There remains

$$\iint_Q (u - v)\chi'(t)\,\mathrm{d}x\,\mathrm{d}t + \chi(0)\int_{\mathbb{R}} (u_0 - v_0)\,\mathrm{d}x = 0, \tag{2.31}$$

which can be written in the form $\int_0^T h\chi' \, dt + \chi(0)h(0) = 0$ with this time $h(t) =: \int_{\mathbb{R}}(u(t) - v(t)) \, dx$ which is a continuous function. We deduce that h is a constant function, that is to say

$$\int_{\mathbb{R}} (u(x,t) - v(x,t)) \, dx = \int_{\mathbb{R}} (u_0(x) - v_0(x)) \, dx.$$

2.8 Comments

Oleĭnik's inequality

We have seen in an exercise that if f is uniformly convex, that is to say, if there exists a number $\alpha > 0$ such that $f'' \geq \alpha$, then the entropy solution satisfies Oleĭnik's inequality

$$u_x \leq \frac{1}{\alpha t}. \tag{2.32}$$

The positive distribution $1/\alpha t - u_x$ is thus a bounded measure for every compact interval and for $t > 0$ and the same is true of u_x. Thus, $u(t) \in \mathrm{BV}(I)$ for every $t > 0$ and every compact interval I, even if u_0 is only in $L^\infty(\mathbb{R})$. There is therefore a smoothing phenomenon which we did not observe in the linear case. In particular, the resolvent operator $S(t)$ is compact as a mapping of $L^\infty_{\mathrm{loc}}(\mathbb{R})$ into itself, and we find again the irreversible character of the entropy formulation of (2.1).

Since f is Lipschitz, $f \circ u(t)$ is also of bounded variation on I; we have in fact

$$\mathrm{TV}(f \circ u(t); I) \leq M \, \mathrm{TV}(u(t)).$$

Thus $f(u)_x \in L^\infty_{\mathrm{loc}}((0,T); M_b(I))$ and, because of (2.1), the same is true of u_t. We deduce that $t \mapsto u(t)|_I$ is not only continuous, but even Lipschitz on $(0,T)$, with values in L^1:

$$\|u(t) - u(s)\|_{L^1(I)} \leq M \int_s^t \mathrm{TV}(u(\tau)) \, d\tau.$$

We can rewrite this result by decomposing u_x into its positive and negative parts: $u_x = u_x^+ - u_x^-$. With $I = (a,b)$ we have

$$\int_I u_x^- \, dx = u(a) - u(b) + \int_I u_x^+ dx \leq 2\|u(t)\| + |I|/\alpha t.$$

From this, on using the maximum principle, we have

$$\|u(t) - u(s)\|_{L^1(I)} \leq M\left(2|t-s|\|u_0\|_\infty + \frac{|I|}{\alpha} \log \frac{t}{s}\right). \tag{2.33}$$

Initial datum with bounded total variation

For a general flux, the property (P6) ensures that an initial datum in $BV(\mathbb{R})$ leads to a solution u whose total variation with respect to x at each instant is bounded by $TV(u_0)$. The preceding calculation remains valid in part but we find

$$\int_{\mathbb{R}} |u(t) - u(s)| \, dx \leq M|t - s| TV(u_0).$$

Uniqueness: the duality method

In the method of Kružkov, the uniqueness is the consequence of a monotonic property. When we consider systems rather than scalar equations, such a property is no longer available and the uniqueness question no longer has a general answer. An alternative argument is needed in approaching the problem. That which was first presented in this spirit is the duality method of Holmgren. Although its generalization to the case of systems is delicate and of limited range it has given several interesting results. Let us look at its application to a scalar question due to Oleĭnik [81].

Being given two entropy solutions u and v of the adjoint equation $u_t + f(u)_x = 0$, their difference $z = u - v$ satisfies the linear equation with variable coefficients $z_t + (hz)_x = 0$ with

$$h(x,t) = H(u,v) := \frac{f(v) - f(u)}{v - u}.$$

The following calculation uses the general solution of the adjoint equation $p_t + hp_x = 0$ with a given *final* condition P at a positive time T. In principle $p(\cdot, t)$ is obtained by forming the composite function $P \circ U$ where $U(t;T)$ is the flow of the differential equation $\dot{x} = h(x,t)$. However, the discontinuities of h prevent the definition of U and deprive us of the solution of the adjoint equation. We get round this obstacle by smoothing h by $h^\varepsilon = h * \rho^\varepsilon$, where $*$ denotes the convolution product with respect to the variable x alone while $\rho^\varepsilon = \rho(x/\varepsilon)/\varepsilon$, with the classical hypotheses on ρ. Being given a smooth function P, we thus denote by p^ε the solution of the Cauchy problem

$$p_t^\varepsilon + h^\varepsilon p_x^\varepsilon = 0, \qquad p^\varepsilon(x,T) = P(x).$$

We assume that the initial values u_0 and v_0 are bounded functions; we denote by J a bounded interval of $\mathbb{R}$ which contains their values. The solutions considered satisfy the following properties:

u and v have values in J,
$u(t) - u_0$ and $v(t) - v_0$ are integrable,

as $\varepsilon \mapsto 0+$, $\|u(t)-u_0\|_{L^1(\mathbb{R})}$ and $\|v(t)-v_0\|_{L^1(\mathbb{R})}$ tend to zero,
if $f'' > \alpha$ on J, then $u_x \le 1/(\alpha t)$ and $v_x \le 1/(\alpha t)$.

In particular, the calculations which follow apply to the solutions obtained as limiting values of the approximate solutions furnished by the parabolic equation $u_t + f(u)_x = \eta u_{xx}$ as η tends to zero.

From now on, we assume that f is uniformly convex on J we denote by α (>0) and β the lower and upper bounds of f'' on J. On $J \times J$, H is bounded and increasing with respect to each argument. In particular h is bounded: $|h(x,t)| \le M$ and hence $|h^\varepsilon| \le M$. As h^ε is Lipschitz with respect to x (uniformly with respect to t but not with respect to ε), we have at our disposal a flow for the differential equation $\dot{x} = h^\varepsilon(x,t)$, which enables us to solve the approximate adjoint equation $p_t^\varepsilon + h^\varepsilon p_x^\varepsilon = 0$ in the class of functions Lip($\mathbb{R}$).

If $v_0 - u_0$ is integrable so also is $z(t)$ by hypothesis. Let us write Green's formula on $\mathbb{R} \times (\tau, T)$, with $0 < \tau < T$:

$$\begin{aligned} 0 &= \int_\tau^T \int_{\mathbb{R}} \left(p^\varepsilon(z_t + (hz)_x) + z\big(p_t^\varepsilon + h^\varepsilon p_x^\varepsilon\big)\right) \mathrm{d}x\, \mathrm{d}t \\ &= \int_{\mathbb{R}} P(x) z(x,T)\, \mathrm{d}x - \int_{\mathbb{R}} p^\varepsilon(x,\tau) z(x,\tau)\, \mathrm{d}x + \int_\tau^T \int_{\mathbb{R}} z(h^\varepsilon - h) p_x^\varepsilon\, \mathrm{d}x\, \mathrm{d}t. \end{aligned}$$

As p^ε is constant along the integral curves of h^ε this becomes

$$\left| \int_{\mathbb{R}} P(x) z(x,T)\, \mathrm{d}x \right| \le \|P\|_\infty \|z(\tau)\|_1 + \int_\tau^T \int_{\mathbb{R}} |z|\, |h^\varepsilon - h|\, |p_x^\varepsilon|\, \mathrm{d}x\, \mathrm{d}t.$$

To exploit this inequality, we establish an estimate for p_x^ε which does not depend on ε; this is possible only with the hypothesis of genuine non-linearity made above. By Taylor's formula

$$h_x = H_u(\bar{u}, \bar{v}) u_x + H_v(\bar{u}, \bar{v}) v_x = \frac{1}{2} f''(u_1) u_x + \frac{1}{2} f''(u_2) v_x \le \beta/\alpha t,$$

which implies that $h_x^\varepsilon = h_x * \rho^\varepsilon \le \beta/(\alpha t)$. Let us write $q^\varepsilon := p_x^\varepsilon$, which satisfies

$$(\partial_t + h^\varepsilon \partial_x) \log|q^\varepsilon| = -h_x^\varepsilon \ge -\beta/\alpha t.$$

The function $|q^\varepsilon| t^{\beta/\alpha}$ increases along the characteristics of the modified adjoint problem. Therefore,

$$|p_x^\varepsilon(x,t)| \le \left(\frac{T}{t}\right)^{\beta/\alpha} \|P_x\|_\infty.$$

Finally,

$$\left| \int_{\mathbb{R}} P(x) z(x,T)\, \mathrm{d}x \right| \le \|P\|_\infty \|z(\tau)\|_1 + \int_\tau^T \|P_x\|_\infty \left(\frac{T}{t}\right)^{\beta/\alpha} \mathrm{d}t \int_{\mathbb{R}} |z|\, |h^\varepsilon - h|\, \mathrm{d}x.$$

The function $(T/t)^{\beta/\alpha}|z|\,|h^{\varepsilon} - h|$ is bounded above by the integrable function $2M(T/\tau)^{\beta/\alpha}|z|$ and converges almost everywhere to zero when ε tends to zero. Its integral also converges to zero as a result of the theorem of dominated convergence. There remains

$$\left|\int_{\mathbb{R}} P(x)z(x,T)\,\mathrm{d}x\right| \le \|P\|_{\infty}\|z(\tau)\|_1.$$

Making τ tend to zero we obtain

$$\left|\int_{\mathbb{R}} P(x)z(x,T)\,\mathrm{d}x\right| \le \|P\|_{\infty}\|v_0 - u_0\|_1,$$

for every bounded Lipschitz function P, which is equivalent to saying that $\|v(T) - u(T)\|_1 \le \|v_0 - u_0\|_1$. This implies uniqueness.

Remark For a system, deriving an estimate for p_x^{ε} is the delicate point. In addition, the adjoint problem not being a transport equation, the constant which we obtain in the eventual upper bound

$$\left|\int_{\mathbb{R}} p^{\varepsilon}(x,t)z(x,\tau)\,\mathrm{d}x\right| \le C\|P\|\cdot\|z\|$$

is, in general, strictly greater than 1, with the result that $\|v(T) - u(T)\|_1 \le C\|v_0 - u_0\|_1$: the semi-group of a system is not contracting in $L^1(\mathbb{R})$. This point is made precise by Temple [104].

2.9 Exercises

2.1 We suppose that f is uniformly convex, that is that $f''(s) \ge \alpha$, where α is a strictly positive constant. Show that the classical solution satisfies $u_x < 1/\alpha t$.

2.2 In the case of the road traffic model, what comparison can we make between the speed of the waves $c(\rho)$ and that of vehicles $V(\rho)$?

2.3 We consider a scalar conservation law in several space dimensions:

$$u_t + \sum_{i=1}^{d} A_i(u)_{x_i} = 0.$$

We note that $u_t + \mathrm{div}_x A(u) = 0$, and we suppose that the vector field $u \mapsto A(u)$ is smooth. We write $a = A'$.

(1) Let u be a classical solution of this conservation law in a domain $\mathbb{R}^d \times [0,T)$ and u_0 its initial value. Show that u is constant along the characteristics defined by $\mathrm{d}x/\mathrm{d}t = a(u(X(t),t))$ and these are straight lines.

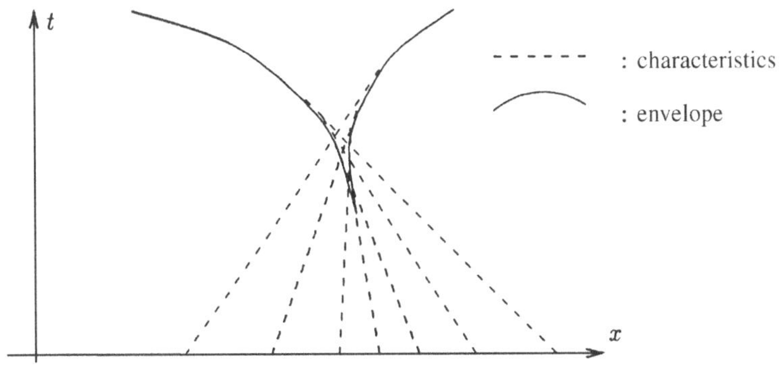

Fig. 2.6: Generic blow-up by a cusp.

(2) Let $q = \mathrm{div}_x(a(u))$. Show that along the characteristics q satisfies the differential equation $\mathrm{d}q/\mathrm{d}t + q^2 = 0$. Deduce that if there exists a point $y \in \mathbb{R}^d$ such that $\mathrm{div}_x(a(u_0))(y) < 0$, then T is finite, more precisely $T \le T^* =: -(\inf_x \mathrm{div}_x(a(u_0)))^{-1}$.

(3) Conversely, show that, if $1 + T \inf_x \mathrm{div}_x(a(u_0)) \ge 0$, then there exists a classical solution of the Cauchy problem on the domain $\mathbb{R}^d \times [0, T)$.

2.4 We consider the Burgers equation ($f(u) = \frac{1}{2}u^2$) with given initial condition u_0 of class $\mathscr{C}^1$ and with non-empty compact support. In the formula $T = -(\inf_{\mathbb{R}} u_0)^{-1}$, the lower bound is thus attained and $T < \infty$. We suppose that it is attained at a single point y_0 and that $u_0'''(y_0) > 0$ (we have $u_0''(y_0) = 0$ and $u_0'''(y_0) \ge 0$ *a priori*). We write $x_0 =: y_0 + Tu_0(y_0)$.

(1) Show that $u(T)$ is continuous on $\mathbb{R}$, and of class $\mathscr{C}^1$ outside of x_0. Prove that $\lim_{x \to x_0} u_x(x, T) = -\infty$.

(2) Show that $(u(T) - u(x_0, T))^3$ is of class $\mathscr{C}^1$ on $\mathbb{R}$ and that its derivative at x_0 has the value $-6(T^4 u_0'''(y_0))^{-1}$. *Hint*: $(y, 0)$ being the base of a characteristic which ends in (x, T), find an equivalent of $x - x_0$ as a function $y - y_0$. Figure 2.6 illustrates this generic behaviour.

2.5 Let f be a function which is not affine. To fix the ideas, there are given three numbers $v < w < z$, $w = av + (1-a)z$, such that $f(w) < af(v) + (1-a)f(z)$. Using either one elementary discontinuity or two, construct two piecewise constant solutions for the Cauchy problem in which the given initial condition is $u_0(x) = v$ if $x < 0$, and $u_0(x) = z$ if $x > 0$. Adapt the question and the solution if $f(w) <$ is replaced by $f(w) >$.

2.6 We assume that f is convex. Show that a discontinuity is admissible if and only if it satisfies *one* entropy inequality

$$[F(u)] \le \frac{\mathrm{d}x}{\mathrm{d}t}[E(u)]$$

for at least one strictly convex entropy.

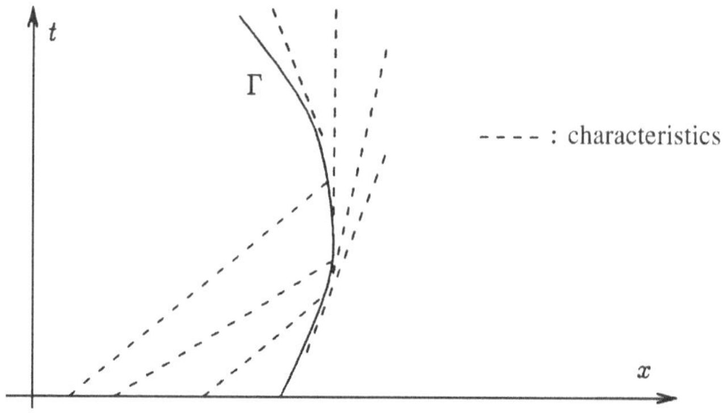

Fig. 2.7: Semi-characteristic shock.

2.7 **Weak shocks.** When $[u] \to 0$, the speed $\sigma = [f(u)]/[u]$ of a shock is of the form $\sigma = c(u_\pm) + O([u])$ from Taylor's formula.

(1) Show that in fact,

$$\sigma = c\left(\frac{1}{2}(u_+ + u_-)\right) + O([u]^2).$$

(2) Find an equivalent of the rate of dissipation of entropy $[F - \sigma E]$ when $[u] \to 0$ $(F' = cE')$ by supposing that $E'' > 0$ and $f'' > 0$.
Solution: $[F - \sigma E] \sim \kappa [u]^3$ where κ is calculable.

(3) Construct an example in which this equivalence is an equality.

2.8 Let $a < b < c$ be three real numbers. We assume that (a, b) and (b, c) are entropic discontinuities of the equation $u_t + f(u)_x = 0$, with speeds σ_1 and σ_2. Using Lax's inequality show that $\sigma_2 \leq \sigma_1$. Deduce that (a, c) is an entropic discontinuity.

2.9 Let u be an entropic solution of (2.1) which is smooth except along a curve $\Gamma: t \mapsto (X(t), t)$ of class $\mathscr{C}^1$, along which there is a semi-characteristic shock to the right: $f'(u_+) = \mathrm{d}x/\mathrm{d}t$. To fix ideas, we assume that the shock is decreasing: $u_+ < u_-$. We also impose the generic condition $f''(u_+) \neq 0$.

(1) Show that Γ is the envelope of a family of straight line characteristics and that its concavity is turned towards the left. Deduce that the continuous extension of u to the left side of Γ is not $\mathscr{C}^1$ (see Fig. 2.7).

(2) Show that $f''(u_+) < 0$ and that there exists a local diffeomorphism G which depends only on f such that along Γ, $u_- = G(u_+)$.

(3) On the other hand, deduce that $t \mapsto u_+(t)$ is of class $\mathscr{C}^1$.

(4) We can then differentiate the Rankine–Hugoniot condition. Show that

$$(f'(u_+) - f'(u_-))^2 \partial_x u_- = \frac{d^2 X}{dt^2}(u_+ - u_-) \geq 0.$$

(5) Show that if u_0 is monotonic, then the entropy solution, if it is piecewise smooth, does not behave as a semi-characteristic shock (verify that we can reduce this case to the one treated above).

2.10 **Converse case.** We consider an initial condition $u(x, 0) = b(x)$ where $b \in \mathscr{C}^1(\mathbb{R}^*)$, and b and b' are bounded, having limits to the left at zero. We assume that $b \equiv 0$ for $x > 0$ and that $b_- = b(0-) > 0$, $b'(0-) > 0$. Finally, we suppose that $(b_-, 0, \sigma_0)$ is a semi-characteristic shock with $\sigma_0 = f'(0)$ and $f''(0) < 0$. Show that there exists $T > 0$ such that the entropy solution is smooth off a curve $\Gamma : t \mapsto (X(t), t)$ issuing from the origin, along which a semi-characteristic shock takes place. Using the method of characteristics to the left side of Γ, derive an ordinary differential equation for $X(t)$.

2.11 **N-wave.** We consider the Burgers equation $u_t + (\frac{1}{2}u^2)_x = 0$.

(1) Let

$$u(x, t) = \begin{cases} x/t, & |x| < \sqrt{t}, \\ 0, & |x| > \sqrt{t}. \end{cases}$$

Show that u is a weak solution of (2.1) for the given initial condition $u_0 \equiv 0$.

(2) Show that u satisfies Oleĭnik's condition along the two curves of discontinuity.

(3) Explain why u is not the entropy solution of a Cauchy problem.

2.12 Show that in the solution of the Riemann problem, we are necessarily in one of the following cases.

There is no discontinuity, the solution is a rarefaction between the constant states u_L and u_R.

The solution is a shock, possibly a (semi-)characteristic shock.

The solution is a contact discontinuity.

The solution involves one or several rarefactions and one or several discontinuities. If there is a discontinuity at $\xi = d(u_L)$, it is semi-characteristic on the right, or characteristic, or is a contact discontinuity. If there is a discontinuity at $\xi = d(u_R)$, it is semi-characteristic on the left, or characteristic, or is a contact discontinuity. The other discontinuities are characteristics or are contact discontinuities.

2.13 We consider the equation $u_t + f(u)_x = 0$ with $f(u) = (u^2 - 1)^2$. Solve the Cauchy problem for the following initial condition:

$$u_0(x) = \begin{cases} -1, & x < -1, \\ a, & -1 < x < 1, \\ 1, & x > 1, \end{cases}$$

where $a \in (\frac{1}{3}, 1)$.

2.14 (See [41]) Let E be a strictly convex entropy of flux F. We introduce the convex conjugate function to E by

$$E^*(\lambda) = \sup_{s \in \mathbb{R}} (s\lambda - E(s)).$$

Let $u = v(x/t)$ be an entropic solution of a Riemann problem.

(1) Show that $F(v)_\xi \leq \xi E(v)_\xi$.
(2) Deduce that, for all real λ, the function $\xi \mapsto F(v) - \lambda f(v) - \xi(E(v) + E^*(\lambda) - \lambda v)$ is decreasing.
(3) Using the property, show anew that v satisfies Oleĭnik's inequality.
(4) We assume that $\lim_{s \to \infty} |s|^{-1} E(s) = +\infty$. We know then that $E^{**} = E$ [19]. Deduce the inequality

$$E\left(\frac{f(v(\xi)) - f(v(\tau)) + \tau v(\tau) - \xi v(\xi)}{\tau - \xi}\right)$$
$$\leq \frac{F(v(\xi)) - F(v(\tau)) + \tau E(v(\tau)) - \xi E(v(\xi))}{\tau - \xi}.$$

for all $\xi < \tau$.
(5) Deduce that, if v is differentiable,

$$E(v + \xi v_\xi - f(v)_\xi) \leq E(v) + \xi E(v)_\xi - F(v)_\xi.$$

Show that this inequality also implies that $\xi v_\xi = f(v)_\xi$.

2.15 Let f and g be two regular functions. We denote the resolvent semi-group of (2.1) by $S_f(t)$, that is, the mapping $u_0 \mapsto u(\cdot, t)$ which is defined from L^∞ into itself. By replacing f by g, we also consider S_g.

(1) If f and g are convex, and if u_0 is an initial condition in a Riemann problem, show that $S_f(t) \circ S_g(s) u_0 = S_g(s) \circ S_f(t) u_0$, for all $s, t > 0$.
(2) If f is strictly convex and g strictly concave, show that to the contrary $S_f(t) \circ S_g(s) u_0 \neq S_g(s) \circ S_f(t) u_0$.

(3) More generally, show that if $S_f(t)$ and $S_g(s)$ commute, then:
  (i) At each point f'' and g'' are of the same sign.
  (ii) The semi-characteristic shocks are the same for the two equations $u_t + f(u)_x = 0$ and $u_t + g(u)_x = 0$.

(4) If $f'' > 0$ on $(0, +\infty)$ and $f'' < 0$ on $(-\infty, 0)$, show that the fluxes g for which $S_f(t)$ and $S_g(s)$ commute obey a second order linear differential equation.

(5) Solve this equation when $f(u) = u^3$.

2.16 Let f and g be two convex fluxes. Again taking the notation of the preceding exercise, show by using Lax's formula that

$$S_f(s) \circ S_g(t) = S_g(t) \circ S_f(s).$$

2.17 We assume that f is strictly convex and that $\lim_{s\to\infty}(f(s)/|s|) = +\infty$. We are given that $u_0 \in L^\infty(\mathbb{R})$.

(1) Show that, for all $t > 0$ and all $x > 0$, the lower bound of $v_0(y)+tg((x-y)/t)$ is attained at least one point $y \in \mathbb{R}$.

(2) Let $t > 0$ and $x_1, x_2 \in \mathbb{R}$ with $x_1 < x_2$. We denote by y_i a point at which $v_0(y) + tg((x_i - y)/t)$ attains its minimum. Using the convexity of g, show that $y_1 < y_2$.

(3) Deduce that, t being fixed, y is unique except for a set of values of x, at most denumerable. To what do these exceptional values correspond?

2.18 Starting from Lax's formula establish again Oleĭnik's condition, if f is convex, or from the dual formula if u_0 is increasing.

2.19 We consider a conservation law $u_t + f(u)_x = 0$ where f'' does not vanish. Let $u_0 \in L^1(\mathbb{R}) \cap L^\infty(\mathbb{R})$ and u be the entropy solution of the Cauchy problem. As a way of simplifying the calculations, we assume that u is of class $\mathscr{C}^1$ off the shock curves which are assumed to form a finite family of smooth curves.

(1) Let $T > 0$ and let $x_1 < x_2$ be two real numbers. Let $(y_j, 0)$ be the base of a characteristic passing through (x_j, T). By integrating the conservation law over a suitable domain, show that

$$T(F \circ u(x_2, T) - F \circ u(x_1, T)) = \int_{y_1}^{y_2} u_0(x)\,\mathrm{d}x - \int_{x_1}^{x_2} u(x, T)\,\mathrm{d}x,$$

where $F(s) := f(s) - sf'(s)$.

(2) Deduce that

$$\mathrm{TV}(F \circ u(\cdot, T)) \le \frac{2}{T}\|u_0\|_1.$$

(3) Assuming that $f(0) = 0$, show that

$$\|F \circ u(\cdot, T)\|_\infty \leq \frac{2}{T}\|u_0\|_1.$$

2.20 **Application**: The Burgers equation. We assume that u_0 is bounded and with compact support: supp $u_0 \subset [a, b]$.

(1) Show that $\|u(t)\|_\infty \leq 2(t^{-1}\|u_0\|_1)^{1/2}$. Deduce that

$$\text{supp}\, u(t) \subset [a - 2(t\|u_0\|_1)^{1/2}, \quad b + 2(t\|u_0\|_1)^{1/2}].$$

(2) Using Oleĭnik's inequality ($x < y$ implies $u(y, t) - u(x, t) \leq (y - x)/t$, (2.32)), deduce that

$$\text{TV}(u(t)) \leq 2\frac{b-a}{t} + 8\left(\frac{\|u_0\|_1}{t}\right)^{1/2}.$$

2.21 Let f be an increasing function (with $f' > 0$) of class $\mathscr{C}^2$. Define a concept of an entropy solution of the mixed problem

$$\begin{aligned} u_t + f(u)_x &= 0, && (x, t) \in \mathbb{R}^+ \times \mathbb{R}^+, \\ u(x, 0) &= u_0(x), && x \in \mathbb{R}^+, \\ u(0, t) &= 0, && t \in \mathbb{R}^+. \end{aligned}$$

Show that this entropy solution exists for all $u_0 \in L^\infty(\mathbb{R}^+)$.

2.22 We suppose that f is convex and that, more precisely, $f'' \geq \alpha$, where α is a strictly positive constant. Show that each approximate solution u_ε in the semi-group method satisfies $(u_\varepsilon)_x \leq 1/\alpha t$. Deduce that the same is true for the entropy solution in a distributional sense which should be made precise.

2.23 We are given $u_0 \in L^\infty(\mathbb{R})$ periodic, of period L.

(1) Show that the entropy solution of the Cauchy problem is equally periodic with respect to x, with the same period (use uniqueness).
(2) Show that the average of $u(t)$ over a period is constant:

$$L^{-1}\int_0^L u(x, t)\,\mathrm{d}x = L^{-1}\int_0^L u_0(x)\,\mathrm{d}x.$$

(3) Suppose that, in addition, $\alpha =: \inf_{x\in\mathbb{R}} f'' > 0$. With the help of Oleĭnik's inequality, show that

$$\sup_{x\in\mathbb{R}} u(x, t) - \inf_{x\in\mathbb{R}} u(x, t) \leq L/t\alpha.$$

Deduce that $u(t)$ converges uniformly to the mean value of u_0.

(4) *Example*: Solve explicitly the Cauchy problem for the Burgers equation with $u_0(x) = x - E[x]$, where $E[x]$ denotes the integral part of x.

2.24 We consider a scalar conservation law in spatial dimension $d = 2$, but whose flux has only a single component:

$$\frac{\partial u}{\partial t} + \frac{\partial}{\partial x_1} f(u) = 0,$$
$$u(x_1, x_2, 0) = a(x_1, x_2).$$

We refer to the comment on Kružkov's theorem which concerns the scalar conservation laws in more than one space dimension. The initial function a is bounded and integrable, and u is the entropy solution.

(1) As $u \in \mathscr{C}(\mathbb{R}_t^+; L^1(\mathbb{R}^2))$, we can speak of the integrable function $u(t)$ for each value of t. Show that, for all $t > 0$ and almost all $x_2 \in \mathbb{R}$, we have

$$\int_{\mathbb{R}} u(x_1, x_2, t)\, \mathrm{d}x_1 = \int_{\mathbb{R}} a(x_1, x_2)\, \mathrm{d}x_1.$$

Denote that value by $m(x_2)$.

(2) We recall that, for every measurable function F from $\mathbb{R}^p$ into $\mathbb{R}$, the total variation of F is

$$\mathrm{TV}_p(F) = \sup_{\xi \in S^{p-1}} \lim_{h \to 0} \frac{1}{|h|} \int_{\mathbb{R}^p} |F(y + h\xi) - F(y)|\, \mathrm{d}y.$$

show that, for all $t > 0$,

$$\mathrm{TV}_2(u(\cdot, t)) \geq \mathrm{TV}_1(m).$$

(3) Compare with Exercise 2.19.

(4) Generalise to the spatial dimension $d \geq 2$.

(5) Construct an example, with $d = 2$, $f(s) = \frac{1}{2}s^2$ and $a(\cdot, x_2)$ an odd function for every x_2 and such that $m \equiv 0$ (because of oddness); nevertheless $u(t)$ should satisfy

$$\liminf_{t \to +\infty} \mathrm{TV}_2(u(t)) > 0.$$

Use could be made of N-waves.

3

Linear and quasi-linear systems

The object of this chapter is to derive the algebraic and geometrical properties which ensure that the Cauchy problem for a first order system of conservation laws is well-posed. In fact, we consider two classes. First of all we consider the quasi-linear systems of the first order, which are of the form

$$\partial_t u + \sum_{\alpha=1}^{d} A^\alpha(u)\partial_{x_\alpha} u = b(u). \tag{3.1}$$

The vector field b is defined and smooth on an open domain $\mathcal{U} \in \mathbb{R}^n$. Similarly the mappings $u \mapsto A^\alpha(u)$ are defined and smooth on $\mathcal{U}$, with values in the space of matrices $M_n(\mathbb{R})$.

The second class, contained in the preceding, will be that of systems of conservation laws

$$\partial_t u + \sum_{\alpha=1}^{d} \partial_{x_\alpha} f^\alpha(u) = b(u). \tag{3.2}$$

A system of the form (3.2) is clearly quasi-linear, with $A^\alpha(u) = \mathrm{d}_u f^\alpha(u)$, where d_u denotes differentiation with respect to u.

Let us consider the Cauchy problem for one or other of these systems:

$$u(x, 0) = u_0(x, 0). \tag{3.3}$$

If u_0 is a constant, the obvious solution is a function of t alone, which satisfies the ordinary differential equation $u' = b(u)$. Let us look at the case of a given initial condition of the form $u_0^\varepsilon = u_0 + \varepsilon v_0(x)$. As we wish that the solution depends continuously on the given initial conditions, we pay attention to a solution of the form $u^\varepsilon(x, t) = u(t) + \varepsilon v(x, t) + O(\varepsilon^2)$ on a bounded time interval. The corrector

v is a solution of the linear problem

$$\partial_t v + \sum_{\alpha=1}^{d} A^\alpha(u(t))\partial_{x_\alpha} v = \mathrm{d}b(u(t)) \cdot v, \tag{3.4}$$

$$v(x, 0) = u_0(x). \tag{3.5}$$

A necessary condition for the asymptotic expansion u^ε to be correct is certainly that v exists! In fact, as the remainder of order ε^2 has to be determined among other things with the help of v, it is important that v has sufficient smoothness, at least that of $v_0(x)$. We begin, therefore, by considering the Cauchy problem for a linear system of the first order whose general coefficients depend on the time. The right-hand side is the least important since we can make it as small as we please by changing the time variable $t \mapsto \eta t$. Thus we suppose that $b \equiv 0$.

3.1 Linear hyperbolic systems

We now therefore consider the following general system:

$$\partial_t u + \sum_{\alpha=1}^{d} A^\alpha(t)\partial_{x_\alpha} u = 0, \tag{3.6}$$

$$u(x, 0) = u_0(x), \tag{3.7}$$

where the matrices A^α depend on t in a smooth manner.

Although it is possible to study the Cauchy problem for (3.6) in a space of smooth functions, for which the partial derivatives have the usual sense, it is simpler and more general to consider the weak solutions. A weak solution of (3.6) is a tempered distribution u, that is to say an element of the dual of the Schwartz class (see below) which satisfies the conditions

$$\left\langle u, \partial_t \varphi + \sum_{\alpha=1}^{d} A^{\alpha\mathrm{T}} \partial_{x_\alpha} \varphi \right\rangle_{\mathscr{S}',\mathscr{S}} + \int_{\mathbb{R}^d} u_0(x)\varphi(x, 0)\,\mathrm{d}x = 0,$$

$$\forall \varphi \in \mathscr{S}(\mathbb{R} \times \mathbb{R}). \tag{3.8}$$

Since the Fourier transform $\mathscr{F}$ with respect to x of (3.6) leads to a linear ordinary differential equation, the natural body of a study of the Cauchy problem is the space $(L^2(\mathbb{R}^d))^n$ or every other space which simply enables the definition of $\mathscr{F}$ and its inverse, for example a Sobolev space $H^s(\mathbb{R}^d) = W^{s,2}$. We notice that the spaces $W^{s,p}(\mathbb{R}^d)$ are in general inappropriate for $\mathscr{F}$ as $\mathscr{F}$ sends $L^p(\mathbb{R}^d)$ into $L^q(\mathbb{R}^d)$ if and only if $p^{-1} + q^{-1} = 1$, and $p \leq 2$, with the result that an isomorphism from L^p to its own dual is possible if and only if $p = 2$. In the case of constant coefficients, we know in fact [4] that the Cauchy problem is not well-posed in L^p for $p \neq 2$, when the matrices A^α do not commute with each other. Obviously in one space dimension that obstruction does not take place.

Fourier analysis

The Fourier transform $\mathscr{F}$ is an isometry of $L^2(\mathbb{R})$ which is defined on the subspace $L^1 \cap L^2$ by the formula

$$\mathscr{F}u(\xi) := (2\pi)^{-n/2} \int_{\mathbb{R}^d} e^{-i\xi\cdot x} u(x)\,dx.$$

Its inverse is the conjugate transformation

$$\bar{\mathscr{F}}u(\xi) = (2\pi)^{-n/2} \int_{\mathbb{R}^d} e^{i\xi\cdot x} u(x)\,dx.$$

We shall use only one of the properties of differentiation $(\mathscr{F}\partial_{x_\alpha} u)(\xi) = i\xi_\alpha \mathscr{F}u(\xi)$, which also furnishes a formula for $\bar{\mathscr{F}}$. It follows that $\mathscr{F}$ is an isomorphism of the Schwartz class $\mathscr{S}(\mathbb{R}^d)$, the space of functions defined on $\mathbb{R}^d$, of class $\mathscr{C}^\infty$ and decreasing rapidly, along with all their derivatives (decreasing at infinity more rapidly than every non-zero rational fraction). By duality, the Fourier transformation extends to an isomorphism of the dual space $\mathscr{S}'(\mathbb{R}^d)$, via the Plancherel formula $\langle \mathscr{F}\varphi, u\rangle_{\mathscr{S}',\mathscr{S}} := \langle \varphi, \bar{\mathscr{F}}u\rangle_{\mathscr{S}}$. If $t \mapsto u(\cdot, t)$ is continuous in $(0, T)$ with values in L^2 (interpreting (3.6) in the sense of distributions) we can apply the operator $\mathscr{F}$. We then have the equivalent differential system, for $v := \mathscr{F}u(\xi, t)$,

$$\partial_t v = -iA(\xi, t)v, \tag{3.9}$$

$$v(\xi, 0) = v_0(\xi) := \mathscr{F}u_0(\xi), \tag{3.10}$$

where $A(\xi, t) := \sum_{1\le\alpha\le d} \xi_\alpha A^\alpha(t)$.

We thus express v with the help of the resolvent: $v(\xi, t) = R(t, 0, \xi)v_0(\xi)$ where $R(\cdot, s; \xi)$ is the matrix solution of the following Cauchy problem:

$$\frac{d}{dt} R(\cdot, s; \xi) = -iA(\xi, \cdot)R(\cdot, s; \xi),$$

$$R(s, s; \xi) = \mathrm{Id}_n.$$

Since $\mathscr{F}$ is an isometry, the Cauchy problem (3.6) is well-posed in L^2 if and only if there exists a constant C_T independent of u_0 and such that

$$\sup_{t\in(0,T)} \|v(t)\|_{L^2} \le C_T \|v_0\|_{L^2}.$$

But since $v_0 \mapsto v(t)$ is just a multiplication operator $v_0(\xi) \mapsto R(t, 0, \xi)v(\xi)$, this is equivalent to saying that

$$\sup_{\substack{t\in(0,T)\\ \xi\in\mathbb{R}^d}} \|R(t, 0, \xi)\| \le C_T. \tag{3.11}$$

In this inequality, the constant C_T depends on the matrix norm chosen, but the fact that it is finite is independent of this. Making the change of variables $(t, \xi) \mapsto (t/a, a\xi)$, the system is transformed into

$$\frac{\mathrm{d}}{\mathrm{d}\tau} R = -\mathrm{i}A(\eta, a\tau)R,$$

which shows that $R(a\tau, 0, \eta/a) \to \exp(-\mathrm{i}\tau A(\eta, 0))$ when $a \to 0$. We deduce therefore that a necessary condition for (3.11) is

$$\sup_{\substack{\tau\in\mathbb{R}\\ \eta\in\mathbb{R}^d}} \|\exp(-\mathrm{i}\tau A(\eta))\| \le C_T, \tag{3.12}$$

where $A(\eta)$ stands for the restriction of A to the initial time or, similarly, to any other instant.

Definition 3.1.1 We say that the linear system with constant coefficients

$$\partial_t u + \sum_{1\le i\le d} A^\alpha \partial_{x_\alpha} u = 0 \tag{3.13}$$

is *hyperbolic* if there exists C such that $\sup_{\xi\in\mathbb{R}} \|\exp(-\mathrm{i}A(\xi))\| \le C$.

The first important result is the following.

Theorem 3.1.2 *For a linear system of the first order with constant coefficients, the Cauchy problem is well-posed in L^2 if and only if this system is hyperbolic. For a hyperbolic system, being given $u_0 \in L^2(\mathbb{R}^d)$, there exists one and only one solution of (3.6) in $\mathscr{C}(\mathbb{R}; L^2(\mathbb{R}^d))$.*

Proof From the preceding analysis, we see that the hyperbolicity ensures that the Cauchy problem is well-posed in L^2 and more precisely that $\|u(t)\|_{L^2} \le C\|u_0\|_{L^2}$ for all $t \in \mathbb{R}$, for $R(t, 0, \xi) = \exp(-\mathrm{i}A(t\xi))$. Let us show that in fact $t \mapsto u(t)$ is a continuous map of $\mathbb{R}$ into $\mathbb{R}^d$. Now $|v(\xi, t)| \le C|v_0(\xi)|$, which bounds $|v|$ above by a square-integrable function independent of t. As $t \mapsto v(\xi, t)$ is continuous, the theorem of dominated convergence assures the continuity demanded.

Conversely, suppose that the Cauchy problem is well-posed on $(0, T)$, the solution being of class L^2. Then for t fixed and non-zero, $u_0 \mapsto u(t)$ and thus $v_0 \mapsto$

$v(t)$ are continuous endomorphisms of L^2. As the second is only a multiplication operator by $\exp(-\mathrm{i}tA(\xi))$, we easily calculate its norm, which takes the value $\sup_{\xi\in\mathbb{R}^d}\|\exp(-\mathrm{i}tA(\xi))\|$. It must therefore be bounded. ■

Geometric conditions of hyperbolicity

It is not simple, *a priori*, to verify the property of hyperbolicity for a given system since it demands the calculation of exponentials of matrices depending on d parameters. This calculation can only be carried out by the diagonalisation of each matrix $A(\xi)$, by reason of the formula $\exp(PMP^{-1}) = P(\exp M)P^{-1}$. First of all, hyperbolicity implies

$$\sup_{\xi\in\mathbb{R}^d} \rho(\exp(-\mathrm{i}A(\xi))) < \infty,$$

where $\rho(M)$ denotes the spectral radius of a matrix M. But as

$$\exp(-\mathrm{i}A(m\xi)) = (\exp(-\mathrm{i}A(\xi)))^m$$

for every integer m, it is equivalent to saying that $\rho(\exp(-\mathrm{i}A(\xi))) = 1$ for all $\xi\in\mathbb{R}^d$, that is, since the eigenvalues of $\exp M$ are the exponentials of those of M, that the spectrum $\mathrm{Sp}A(\xi)$ of the matrices $A(\xi)$ is real. In addition, if one of these matrices possesses an eigenvalue λ of which the algebraic and geometrical multiplicities differ one from the other, then there exist two non-zero vectors w and z such that $Aw = \lambda w$ and $Az = \lambda z + w$. We then deduce that

$$\mathrm{e}^{-\mathrm{i}tA(\xi)}z = \mathrm{e}^{-\mathrm{i}t\lambda}(z - \mathrm{i}tw),$$

which contradicts the boundedness condition (3.12). We have therefore

Lemma 3.1.3 *If the system (3.13) is hyperbolic, then the matrix $A(\xi)$ is diagonalisable with real eigenvalues, for all ξ in $\mathbb{R}^d$.*

Although the converse of this lemma turns out to be true (this is the so-called Kreiss matrix theorem), we shall give two proofs of hyperbolicity under more restrictive (but rather natural) conditions. Let us look first of all at the case $d = 1$. Writing $A := A(1)$, we have $A(\xi) = \xi A$. If A is diagonalisable with real eigenvalues, $A = PDP^{-1}$, we have $\exp(-\mathrm{i}A(\xi)) = P(\exp(-\mathrm{i}\xi D))P^{-1}$. Now $\exp(-\mathrm{i}\xi D)$ is a diagonal matrix whose diagonal terms are the complex numbers $\mathrm{e}^{\mathrm{i}\lambda}$ of modulus one where $\lambda\in\mathrm{Sp}(A)$, it is therefore bounded and the same is true of $\exp(-\mathrm{i}A(\xi))$.

In the case $d \geq 2$, we proceed in the same way, but the matrices P and D depend on $\xi : \exp(-\mathrm{i}A(\xi)) = P(\xi)\exp(-\mathrm{i}D(\xi))P^{-1}(\xi)$. The matrix D is homogeneous of degree 1 with respect to ξ and we can choose P to be homogeneous of degree 0. Then

$$\|\exp(-\mathrm{i}A(\xi))\| \le K(\xi)\|\exp(-\mathrm{i}D(\xi))\|,$$

where $K(\xi) := \|P(\xi)\| \cdot \|P(\xi)^{-1}\|$. If the matrix $A(\xi)$ is diagonalisable with real eigenvalues, then again $\|\exp(-\mathrm{i}D(\xi))\| = 1$ and so $\|\exp(-\mathrm{i}A(\xi))\| \le K(\xi)$. From this we have the sufficient condition

Proposition 3.1.4 *We suppose that the matrices $A(\xi)$ are diagonalisable with real eigenvalues, uniformly with respect to ξ, that is that $\xi \mapsto K(\xi) = \|P(\xi)\| \cdot \|P(\xi)^{-1}\|$ is bounded on $\mathbb{R}^d$. Then the system (3.13) is hyperbolic.*

An essential application of this proposition is the following.

Definition 3.1.5 The system (3.13) is *symmetrisable hyperbolic* if there exists a positive definite symmetric matrix S such that the matrices $S^\alpha := SA^\alpha$ are symmetric.

Theorem 3.1.6 *If the system (3.13) is symmetrisable hyperbolic, then it is hyperbolic.*

Proof Let $S(\xi) := \sum_{1\le\alpha\le d} \xi_\alpha S^\alpha$. We have $A(\xi) = S^{-1}S(\xi)$. Let Σ be the positive symmetric square root of S^{-1}. As $\Sigma S(\xi)\Sigma$ is symmetric, it is diagonalisable in an orthonormal basis, that is to say that there exists a matrix $O(\xi) \in O_n(\mathbb{R})$ such that $O(\xi)^{-1}\Sigma S(\xi)\Sigma O(\xi)$ is diagonal and real. Then we can choose $P(\xi) = \Sigma O(\xi)$ and we have $K(\xi) \le K_0 := \|\Sigma\|\|\Sigma^{-1}\|$. ∎

Another favourable case, independent of the preceding one, is that of strictly hyperbolic systems.

Definition 3.1.7 We say that the system (3.6) is *strictly hyperbolic* if the matrices $A(\xi)$ are diagonalisable with real eigenvalues, with constant multiplicities when ξ ranges over $\mathbb{R}^d - \{0\}$.

It comes to the same thing to say that the eigenvalues are continuous functions on $\mathbb{R}^d - \{0\}$, $\xi \mapsto \lambda_j(\xi)$, with

$$\lambda_1(\xi) < \lambda_2(\xi) < \cdots < \lambda_r(\xi).$$

Theorem 3.1.8 *If the system (3.13) is strictly hyperbolic, then it is hyperbolic.*

Proof The key point of the proof is a geometrical lemma, which uses only the continuity of the mapping $\xi \mapsto A(\xi)$.

Lemma 3.1.9 *If the matrices $A(\xi)$ are diagonalisable with eigenvalues of multiplicities independent of ξ for $\xi \neq 0$, then the eigenspaces depend continuously (and even analytically, but that is indifferent to us) on ξ.*

For all $\xi_0 \neq 0$, there thus exist a compact neighbourhood $V(\xi_0)$ and a continuous option, hence bounded, of $P(\xi)$ on $V(\xi_0)$. As the sphere S^{d-1} is compact, it is covered by a finite number of such neighbourhoods, with the result that we can choose a bounded option (but not necessarily continuous) of P on S^{d-1}, that is on $\mathbb{R}^d$ since P is homogeneous of degree zero with respect to ξ.

It remains to prove the lemma. Let $\xi_0 \neq 0$ and ξ_m be a sequence tending to ξ_0. The eigenvalues $\lambda_j(\xi)$ are continuous functions of the coefficients, thus of ξ. Let n_j be the dimension of the sub-eigenspace associated with $E_j(\xi)$. As the grassmannian variety of the sub-spaces of dimension n_j of $\mathbb{R}^n$ is compact, the sequence $E_j(\xi_m)$ takes a limiting value F_j which is included, by continuity, in $E_j(\xi_0)$. Their dimensions being the same, these two sub-spaces are equal. ∎

Example 3.1.10 Let us consider the example of the model below, for which $n = d = 2$:

$$u_t + \begin{pmatrix} 1 & 0 \\ 0 & -1 \end{pmatrix} u_x + \begin{pmatrix} 0 & 1 \\ 1 & 0 \end{pmatrix} u_y = 0. \tag{3.14}$$

We have

$$A(\xi) = \begin{pmatrix} \xi_1 & \xi_2 \\ \xi_2 & -\xi_1 \end{pmatrix}.$$

The eigenvalues are $\pm|\xi|$, that is to say that the speeds of propagation (see below) take the values ± 1. They are independent of the direction, which is not the general case but corresponds to an invariance of the system under the group $O_2(\mathbb{R})$.

Plane waves

The normalised eigenvalues $c_j(\xi) := \lambda_j(\xi)/|\xi|$ must be seen as the speeds of propagation of plane waves in the direction ξ for $\xi \neq 0$. There are two ways in which to see that.

First of all, if $u_0 \in L^2(\mathbb{R}^d)$, the solution of the Cauchy problem is given formally by

$$u(x,t) = (2\pi)^{-\frac{1}{2}n} \int_{\mathbb{R}^d} P(\xi) e^{i(x\cdot\xi I_2 - tD(\xi))} w_0(\xi)\, d\xi,$$

where $w_0(\xi) := P(\xi)^{-1} v_0(\xi)$. Thus, u appears as an infinite sum of one-dimensional solutions

$$(x,t) \mapsto P(\xi) e^{i(x\cdot\xi I_2 - tD(\xi))} w_0(\xi).$$

These can be decomposed in their turn, using the column vectors $r_j(\xi)$ of $P(\xi)$, which are the eigenvectors of $A(\xi)$, into plane waves of the form

$$(x,t) \mapsto a_j(\xi) e^{i(x\cdot\xi - t\lambda_j(\xi))} r_j(\xi).$$

The second approach, more elementary, consists of verifying that, for all $\xi \neq 0$ and every locally integrable function f, the plane wave

$$u(x,t) := f(x\cdot\xi - t\lambda_j(\xi)) r_j(\xi)$$

is a weak solution of (3.13). Since here $u(x,t) = u_0(x - c_j(\xi)t\nu)$ where $\nu = \xi/|\xi|$ is a unit vector, the number $c_j(\xi)$ clearly plays the rôle of a speed of propagation.

Exercises

3.1 Show that every scalar equation ($n = 1$) is hyperbolic.

3.2 Show that if $d = 1$, every hyperbolic system is symmetrisable.

3.3 Assume that the matrices A^α commute with each other: $A^\alpha A^\beta = A^\beta A^\alpha$.

(1) Show that (3.6) is hyperbolic if and only if each one-dimensional system $v_t + A^\alpha v_x = 0$ is hyperbolic.

(2) By a linear change of variables $u \mapsto v := Pu$, show that the system is equivalent to n decoupled transport equations:

$$\partial_t v_i + V_i \cdot \nabla_x v_i = 0.$$

3.4 Show that the system of Maxwell's electromagnetic equations is hyperbolic. Here $n = 6, d = 3$ and u is composed of two vector fields B and E. The equations are

$$B_t + \operatorname{curl} E = 0, \tag{3.15}$$

$$E_t - c^2 \operatorname{curl} B = 0. \tag{3.16}$$

Calculate the speeds of propagation and determine which correspond to plane waves of a physical nature, that is, which satisfy the constraint $\operatorname{div} B = 0$.

3.5 We consider the linear system of isotropic elasticity of small deformations. The unknown is the *displacement* $y(x,t) \in \mathbb{R}^d$. The equations of the second order are

$$\frac{\partial^2 y_i}{\partial t^2} = \alpha \Delta y_i + \beta \partial_i \operatorname{div} y.$$

The parameters $\alpha > 0$, $\beta > 0$ are the *Young's moduli* of the material.

(1) Put the system into a first order form with a constraint,

$$\partial_t u + \sum_{\alpha=1}^{d} A^\alpha \partial_\alpha u = 0, \qquad \sum_{\alpha=1}^{d} B^\alpha \partial_\alpha u = 0.$$

(2) Calculate the plane waves. Deduce that the system is strictly hyperbolic. Calculate the speeds of propagation. *Note*: they are traditionally denoted by $c_S < c_P$ and correspond respectively to shear waves, in which the material vibrates transversely to the direction of propagation, and to pressure waves where the vibration is parallel to this propagation.
(3) Show that the system is symmetric hyperbolic in expressing the conservation of mechanical energy. This is the sum of a kinetic term and of an energy of deformation.

3.6 We suppose that $u_0 \in H^s(\mathbb{R}^d)$, that it to say that $\xi \mapsto (1 + |\xi^2|)^{\frac{1}{2}s} v_0(\xi)$ is square-integrable. Show that the weak solution of a hyperbolic problem (3.6) satisfies $u \in \mathscr{C}(\mathbb{R}; H^s(\mathbb{R}^d)) \cap \mathscr{C}^1(\mathbb{R}; H^{s-1}(\mathbb{R}^d))$.

3.7 (1) Let $\theta \in \mathbb{R}$. Find a matrix M_θ such that $u \mapsto u^\theta$ where

$$u^\theta(x, y, t) := M_\theta u(x \cos\theta + y \sin\theta, -x \sin\theta + y \cos\theta, t)$$

preserves the set of solutions of (3.14).
(2) Show that it is not possible to choose the matrix of the change of basis $P(\xi)$ with the result that $\xi \mapsto P(\xi)$ is continuous on S^1.

3.8 We assume that $u_0 \in H^s(\mathbb{R}^d)$ for $s > 0$ sufficiently large (see Exercise 3.6) and we consider a symmetrisable hyperbolic system.

(1) Show that there exists a number $M > 0$ such that, for all $\xi \in S^{d-1}$ and all $w \in \mathbb{R}^n$, we have

$$|(S(\xi)w, w)| \leq M(Sw, w),$$

where $(\cdot,\cdot)$ denotes the usual scalar product in $\mathbb{R}^n$.
(2) Verify, for the solution of the Cauchy problem, the equality

$$\partial_t(Su, u) + \sum_{1\leq\alpha\leq d} \partial_{x_\alpha}(SA^\alpha u, u) = 0.$$

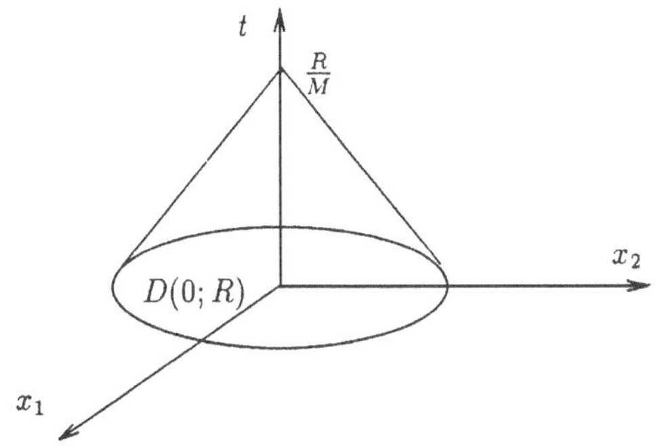

Fig. 3.1: The cone of dependence of a disk $D(0, R)$.

(3) Deduce that, if u_0 is zero for $|x| < R$, then u is zero in the interior of the cone defined by $t > 0$ and $|x| + Mt < R$ (integrate the preceding formula on the cone: see Fig. 3.1).

(4) Express this result in terms of a propagation phenomenon with finite speed.

3.9 We consider a hyperbolic system for which $n = d = 2$.

(1) By a linear change of variable $u \mapsto v := Pu$, reduce to the case where A^1 is a diagonal matrix.

(2) Show that if A^1 is of the form $aI_2, a \in \mathbb{R}$, the system (3.6) may be reduced to the one-dimensional case, with a given initial value depending on a parameter.

(3) We suppose now that A^1 is diagonal but is not of the form aI_2. In calculating the characteristic polynomial of $A^2 + xA^1$, show that either A^2 is diagonal or $a_{12}^2 a_{21}^2 > 0$.

(4) Show then that the system is symmetrisable.

3.10 We consider the system (3.6), where the matrices A^α depend on the time t. We suppose that at each instant, the system is symmetrisable by a matrix $S_0(t)$ which is of class $\mathscr{C}^1$ with respect to t:

$S_0(t)$ is symmetric and positive definite,
$S(\xi, t) := S_0(t)A(\xi, t)$ is symmetric.

(1) Show that $(R^* S_0 R)_t = R^*(\mathrm{d}S_0/\mathrm{d}t)R$, where $R(t, 0, \xi)$ is the resolvent considered above.

(2) Show that there exists a number $c_T > 0$ such that, for all $t \in [0, T]$ and all $\xi \in \mathbb{R}^d$, $\mathrm{Tr}(R^*(\mathrm{d}S_0/\mathrm{d}t)R) \leq c_T \mathrm{Tr}(R^* S_0 R)$.

(3) Deduce that there exists a number $C_T > 0$ such that, for all $t \in [0, T]$ and all $\xi \in \mathbb{R}^d$, $\|R(\xi, t)\| \leq C_T$.

The Cauchy problem for (3.6) is therefore well-posed in $L^2(\mathbb{R}^d)$.

3.11 We suppose that (3.6) is symmetrisable hyperbolic. Let $B \in M_n(\mathbb{R})$. We wish to show that the Cauchy problem for the system

$$\partial_t u + \sum_{\alpha=1}^{d} A^\alpha(t)\partial_{x_\alpha} u = Bu \tag{3.17}$$

is well-posed in $L^2(\mathbb{R}^d)$. We denote by $S(t)$ the solution operator of (3.6), $u_0 \mapsto S(t)u_0 = u(t)$ (see the preceding exercise for the construction of this operator, an endomorphism of $H^s(\mathbb{R}^d)$ (for $s \geq 0$).

(1) Let $u_0 \in L^2(\mathbb{R}^d)$ and $f \in L^1(\mathbb{R}; L^2(\mathbb{R}^d))$. Show that the inhomogeneous Cauchy problem

$$\partial_t u + \sum_{1\leq\alpha\leq d} A^\alpha(t)\partial_{x_\alpha} u = Bf, \tag{3.18}$$

$$u(x,0) = u_0(x) \tag{3.19}$$

possesses a unique solution in $\mathscr{C}(\mathbb{R}; L^2(\mathbb{R}^d))$, given by Duhamel's formula

$$u(t) = S(t)u_0 + \int_0^t S(t-s)Bf(s)\,\mathrm{d}s.$$

We write $u = \mathscr{T} f$.

(2) We construct a sequence $(u^m)_{m\in\mathbb{N}}$ by $u^0(x,t) \equiv u_0(x)$, and $u^{m+1} = \mathscr{T} u^m$. Show that $\mathscr{T}$ restricted to the space $L^1(0,T; L^2(\mathbb{R}))$ is a contraction mapping provided that $T > 0$ is sufficiently small.

(3) Deduce that the sequence $(u^m)_{m\in\mathbb{N}}$ converges in $\mathscr{C}(0,T; L^2(\mathbb{R}^d))$ and that its limit is a weak solution of (3.17) in the band $\mathbb{R}^d \times (0,T)$.

(4) Making use of the fact that the number T does not depend on u_0 show that (3.17) possesses a weak solution on $\mathbb{R}^d \times \mathbb{R}$.

(5) Show that the mapping $u_0 \mapsto u$ is continuous in $L^2(\mathbb{R}^d)$ in $\mathscr{C}(0,T; L^2(\mathbb{R}^d))$, u being the solution of (3.17).

(6) Show that there exists a constant C, depending only on S_0 and B, such that

$$\frac{\mathrm{d}}{\mathrm{d}t}\int_{\mathbb{R}^d}(S_0 u, u)\,\mathrm{d}x \leq \|B\|\int_{\mathbb{R}^d}|u|^2\,\mathrm{d}x \leq C\int_{\mathbb{R}^d}(S_0 u, u)\,\mathrm{d}x.$$

(Do it first of all for $u_0 \in H^s(\mathbb{R}^d)$ for s sufficiently large, then deduce the general case with the help of the preceding question).

(7) Deduce that the solution of (3.17) is unique (reduce to the case $u_0 \equiv 0$, then apply Gronwall's lemma).

3.2 Quasi-linear hyperbolic systems

We return to the case of quasi-linear systems, that is to say to systems of the form (3.1). We have seen that a formal analysis of the stability of perturbations of small amplitude, via an asymptotic development with respect to this amplitude, requires the hyperbolicity of the linearised system. That condition is in fact far from being sufficient, but as we have not found one which is truly satisfying, mathematicians have for a long time adopted the following definition. We shall use the notation

$$A(\xi; u) = \sum_{1 \leq \alpha \leq d} \xi_\alpha A^\alpha(u)$$

and more generally $P(\xi; u), \ldots$, for the matrices having been defined in the study of the linear case but which now depend on the state u of the system.

Definition 3.2.1 The quasi-linear system (3.1) is said to be *hyperbolic* if for all $\bar{u}$, the linear system

$$\partial_t u + \sum_{1 \leq \alpha \leq d} A^\alpha(\bar{u}) \partial_{x_\alpha} u = 0$$

is hyperbolic, the matrices $P(\xi; u)$ and their inverses being bounded on every compact set of $S^d \times \mathcal{U}$.

This definition does not assure us that the Cauchy problem for (3.1) is well-posed (one may no longer apply Kreiss' matrix theorem; much more, the well-posedness is no longer a matter of matrices only), even in a space of smooth functions and locally in time. Its popularity comes from the fact that it is invariant under the change of unknown $u \mapsto v = \varphi(u)$. If φ is a diffeomorphism of $\mathcal{U}$ into $\mathcal{V}$, this change of variable transforms a quasi-linear system into another quasi-linear system

$$\partial_t v + \sum_{1 \leq \alpha \leq d} B^\alpha(v) \partial_{x_\alpha} v = 0,$$

where the matrices B^α are conjugate to the matrices A^α:

$$B^\alpha(\varphi(u)) = \mathrm{d}\varphi(u)\, A^\alpha(u)\, (\mathrm{d}\varphi(u))^{-1}.$$

In fact, if $A(\xi; u) = P(\xi, u) D(\xi, u) (P(\xi, u))^{-1}$, we diagonalise $B(\xi, u)$ by the matrix $\mathrm{d}\varphi(u) P(\xi, u)$ which is bounded on every compact set when P is.

However, other changes of the unknown field (the *dependent* variables) can transform a hyperbolic quasi-linear system into a non-hyperbolic quasi-linear system. The basic example is the following, which is clearly hyperbolic in the sense of the definition given above. We have $n = 2$, $d = 1$:

$$\partial_t u + \begin{pmatrix} u_1 & 0 \\ 0 & u_1 \end{pmatrix} \partial_x u = 0. \tag{3.20}$$

We transform this system by $v_1 = u_1$, $v_2 = \partial_x u_2$ and we obtain, on differentiating the second equation of (3.20) with respect to x,

$$\partial_t v + \begin{pmatrix} v_1 & 0 \\ v_2 & v_1 \end{pmatrix} \partial_x v = 0, \tag{3.21}$$

whose matrix is no longer diagonalisable, except for $v_2 = 0$.

3.3 Conservative systems

The systems originating in physics or mechanics form in fact a much more restricted class than those of the first order quasi-linear systems. These are the systems of conservation laws which can be written in the form

$$\partial_t u + \partial_{x_\alpha} f^\alpha(u) = 0, \tag{3.22}$$

where f^α is a smooth vector field defined on $\mathcal{U}$. We can re-write such a system under the quasi-linear form. We shall then have $A^\alpha(u) = \mathrm{d} f^\alpha(u)$ and $A(\xi, u) = \mathrm{d}(\xi \cdot f)(u)$.

The conservation principle which is essential to define the weak solutions beyond the formation of the discontinuities is not preserved by the diffeomorphisms $u \mapsto v = \varphi(u)$ in general. And even when a diffeomorphism φ transforms (3.22) into another conservative system $v_t + \operatorname{div} g(v) = 0$, the notion of a weak solution will be in general modified, which is unacceptable because, even in dimension $d = 1$, the Rankine–Hugoniot conditions $[f(u)] = \sigma[u]$ and $[g(\varphi(u))] = \sigma[\varphi(u)]$ are not equivalent. There are notable exceptions. For example when f is linear, and more generally if all the characteristic fields (see below) are linearly degenerate (*idem*). For a general system, the only transformations which preserve the notion of a weak solution are the affine functions $u \mapsto Au + b$, because then $g(v) = Af(A^{-1}(v - b))$ and $[g(v)] = A[f(u)] = \sigma A[u] = \sigma[v]$. From which comes the importance of an affine theory of the system of laws of conservation. One should consult the thesis of B. Sévennec [93] on this subject. The simplest result in this direction is that of G. Boillat. For this statement we first of all need a new concept.

Definition 3.3.1 A *characteristic field* of a quasi-linear system of the form (3.1) is a mapping $(\xi, u) \mapsto (\lambda, E)$ defined and smooth on an open set $\mathcal{O}$ of $\mathcal{U}$, where λ is an eigenvalue of $A(\xi, u)$, of constant multiplicity, and E the associated eigenspace, whose dimension is the multiplicity of λ.

We have thus excluded the case where λ is associated with a non-trivial Jordan form in the decomposition into characteristic sub-spaces of the matrix $A(\xi, u)$. Clearly $\xi \mapsto \lambda(\xi, u)$ is homogeneous of degree 1. When there is a single spatial dimension we set $\xi = 1$ and a characteristic field is merely a mapping $u \mapsto (\lambda, E)$. Another essential notion is that of *differential eigenform*, that is to say of a left eigenvector field $(\xi, u) \mapsto l$:

$$l(\xi, u)(A(\xi, u) - \lambda(\xi, u)I_n) = 0.$$

Definition 3.3.2 A characteristic field is said to be *linearly degenerate* on $\mathcal{O}$ if the differential of λ is zero on $E = \mathrm{Ker}(A - \lambda I_n)$ when u ranges over $\mathcal{O}$:

$$\mathrm{d}_u\lambda \cdot r \equiv 0, \quad \forall r \in \mathrm{Ker}(A - \lambda I_n), \quad \forall u \in \mathcal{O}.$$

Theorem 3.3.3 (Boillat [3]) *Let us consider a system of conservation laws (3.22). We suppose that $A(\xi, u)$ has an eigenvalue $\lambda(\xi, u)$ whose multiplicity m is a constant greater than or equal to 2. Then the characteristic field $(\lambda, \mathrm{Ker}\,(A - \lambda I_n))$ is linearly degenerate. In addition, $\xi \neq 0$ being given, the affine sub-spaces $u + \mathrm{Ker}\,(A(\xi, u) - \lambda(\xi, u)I_n)$ are the tangent spaces to a family of sub-manifolds of dimension m.*

Obviously, the integral sub-manifolds mentioned in the theorem form a foliation of the open set $\mathcal{O}$ called the *characteristic foliation* associated with λ. The characteristic foliation depends only on the direction of ξ but on neither its sense nor its norm. If the multiplicity of λ has the value 1, then $\mathrm{Ker}\,(A - \lambda I_n)$ is generated by a vector $r(\xi, u)$, with the result that the foliation still exists, formed by the integral curves of the vector field $u \mapsto r$.

Proof Let us fix $\xi \neq 0$. Let $u \mapsto r$ be a smooth field of eigenvectors on $\mathcal{O}$. Since $m > 1$, we can choose a second field of eigenvectors $u \mapsto s$, smooth and linearly independent of the first at every point. Differentiating the relation $(\mathrm{d}_u(\xi \cdot f) - \lambda(u))r(u) \equiv 0$ in the direction s we obtain

$$\mathrm{D}^2(\xi \cdot f)(r, s) - (\mathrm{d}_u\lambda \cdot s)r = (\mathrm{d}f - \lambda)\mathrm{d}_u r \cdot s.$$

Interchanging the rôles of r and s, we have also

$$\mathrm{D}^2(\xi \cdot f)(s, r) - (\mathrm{d}_u\lambda \cdot r)s = (\mathrm{d}f - \lambda)\mathrm{d}_u s \cdot r.$$

As $\mathrm{D}^2 f$ is a symmetric bilinear form, we can eliminate the term $\mathrm{D}^2(\xi \cdot f)(r, s)$ between these two equalities. This gives

$$(\mathrm{d}f - \lambda)\{r, s\} = (\mathrm{d}_u\lambda \cdot s)r - (\mathrm{d}_u\lambda \cdot r)s, \tag{3.23}$$

where $\{r, s\}$ denotes the Poisson bracket of the field vectors r and s.

The right-hand side of equation (3.23) is a vector of $\mathrm{Ker}(d\xi \cdot f - \lambda)$ with the result that $\{r, s\} \in \mathrm{Ker}(d\xi \cdot f - \lambda)^2 = \mathrm{Ker}(d\xi \cdot f - \lambda)$, the second equality being due to the equality of the algebraic and geometric multiplicities of λ. The set of eigenvector fields associated with the eigenvalue λ is thus a Lie algebra. This property, called the *Frobenius integrability condition*, ensures the existence of the foliation whose affine spaces $u + \mathrm{Ker}(A(\xi, u) - \lambda(\xi, u)I_n)$ are the tangent spaces.

Finally, again using (3.23), we find the relation of linear dependence $(d_u\lambda \cdot s)r - (d_u\lambda \cdot r)s = 0$, which implies the nullity of the coefficients. For example, we have $d_u\lambda \cdot r = 0$. The characteristic field is degenerate. ∎

In the light of this theorem, we are led to a finer definition of hyperbolicity in the case of a quasi-linear system. In addition to imposing that the matrices $A(\xi, u)$ are diagonalisable on $\mathbb{R}$ with eigenvalues of constant multiplicities m_j we shall demand, when $m_j \geq 2$, that the corresponding characteristic field be linearly degenerate and that the eigenvector fields form a Lie algebra. This now excludes the pathological example (3.20) of the preceding section.

3.4 Entropies, convexity and hyperbolicity

Physical systems

Most of the systems arising in physics or mechanics are conservative systems and admit a supplementary conservation law of the form $E(u)_t + \mathrm{div}_x F(u) = 0$ where the time component E is a strictly convex function on $\mathcal{U}$. The convexity makes good sense here since it is an affine notion and the only group of transformations of dependent variables u which we accept is the affine group. On the other hand, care must be taken not to apply a non-linear change of variables, even in preserving the conservative nature of the system, for the function $v \mapsto E(\varphi^{-1}(v))$ need not be convex if E is.

Definition 3.4.1 We say that a real function $u \mapsto E$ is an *entropy* of the system (3.1) if there exists a mapping $u \mapsto F(u)$ with values in $\mathbb{R}^d$, called the entropy flux, such that every classical solution of (3.1) satisfies the equality $E(u)_t + \mathrm{div}_x F(u) = 0$.

The entropy–entropy-flux pairs are thus the solutions of the linear first order equations in $\mathcal{U}$,

$$\frac{\partial F^\alpha}{\partial u_j} = \sum_{1 \leq i \leq n} a_{ij}^\alpha \frac{\partial E}{\partial u_i}, \quad \forall 1 \leq j \leq n, \ \ 1 \leq \alpha \leq d. \tag{3.24}$$

The entropies and their convexity have an essential rôle in the theory of hyperbolic systems of conservation laws. In particular, the mere convexity assures the hyperbolicity.

Theorem 3.4.2 (26, 35) *If a conservative system (3.22), whose state $u(x, t)$ takes its values in a convex domain $\mathcal{U}$, possesses a strictly convex entropy in the sense that $D_u^2 E$ is positive definite at every point, then matrices $A(\xi, u)$ are symmetrisable: there exist positive definite symmetric matrices $S(u)$ (in fact $S(u) = D_u^2 E$) such that the matrices $SA(\xi, u)$ are symmetric.*

From Theorem 3.1.6, such a system is thus hyperbolic, strictly hyperbolic where the eigenvalues are of constant multiplicities. We thus shall adopt the following definition.

Definition 3.4.3 A *physical system* is a system of conservation laws whose state $u(x, t)$ takes its values in a convex domain $\mathcal{U}$ of $\mathbb{R}^n$ and which possesses a strongly convex entropy on $\mathcal{U}$.

Proof of theorem

We introduce the conjugate convex function $E^*(q) := \sup_{u\in\mathcal{U}}(q\cdot u - E(u))$, defined on the range of $d_u E$, and we make the change of variables $q := d_u E(u)$ whose inverse is $u = d_q E^*(q)$. We have $E^*(q) = q \cdot u - E(u)$. The matrix $S(u) = D^2 E(u)$ is positive definite symmetric and we have $Su_t = q_t$.

We rewrite the equations (3.24):

$$\frac{\partial F^\alpha}{\partial u_j} = \sum_{1\le i\le n} q_i \frac{\partial f_i^\alpha}{\partial u_j}.$$

Writing $g(q) = f(u(q))$, we have also

$$\frac{\partial f_i^\alpha}{\partial u_j} = \sum_{1\le k\le n} \frac{\partial g_i^\alpha}{\partial q_k}\frac{\partial q_k}{\partial u_j} = \sum_{1\le k\le n} \frac{\partial g_i^\alpha}{\partial q_k}\frac{\partial^2 E}{\partial u_j \partial u_k},$$

and, similarly, with $H(q) := F(u(q))$,

$$\frac{\partial F^\alpha}{\partial u_j} = \sum_{1\le k\le n} \frac{\partial H^\alpha}{\partial q_k}\frac{\partial q_k}{\partial u_j} = \sum_{1\le k\le n} \frac{\partial H^\alpha}{\partial q_k}\frac{\partial^2 E}{\partial u_j \partial u_k}.$$

The equation which governs the entropies and their fluxes can thus be written

$$S d_q H = S \sum_{1\le i\le n} q_i d_q g_i.$$

Since S is invertible, this becomes $d_q H = \sum_{1\le i\le n} q_i d_q g_i = d_q(\sum_{1\le i\le n} q_i g_i) - g$. Finally, we have

$$g_k^a = \frac{\partial h^\alpha}{\partial q_k},$$

where

$$h^\alpha = \left(\sum_{1 \le i \le n} \frac{\partial E}{\partial u_i} f_i^\alpha - F^\alpha \right) \circ u.$$

The system (3.22) thus has the equivalent form (even when this concerns the weak solutions)

$$\partial_t (\mathrm{d}_q E^*) + \mathrm{div}_x (\mathrm{d}_q h) = 0. \tag{3.25}$$

Finally, in quasi-linear form, for the classical solutions, it is symmetric:

$$S^{-1} \partial_t q + \sum_{1 \le \alpha \le d} \mathrm{D}^2 h^\alpha \partial_{x_\alpha} q = 0.$$

In particular, $SA^\alpha = S(D^2 h^\alpha) S$ is symmetric. ∎

Exercises

3.12 For a hyperbolic linear system $u_t + Au_x = 0$, find all the entropies. Determine those which are convex.

3.13 For a decoupled system of scalar conservation laws $\partial_t u_i + \partial_x f_i(u_i) = 0$, $1 \le i \le n$, find all the entropies and determine those which are convex.

3.14 Let $H : \mathcal{U} \to \mathbb{R}$ be a smooth function. Find a non-trivial entropy (i.e., non-affine) for the system

$$\partial_t u_i + \partial_x \frac{\partial H}{\partial u_i} = 0, \quad 1 \le i \le n.$$

3.15 Let $\mathcal{U} = (0, +\infty) \times \mathbb{R}^{n-1}$ and $H : \mathcal{U} \to \mathbb{R}$ be a smooth function. Find a list of entropies of the form $E^g(u) = E_0(u) g(q_1, \dots, q_{n-1})$, parametrised by smooth functions g of $n - 1$ variables, for the Keyfitz and Kranzer system:

$$\partial_t u_i + \partial_x (H(u) u_i) = 0.$$

3.16 (Converse to Theorem 3.3.3) Let $u_t + f(u)_x = 0$ be a strictly hyperbolic system whose one eigenvalue λ is linearly degenerate. Let $v := (u, z) \in \mathcal{U} \times \mathbb{R}$ be new dependent variables. We consider the augmented system $v_t + g(v)_x = 0$, defined by

$$u_t + f(u)_x = 0,$$
$$z_t + (\lambda(u) z)_x = 0.$$

Show that this system is hyperbolic and has the same propagation speeds as the preceding system, the multiplicity of λ being augmented by unity.

3.17 This problem occurs in spatial dimension $d = 1$. We suppose that the eigenvalues of $\mathrm{d}f(u)$ are real and strictly positive for all $u \in \mathcal{U}$ and that f is *proper*, that is to say that $\lim_{\mathrm{d}(u;\partial\mathcal{U})\to 0} |f(u)| = \infty$.

(1) Show that the mapping $u \mapsto v := f(u)$ is invertible from $\mathcal{U}$ into $\mathbb{R}^n$. We write $g = f^{-1}$. (Optional question, alternatively pass directly to the following question.)
(2) Using this change of dependent variables, we write $G(v) := F(u)$ and $H(v) := E(u)$. Show that $\mathrm{d}_v G(v) = \mathrm{d}_u E(u)$.
(3) Show that G is an entropy of flux H, of the system $v_s + (g(v))_y = 0$, and that G is strictly convex if and only if E is strictly convex. *Hint*: Verify that $D_v^2 G = \mathrm{D}_u^2 E \cdot \mathrm{d}_v g$ and deduce that the list of signs of eigenvalues of $\mathrm{d}g$ is equal to the signature of $D_v^2 G$ if $D_u^2 E > 0$.
(4) Deduce that the system is hyperbolic in the spatial direction x, that the notions of weak and entropy solutions are the same, that they respectively occur in $\mathbb{R}\times\mathbb{R}^+$ with t for time variable, or in $\mathbb{R}^+\times\mathbb{R}$ with x for time variable.

3.18 We consider gas dynamics in dimension 1 in its lagrangian representation

$$\begin{aligned} v_t &= u_x, \\ u_t + (p(v,e))_x &= 0, \\ \left(e + \frac{1}{2}u^2\right)_t + (up)_x &= 0. \end{aligned}$$

We denote by $S(v, e)$ a smooth function such that $S_v = pS_e$ and $S_e > 0$. We put $T = S_e^{-1}$.

(1) Show that the system is hyperbolic if and only if $pp_e - p_v > 0$. We shall then write $c(v, e) = \sqrt{(pp_e - p_v)}$.
(2) Show that for every numerical function g, $E = g \circ S$ is an entropy.
(3) We define the differential form $\alpha = p\,\mathrm{d}v + \mathrm{d}e = T\mathrm{d}S$. Show that

$$T^2\mathrm{D}^2E = (g'' - T_e g')\alpha^2 - Tg'(c^2\mathrm{d}v^2 - 2p_e\alpha\,\mathrm{d}v + \mathrm{d}u^2),$$

where D^2E is the hessian form of E in the variables $(v, u, \varepsilon := e + \frac{1}{2}u^2)$.
(4) Deduce that E is a convex entropy (with respect to $(v, u, e + \frac{1}{2}u^2)$) if and only if

i. $g' \le 0$,
ii. $c^2 g'' \ge T^2(p_v S_{ee} - p_e S_{ve})g'$.

(5) For a perfect gas ($P(v, e) = (\gamma - 1)e/v$ where $\gamma > 1$ is a constant), we can choose $T = e$ and $S = (\gamma - 1)\log v + \log e$. Show that E is convex if and only if $g' \le 0$ and $\gamma g'' + g' \ge 0$. In particular, $-S$ is itself a convex entropy.

3.19 We keep the notation of the preceding exercise. The aim of the present one is to determine all the entropies of gas dynamics in lagrangian coordinates in dimension 1. Let E be a general entropy and F its flux.

(1) Show that p and S are independent functions. Henceforth, we shall express E and F as functions of (u, p, S).
(2) Write down the equations satisfied by the pair (E, F). Show that E decomposes in the form $E = \varepsilon(p, S) + a(p, u)$.
(3) We suppose that c and p are two independent functions, that is to say that $c_S \neq 0$. Show that $\varepsilon_p + c\varepsilon_{pS}/(2c_S) = -a_p$ and deduce that we can choose to decompose E into the sum $E = \varepsilon(p, S) + a(u)$ and that then F is of the form $pa'(u) + h(p)$.
(4) Show that a'' is a constant. Deduce that E is a linear combination of two entropies, one which we knew already and an entropy which depends only on (p, S).
(5) We are therefore led to set up the list of entropies of the form $E(p, S)$. Show that then F depends only on u and is affine. Again, using an entropy already known, this leads to the case $F \equiv 0$.
(6) Show that then E is of the form $g \circ S$.
(7) In the (non-realistic) case where c is a function of p alone, show that the system decouples, at least for smooth solutions, into two independent systems, one governing S, the other governing the pair (u, p). This latter, consisting of two equations only, possesses many entropies as we shall see in Chapter 9, those evidently not being of the form $g \circ S$.

3.5 Weak solutions and entropy solutions

The hyperbolic quasi-linear systems having a complexity at least as large (it is in fact greater) as that of scalar equations, classical solutions only exist in general during a finite time, after which they give place to less smooth solutions, typically bounded measurable ones. Those satisfy partial differential equations in the sense of distributions, which can only be written properly for systems of conservation laws (in fact a product $A^\alpha \partial_{x_\alpha} u$ does not have a sense for $u \in L^\infty(\omega_{x,t})$).

Without repeating the analysis made in the preceding chapter, we define a notion of weak solution which translates well that which a conservation law wants to communicate to a physical level when u is piecewise continuous.

Definition 3.5.1 Let ω be an open set of $\mathbb{R}^{d+1}$ and $u \in L^\infty(\omega)^n$. We say that u is a *weak solution* of (3.2) in ω, if for all $\Phi \in \mathscr{D}(\omega)^n$, we have

$$\iint_\omega \left(u \cdot \partial_t \Phi + \sum_{1 \le \alpha \le d} f^\alpha(u) \cdot \partial_{x_\alpha} \Phi + b(u) \cdot \Phi \right) \mathrm{d}x \, \mathrm{d}t = 0.$$

We note that we have taken vector-valued test functions. By choosing $\Phi = (0, \ldots, 0, \varphi_j, 0, \ldots, 0)^{\mathrm{T}}$ where φ_j ranges over $\mathscr{D}(\omega)$ and j over $\{1, \ldots, n\}$, this reduces to writing that $(u_j, f_j^1(u), \ldots, f_j^d(u))$ satisfies the jth conservation law in the sense of distributions. For the Cauchy problem, we have

Definition 3.5.2 Let T be a positive real number, $u_0 \in L^\infty(\mathbb{R}^d)^n$ and $u \in L^\infty(\mathbb{R}^d \times (0, T))^n$. We say that u is a *weak solution of the Cauchy problem*

$$\left.\begin{aligned} \partial_t u + \sum_{1\le\alpha\le d} \partial_{x_\alpha} f^\alpha(u) &= b(u), \quad (x,t) \in \mathbb{R}^d \times (0, T), \\ u_0(x, 0) &= u_0(x), \quad x \in \mathbb{R}^d, \end{aligned}\right\} \tag{3.26}$$

if, for all $\Phi \in \mathscr{D}(\mathbb{R}^d \times (-\infty, T))^n$, we have

$$\iint_{\mathbb{R}^d\times(0,T)} \left(u \cdot \partial_t \Phi + \sum_{1\le\alpha\le d} f^\alpha(u) \cdot \partial_{x_\alpha} \Phi + b(u) \cdot \Phi \right) \mathrm{d}x\, \mathrm{d}t$$
$$+ \int_{\mathbb{R}^d} u_0(x) \cdot \Phi(x, 0)\, \mathrm{d}x = 0.$$

However, as in the scalar case, the class of weak solutions is not appropriate because the solution of the Cauchy problem is not in general unique, whereas the models considered are conceived in a deterministic setting. We thus must introduce a new admissibility criterion to select, from all weak solutions, that which is stable from the physical or mathematical point of view, hoping that there exists only one. The only criterion of general power, whose application is not restricted to piecewise smooth solutions, is Lax's entropy condition, which, for physical systems, can be written:

Definition 3.5.3 We consider a physical system whose strongly convex entropy and its flux are denoted by E and F. Being given u and u_0 as above, u being a weak solution of the Cauchy problem, we say that u satisfies *Lax's entropy condition* if, for all $\varphi \in \mathscr{D}^+(\mathbb{R}^d \times (-\infty, T))$, we have

$$\iint_{\mathbb{R}^d\times(0,T)} (E(u)\partial_t\varphi + F(u) \cdot \nabla_x \varphi + \mathrm{d}E(u) \cdot b(u)\varphi)\, \mathrm{d}x\, \mathrm{d}t$$
$$+ \int_{\mathbb{R}^d} E(u_0(x))\varphi(x, 0)\mathrm{d}x \ge 0.$$

The entropy condition implies, when we take only test functions with compact support in $\mathbb{R}^d \times (0, T)$, the inequality (in the sense of distributions)

$$\partial_t E(u) + \operatorname{div}_x F(u) \le 0.$$

As we have seen in the scalar case (see the exercise on the N-wave), the entropy condition is, in general, strictly more precise than this single inequality.

The above definitions are written in the most general framework, but the hypothesis $u \in L^\infty$ can often be weaker. For example, in the linear case, we know that for the majority of systems, the Cauchy problem is not well-posed in L^∞, but that it is in L^2_{loc}. Natural conditions are to suppose that u and $E(u)$ are in $\mathscr{C}((0,T); L^1_{\mathrm{loc}}(\mathbb{R}^d))$ and $f(u) \in L^1_{\mathrm{loc}}(\mathbb{R}^d \times (0,T))$. However, without the growth of u at infinity being controlled, there is a risk of an unavoidable blow-up in finite time of the entropy solution if, because of the non-linearity, a propagation speed is unbounded. It is thus prudent to restrict the study of the Cauchy problem to solutions satisfying $u \in \mathscr{C}((0,T); L^1_{\mathrm{loc}}(\mathbb{R}^d)) \cap L^\infty(\mathbb{R}^{d+1})$.

The Rankine–Hugoniot condition

We are now going to interpret the weak formulation for a discontinuous but piecewise smooth solution. To be precise, we consider an open set ω of $\mathbb{R}^{d+1}$ which a regular hypersurface Σ separates into two connected components ω_+ and ω_-. The field of unit vectors normal to Σ and directed towards ω_+ is denoted by ν. The field of unit vectors normal to $\partial\omega_\pm$ and pointing outwards is denoted by $\nu^\pm$. Along Σ, we have $\nu^- = \nu = -\nu^+$ (see Fig. 3.2). We assume that u is a function defined on ω with values in $\mathscr{U}$ whose restrictions to ω_+ and ω_- are of class $\mathscr{C}^1$ and can be extended by continuity to Σ. Their traces on Σ are denoted by u_+ and u_- ; u is genuinely discontinuous as long as $u_+ \neq u_-$.

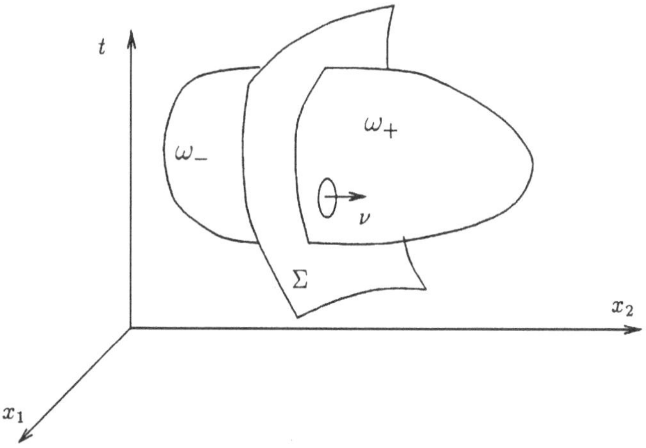

Fig. 3.2: Surface of discontinuity and unit normal.

Let us suppose that u is a weak solution of (3.2) in ω. Then for every test function $\Phi \in \mathscr{D}(\omega)^n$, using Green's formula we have that

$$0 = \iint_\omega \left(u \cdot \partial_t \Phi + \sum_{1 \le \alpha \le d} f^\alpha(u) \cdot \partial_{x_\alpha} \Phi + b(u) \cdot \Phi \right) \mathrm{d}x \, \mathrm{d}t$$

$$= \left(\iint_{\omega_-} + \iint_{\omega_+} \right) \Phi \cdot \left(b(u) - u_t - \sum_{1 \le \alpha \le d} \partial_{x_\alpha} f^\alpha(u) \right) \mathrm{d}x \, \mathrm{d}t$$

$$+ \int_{\partial \omega_-} \left(\nu_0^- u + \sum_{1 \le \alpha \le d} \nu_\alpha^- f^\alpha(u) \right) \cdot \Phi \, \mathrm{d}s(x,t)$$

$$+ \int_{\partial \omega_+} \left(\nu_0^+ u + \sum_{1 \le \alpha \le d} \nu_\alpha^+ f^\alpha(u) \right) \cdot \Phi \, \mathrm{d}s(x,t).$$

When Φ ranges over $\mathscr{D}^+(\omega_+)$, we deduce from

$$0 = \iint_{\omega_+} \Phi \cdot \left(b(u) - u - \sum_{1 \le \alpha \le d} \partial_{x_\alpha} f^\alpha(u) \right) \mathrm{d}x \, \mathrm{d}t,$$

that the conservation law holds at each point of ω_+, because $b(u) - u_t - \sum_{1 \le \alpha \le d} \partial_{x_\alpha} f^\alpha(u)$ is continuous on ω_+. Similarly in ω_-. Finally the integrals on ω_+ and on ω_- are zero for every test function in the above formula. It only remains to show that the boundary integrals on $\partial\omega_+$ and $\partial\omega_-$ are zero. These simplify for two reasons. First of all because Φ is zero on $\partial\omega$, with the result that these integrals reduce to the domain Σ. Then, because on Σ, $\nu_+ = -\nu_-$. Finally, with the usual notation $[u] = u_+ - u_-$,

$$0 = \int_\Sigma \left(\nu_0[u] + \sum_{1 \le \alpha \le d} \nu_\alpha [f^\alpha(u)] \right) \cdot \Phi \, \mathrm{d}s(x,t).$$

When Φ ranges over $\mathscr{D}(\omega)^n$, the traces of Φ on Σ run through a sub-space dense in $\mathscr{C}^1(\Sigma)$. Since $\nu_0 u + \sum_{1 \le \alpha \le d} \nu_\alpha f^\alpha(u)$ is continuous on Σ, we therefore obtain at each point the Rankine–Hugoniot condition:

$$\nu_0[u] + \sum_{1 \le \alpha \le d} \nu_\alpha [f^\alpha(u)] = 0. \tag{3.27}$$

Conversely, if u is a classical solution of (3.2) in ω_+ and ω_- and if u satisfies the Rankine–Hugoniot condition along Σ, the same calculation, carried out in the reverse order, shows that u is a weak solution of (3.2) in ω.

Let M_α be the Lipschitz constant of f^α on the interval with extremities u_- and u_+. We deduce from (3.27) the inequality

$$|\nu_0| \, \|[u]\| \le \sum_{1\le\alpha\le d} |M_\alpha \nu_\alpha| \, \|[u]\|.$$

This shows that, if u is genuinely discontinuous, then $|\nu_0| \le \sum_{1\le\alpha\le d} |M_\alpha \nu_\alpha|$. As ν is a unit vector, it follows that $(\nu_1, \dots, \nu_d)$ does not vanish: the hypersurface Σ is never tangent to the horizontal space $\{0\} \times \mathbb{R}^d$, it is a space-like hypersurface. As in the one-dimensional case, we introduce a unit vector

$$n = \frac{(\nu_1, \dots, \nu_d)}{|\nu|}, \qquad |\nu| = \sqrt{(\nu_1^2 + \cdots + \nu_d^2)},$$

and a number

$$V = -\frac{\nu_0}{|\nu|}.$$

The field n is a field of unit vectors normal to the section $\Sigma_t = \Sigma \cap (\{t\} \times \mathbb{R}^d)$ and V is the normal speed of the displacement of Σ_t with respect to the time.

Rewriting the Rankine–Hugoniot condition with the help of V and of n, we have

$$n \cdot [f(u)] = V[u]. \tag{3.28}$$

The above formula is vectorial. There is no ambiguity in the scalar product between the vector n and the tensor $[f(u)]$ as the dimensions of these, d and p, are in general distinct. Since n is a unit vector, we observe that the speed of propagation of a discontinuity is dominated by the Lipschitz constant of f on the interval $[u_-, u_+]$, which confirms the general idea that the hyperbolic systems are associated with phenomena of propagation with finite speeds.

In what concerns Lax's entropy condition, the same calculation as above, carried out with positive test functions, shows that a weak solution u of (3.2) is an entropy solution if and only if

$$n \cdot [F(u)] \le V[E(u)]. \tag{3.29}$$

Note that the significance of this inequality does not depend on the direction of n. In fact, if we replace n by $-n$ this results in interchanging ω_+ and ω_-, hence replacing $[E(u)]$ by $-[E(u)]$ and $[F(u)]$ by $-[F(u)]$.

Reversibility

If u is a weak solution of (3.2), then $v(x,t) := u(x,-t)$ is also a weak solution of the system $v_t - \sum_{1\le\alpha\le d} \partial_{x_\alpha} f^\alpha(v) = 0$. On the other hand, in conserving the same field of unit vectors n, the speed V and the flux F change sign. It follows that u

and v cannot satisfy their entropy conditions simultaneously unless the inequality (3.29) is an equality. Thus as long as (3.29) is strict, the solution u is irreversible.

3.6 Local existence of smooth solutions

The method of characteristics allowed us to show the existence of a smooth solution in a band $(0, T) \times \mathbb{R}$ for a scalar equation when the given initial condition is itself smooth. This result remains true for sufficiently general systems, but as the method of characteristics is not transposable to this case,[1] we must turn back to the energy estimates in the Sobolev spaces H^s. By this method, we can treat only symmetrisable systems. The regularity demanded for the given initial condition grows with the dimension. Typically, we need to show that if $u \in H^s(\mathbb{R}^d)$, then the non-linear terms of the form $g(u)\nabla_x u$ are bounded. From the Sobolev injection theorems and the Gagliardo–Nirenberg inequality, this reduces to requiring that $s > 1 + \frac{1}{2}d$. The following theorem expresses that the Cauchy problem for a symmetrisable system is locally well-posed for $s > 1 + \frac{1}{2}d$ [28].

Theorem 3.6.1 (L. Gårding, J. Leray) *Let s be a real number, $s > 1 + \frac{1}{2}d$. Let $\mathcal{U}_1$ be an open set relatively compact in $\mathcal{U}$ and $u_0 \in H^s(\mathbb{R}^d)$ with values in $\mathcal{U}_1$.*

Then there exists a time $T > 0$ such that the symmetric hyperbolic system

$$S_0(u)\partial_t u + \sum_{1 \le \alpha \le d} S_\alpha(u)\partial_{x_\alpha} u = b(u)$$

has a classical solution $u \in \mathcal{C}^1([0, T] \times \mathbb{R}^d)$ satisfying the initial condition

$$u(x, 0) = u_0(x).$$

In addition, $u \in \mathcal{C}([0, T]; H^s(\mathbb{R}^d)) \cap \mathcal{C}^1([0, T]; H^{s-1}(\mathbb{R}^d))$ and this solution is unique.

Remark The time T obtained in the proof of the theorem depends *a priori* on $\|u_0\|_s$, on $\mathcal{U}_1$ and on the smooth functions S_0, S_α and b defined on $\mathcal{U}$. When s is an integer, the norm on $H^s(\mathbb{R}^d)$ is defined classically by

$$\|v\|_s := \left(\sum_{|\gamma| \le s} \int_{\mathbb{R}^d} |D^\gamma v|^2 \, dx \right)^{\frac{1}{2}},$$

[1] In fact, it remains effective in dimension $d = 1$, even for systems. See a proof in Courant and Hilbert [12], pp. 476–8.

where $\gamma = (\gamma_1, \ldots, \gamma_d)$ denotes a positive multi-integer of length $|\gamma| = \gamma_1 + \cdots + \gamma_d$ and D^γ is the differential operator

$$\left(\frac{\partial}{\partial x_1}\right)^{\gamma_1} \cdots \left(\frac{\partial}{\partial x_d}\right)^{\gamma_d}.$$

Indications about the proof

The proof of Theorem 3.6.1 makes use of an iterative scheme in which each iteration consists of solving a linear system with variable coefficients but of class $\mathscr{C}^\infty$. We choose for this a sequence $(u_0^k)_{k\geq 0}$ of functions of class $\mathscr{C}^\infty$ such that the series $\sum_k \|u_0^{k+1} - u_0^k\|_s$ converges, the limit of the sequence being u_0. Since, by hypothesis, $H^s(\mathbb{R}^d)$ is included in $\mathscr{C}^1(\mathbb{R}^d)$, we may suppose that each element of this sequence has values in $\mathscr{U}_2$, a neighbourhood of $\mathscr{U}_1$ relatively compact in $\mathscr{U}$. If $u^k \in \mathscr{C}^\infty([0, T_k] \times \mathbb{R}^d)$ takes its values in $\mathscr{U}$, the linear system

$$\left.\begin{aligned} S_0(u^k)\partial_t u^{k+1} + \sum_{1\leq\alpha\leq d} S_\alpha(u^k)\partial_{x_\alpha} u^{k+1} &= b(u^k), \\ u^{k+1}(x,0) &= u_0^k(x) \end{aligned}\right\} \tag{3.30}$$

makes sense and possesses a solution of class $\mathscr{C}^\infty$ in the band $[0, T_k] \times \mathbb{R}^d$. This is a consequence of the linear theory (which we shall not develop in this work because of space constraints). We call $T_{k+1} \leq T_k$, the maximal time in which u^{k+1} takes its values in $\mathscr{U}$.

The aim of these estimates is two-fold: on the one hand to control the distance of $u^k(x, t)$ from the boundary of $\mathscr{U}$, since the coefficients of the system can reach singular values at the boundary of this domain (leading for example to an infinite propagation speed), on the other hand to control the norm of u^k in $\mathscr{C}^1$ (which can only be done by passing through H^s) so as to be able to pass to the limit in the products $S_0(u^k)\partial_t u^{k+1}$, etc. In both cases, it is obviously essential to show that the sequence of times of existence T_k is bounded below by a number $T > 0$, which is the number T stated in the theorem.

As frequently happens in the study of non-linear (and also linear) partial differential equations, the estimates for the scheme are adapted from an *a priori* estimate obtained on the equation we seek to solve, when we assume that it possesses a solution which is sufficiently smooth. It is there that the deep idea remains, the rest is essentially a matter of technique. For simplicity, we shall suppose that $\mathscr{U} = \mathbb{R}^d$, which reduces the first estimate to a control of $|u^k|_\infty$, the norm of u^k in L^∞, which itself is bounded by $\|u^k\|_{s,T} := \sup_{0\leq t\leq T} \|u^k(t)\|_s$. Finally, there is only a single important estimate, that of $\|u^k\|_{s,T}$. We shall assume also that s is an integer, so that we shall avoid having to manipulate with fractional derivatives in the energy estimates.

In fact, in the case of the iterative scheme, a supplementary difficulty appears, it must be shown that the whole sequence $(u^k)_{k\geq 0}$ converges. For that, we shall first establish a bound on $\|u^{k+1} - u^k\|_{0,T}$ by using that on $\|u^k\|_{s,T}$.

A priori estimate of $\|u\|_{s,T}$

As we have said above, we work directly on a smooth solution u of the problem to be solved. The calculations carried out below do not constitute a proof of Theorem 3.6.1. In addition, the right-hand side $b(u)$ plays a minor rôle in the theory and we shall suppose it to be identically zero.

For an integer $k \geq 0$, let us denote by v_k the list of derivatives of u of order k:

$$v_k = (D^\gamma u)_{|\gamma|=k}.$$

For a monomial

$$M(v_1, \ldots, v_k) = \prod_{j=1}^{k} v_j^{\beta_j},$$

we denote its *weight* by

$$p(M) = \sum_{j=1}^{k} j|\beta_j|.$$

Differentiating the system k times with respect to the spatial variables, we obtain a system, linear with respect to derivatives of higher order:

$$A_0(u)\partial_t v_k + \sum_{1\leq\alpha\leq d} A_\alpha(u)\partial_{x_\alpha} v_k = P(u; v_1, \ldots, v_k), \tag{3.31}$$

where $P(u; v_1, \ldots, v_k)$ is a polynomial homogeneous in weight, of weight $k+1$. In the energy method we use as a matter of fact a norm equivalent to the usual norm on $(L^2)^n$, which depends on the solution and on the time, namely

$$[w(t)] := \left(\int_{\mathbb{R}^d} w^* A_0(u(t)) w \, dx\right)^{\frac{1}{2}}.$$

So as not to overburden the notation, we have chosen not to mention the dependence of this norm on $u(t)$, but we hope that the context will recall it clearly. The equivalence of $[\cdot]$ and the usual norm $\|\cdot\|_0$ of $(L^2)^n$ is not in general uniform with respect to u, in any case if A_0 or A_0^{-1} is not bounded as a function of u. Precisely, there exists an increasing numerical function $C \geq 1$, such that

$$C(|u(t)|_\infty)^{-1}\|w\|_0 \le [w] \le C(|u(t)|_\infty)\|w\|_0.$$

Taking the scalar product of (3.31) with v_k, we obtain

$$\begin{aligned}\partial_t\left(\frac{1}{2}v_k^* A_0(u)v_k\right) + \sum_{1\le\alpha\le d}\partial_{x_\alpha}\left(\frac{1}{2}v_k^* A_\alpha(u)v_k\right)\\ = v_k\cdot P + \frac{1}{2}v_k^*\left(\partial_t A_0(u) + \sum_{1\le\alpha\le d}\partial_{x_\alpha}A_\alpha(u)\right)v_k\\ = Q(u; v_1,\dots,v_k) \qquad (3.32)\end{aligned}$$

where $Q(u;\dots)$ is a homogeneous polynomial in weight, of weight $2k+1$.

Integrating (3.32) over a ball of radius R and admitting that v_k tends to zero at infinity sufficiently fast that the boundary integrals of $v_k^* A_\alpha(u)v_k$ tend to zero when R tends to infinity,[2] we deduce that

$$\frac{\mathrm{d}}{\mathrm{d}t}\left(\frac{1}{2}[v_k]^2\right) = \int_{\mathbb{R}^d} Q(u; v_1,\dots,v_k)\,\mathrm{d}x. \qquad (3.33)$$

Let us for the moment accept the following lemma.

Lemma 3.6.2 *If $Q(u; v_1,\dots,v_k)$ is a polynomial, homogeneous in weight with respect to v, of weight $2k+1$, then there exists a numerical function C_k such that, for all $u \in H^k(\mathbb{R}^d)^n$ we have*

$$\int_{\mathbb{R}^d} Q\left(u; \mathrm{d}_x u,\dots,\mathrm{D}_x^k u\right)\mathrm{d}x \le C_k(|u|_\infty, |\mathrm{d}_x u|_\infty)\left\|\mathrm{D}_x^k u\right\|_0^2.$$

Applying the lemma to the formula (3.33), then summing from $k=0$ to $k=s$, we obtain

$$\frac{\mathrm{d}}{\mathrm{d}t}\left(\frac{1}{2}[u]_s^2\right) = \left(\max_{0\le k\le s} C_k\right)\|u\|_s^2, \qquad (3.34)$$

where we have defined

$$[w]_s := \left(\sum_{0\le k\le s}[\mathrm{D}_x^k w]^2\right)^{\frac{1}{2}}.$$

[2] Each approximation u_0^k to the given initial function u_0 is with compact support, with the result that the approximate solution u^k is with compact support with respect to x. In practice, the functions v_k are therefore with compact support.

The expression $[\cdot]_s$ is a norm on $H^s(\mathbb{R}^d)^n$, equivalent to the norm $\|\cdot\|_s$ but this equivalence depends on $u(t)$ in the same manner as for $[\cdot]$:

$$C(|u(t)|_\infty)^{-1}\|w\| \leq [w]_s \leq C(|u(t)|_\infty)\|w\|.$$

We can thus simplify the inequality (3.34), to reduce it to

$$\frac{\mathrm{d}}{\mathrm{d}t}[u] \leq c_1(|u|_\infty, |\mathrm{d}_x u|_\infty)\|u\|_s, \tag{3.35}$$

where c_1 is an explicit numerical function.

Integrating (3.35) from 0 to t, we find that

$$[u(t)]_s \leq \int_0^t c_1(|u(\tau)|_\infty, |\mathrm{d}_x u(\tau)|_\infty)\|u(\tau)\|_s \mathrm{d}\tau + [u_0]_s,$$

that is to say,

$$\|u(t)\|_s \leq C(|u(t)|_\infty) \times \left\{\int_0^t c_1(|u(\tau)|_\infty, |\mathrm{d}_x u(\tau)|_\infty)\|u(\tau)\|_s \mathrm{d}\tau + [u_0]_s\right\}. \tag{3.36}$$

This inequality is more precise than we need immediately, but it will serve us in the sequel to establish a characterisation of the maximal time of existence of the classical solution.

For the moment, we make use of a rough upper bound, which expresses that $H^s(\mathbb{R}^d)$ is included in $\mathscr{C}^1(\mathbb{R}^d)$; let $|u|_\infty + |\mathrm{d}_x u|_\infty \leq c_2\|u\|_s$, where c_2 is a constant. From (3.36) we then derive

$$\|u(t)\|_s \leq C(c_2\|u(t)\|_s)\left\{\int_0^t c_3(\|u(\tau)\|_s)\|u(\tau)\|_s \mathrm{d}\tau + [u_0]_s\right\}, \tag{3.37}$$

where c_3 is an explicit numerical function.

Let us introduce the numbers $R := C(c_2\|u_0\|_s)[u_0]_s$ and c_4, the supremum of the expression $C(c_2 z)c_3(z)z$ when z ranges over the interval $[0, R+1]$. We have $\|u_0\|_s \leq R$. The inequality (3.37) then ensures that $\|u(t)\| \leq R+1$ provided that $0 \leq t \leq T$ where $T := R/c_4$.

Lemma 3.6.3 *There exist a time $T > 0$ and a real number $L > 0$ such that every smooth solution of the Cauchy problem satisfies*

$$\sup_{0\leq t\leq T} \|u(t)\|_s \leq L.$$

The numbers L and T depend both on the system considered and on the initial norm $\|u_0\|_s$, $s > 1 + \frac{1}{2}d$.

Corollary 3.6.4 *With the notation of the preceding lemma, there exists a number L_1 such that every regular solution of the Cauchy problem satisfies*

$$\sup_{0\le t\le T}\left\|\frac{\partial u}{\partial t}\right\|_{s-1}\le L_1.$$

Proof We recall the formula (3.31) for $k\le s-1$, which shows that

$$\mathrm{D}_x^k(u_t)=A_0(u)^{-1}\left(P_k-\sum_{1\le\alpha\le d}A_\alpha\partial_{x_\alpha}v_k\right).$$

The same calculation which previously showed that $A_0^{-1}P_k$ is bounded in $L^\infty(0,T;L^2(\mathbb{R}^d))$ is still valid and even trivial for $A_0^{-1}A_\alpha\partial_{x_\alpha}v_k$. ∎

Proof of Lemma 3.6.2

The proof of this lemma is based on the Gagliardo–Nirenberg inequality. If r is a positive real number, i an integer between 0 and r and if z belongs to $L^\infty\cap H^r$, then $\mathrm{D}_x^i z\in L^{2r/i}$ with

$$\left|\mathrm{D}_x^i z\right|_{2r/i}\le C_{i,r}|z|_\infty^{1-i/r}\|z\|_r^{i/r}, \tag{3.38}$$

where $|\cdot|_q$ denotes the usual norm of L^q. Applying this inequality to v_1 and $i=j-1$ for $j\le r+1$, we have

$$|v_j|_{2r/(j-1)}\le C_{j-1,r}|v_1|_\infty^{1-(j-1)/r}\|u\|_{r+1}^{(j-1)/r}.$$

Thus, for $u\in W^{1,\infty}\cap H^k$, $k\ge j$, we have $v_j\in L^{p_j}$ for all p_j comprised between 2 and $2(k-1)/(j-1)$. A monomial of the form

$$Q:=\prod_{1\le j\le k}v_j^{\gamma_j},$$

of weight $2k+1$, thus belongs to L^p for p satisfying $a\le p\le b$, where

$$\frac{1}{a}=\frac{1}{2}\sum_j|\gamma_j|,\quad \frac{1}{b}=\frac{1}{2}\sum_j\frac{j-1}{k-1}|\gamma_j|.$$

In addition, we have

$$|v_j|_{p_j}\le C|v_1|_\infty^{1-2/p_j}\|u\|_k^{2/p_j}.$$

We thus have $|Q|_p \leq C|v_1|_\infty^\sigma \|u\|_k^\theta$, where

$$\theta := \sum_j \frac{2}{p_j}|\gamma_j| = \frac{2}{p}, \quad \sigma := \sum_j |\gamma_j| - \theta.$$

The lemma then comes from the inequalities $a \leq 1 \leq b$, which we prove now. We have

$$\sum_{1\leq j\leq k} |\gamma_j| \geq \sum_{1\leq j\leq k} \frac{j}{k}|\gamma_j| = 2 + \frac{1}{k},$$

which indicates that $\sum_{1\leq j\leq k} |\gamma_j| \geq 3$ (because this is an integer). We deduce that $a \leq 2/3$, as well as $b \geq 1$ (which achieves the proof of the lemma) by reason of the formula

$$\frac{1}{b} = 1 + \frac{1}{2(k-1)}\left(3 - \sum |\gamma_j|\right).$$

Convergence of the iterative scheme

Returning to the iterative scheme, we shall accept that the *a priori* estimates established in §3.6 remain valid for the approximate solutions introduced there, even if it entails that the time T of existence common to all the solutions u^m is a little smaller (but, however, strictly positive). There are thus a time T_1 and a number R_1 such that, for all $m \geq 1$ with $s > 1 + \frac{1}{2}d$, we have

$$\sup_{1\leq t\leq T_1} \|u^m(t)\|_s \leq R_1, \tag{3.39}$$

$$\sup_{0\leq t\leq T_1} \|\partial_t u^m(t)\|_{s-1} \leq R_1. \tag{3.40}$$

We are going first to show the convergence of the iterative scheme in $L^2(\mathbb{R}^d)$ on a time interval eventually smaller, then we conclude in $H^r(\mathbb{R}^d)$ for all $0 \leq r < s$ by interpolation. For a start, let us define the difference $z^m := u^{m+1} - u^m$ and form the difference of two successive equations of the scheme:

$$A_0(u^m)\partial_t z^m + \sum_{1\leq\alpha\leq d} A_\alpha(u^m)\partial_{x_\alpha} z^m = F_m, \tag{3.41}$$

where

$$\begin{aligned} F_m &= (A_0(u^{m-1}) - A_0(u^m))\partial_t u^m + \sum_\alpha((A_\alpha(u^{m-1}) - A_\alpha(u^m))\partial_{x_\alpha} u^m) \\ &= R(u^{m-1}, u^m, d_{t,x}u^m, z^{m-1}), \end{aligned}$$

R being linear with respect to its last argument because of Taylor's formula (mean value theorem). Thus

$$|F_m|_2 \le C(R_1)|z^{m-1}|_2.$$

Multiplying (3.41) by z^m, it becomes

$$\partial_t\left(\frac{1}{2}z^m A_0(u^m)z^m\right) + \sum_\alpha \partial_{x_\alpha}\left(\frac{1}{2}z^m A_\alpha(u^m)z^m\right)$$
$$= z^m \cdot F_m + \frac{1}{2}z^m(\partial_t A_0(u^m) + \sum_\alpha \partial_{x_\alpha} A_\alpha(u^m))z^m.$$

Integrating again over $\mathbb{R}^d$ and assuming that z^m tends to zero at infinity rapidly enough for the integrals of $\partial_{x_\alpha}(z^m A_\alpha(u)z^m)$ to be null (same remark as previously), it becomes, for $0 \le t \le T_1$,

$$\frac{\mathrm{d}}{\mathrm{d}t}\int_{\mathbb{R}^d} z^m A_0(u)z^m\,\mathrm{d}x \le C(R_1)|z^m|_2|z^{m-1}|_2 + |\partial_t A_0(u^m) + \sum_\alpha \partial_{x_\alpha} A_\alpha(u^m)|_\infty |z^m|_2^2$$
$$\le C_1(R_1)[z^m][z^{m-1}] + C_2(R_1)[z^m]^2,$$

which immediately reduces to

$$2\frac{\mathrm{d}}{\mathrm{d}t}[z^m] \le C_1(R_1)[z^{m-1}] + C_2(R_1)[z^m].$$

Integrating from 0 to $T^* \le T_1$ and writing $y_m := \sup_{0\le t\le T^*}[z^m(t)]$, we have

$$y_m \le \delta y_{m-1} + \beta_m, \tag{3.42}$$

where $\delta = C_2(R_1)T^*\mathrm{e}^{C_1(R_1)T^*}$ and $\beta_m = \mathrm{e}^{C_1T^*}[z^m(0)]$. We then choose T^* to be such that $0 < \delta < 1$; this is possible and we obtain

$$\sum_m y_m \le \frac{1}{1-\delta}\sum_m \beta_m. \tag{3.43}$$

Lemma 3.6.5 *The sequence* $(\beta_m)_{m\ge 0}$ *has a finite sum.*

Proof Since $u_0 \in H^s$ with $s > 1+\frac{1}{2}d$, we have $u_0 \in \mathscr{C}^1$. We choose $u_0^m = u_0 * j_m$, the convolution product of u_0 with a smoothing function $j_m(x) := 2^{md} j(2^m x)$ where $j \in \mathscr{D}(\mathbb{R}^d)$ and $\int_{\mathbb{R}^d} j\,\mathrm{d}x = 1$. We have $\int_{\mathbb{R}^d}(j_1 - j_0)\,\mathrm{d}x = 0$ with the result that we can write $j_1 - j_0$ in the form $\operatorname{div} p$, p being of class $\mathscr{C}^\infty$ and with compact support. Let $p_m(x) = 2^{(m-1)(d-1)}p(2^{m-1}x)$. We have $j_m - j_{m-1} = \operatorname{div} p_m$ and

$|p_m|_1 = 2^{-m}C$. Thus

$$\begin{aligned} z^m(0) &= -p_m * d_x u_0, \\ |z^m(0)|_2 &\le |p_m|_1 |d_x u_0|_2 \le |p_m|_1 \|u_0\|_1 \\ &\le 2^{-m}\|u_0\|_1 \le 2^{-m}\|u_0\|_s, \end{aligned}$$

which shows clearly that the sequence has a finite sum. ■

We deduce from the lemma and (3.43) that the sequence $(y_m)_{m\ge 0}$ equally has a finite sum, that is to say that u^m converges at least in $L^\infty(0, T^*; L^2(\mathbb{R}^d))$ since the norms $[\cdot]$ and $|\cdot|_2$ are equivalent on $L^2(\mathbb{R}^d)$ uniformly for t varying from 0 to T^*. If u is the limit of this sequence, then $u \in L^\infty(0, T^*; L^2(\mathbb{R}^d))$ by Fatou's lemma. In addition by an interpolation lemma between $L^2 = H^0$ and H^s (see [1]), we have for all $0 \le r \le s$

$$\|u - u^m\|_r \le |u - u^m|_2^{1-r/s} \|u - u^m\|^{r/s},$$

the right-hand side of which tends to zero from the preceding argument and Lemma 3.6.2. Thus, the sequence $(u^m)_{m\ge 0}$ tends to u in $L^\infty(0, T^*; H^r(\mathbb{R}^d))$ for all $r < s$.

Finally, the convergence of the equations of the iterative scheme takes place, in a uniform manner, with the result that u is clearly a regular solution of the Cauchy problem.

Remarks (1) In fact, the approximate solutions are continuous with respect to the time with values in H^{s-1}, hence with values in H^r for all $r < s$ (again the argument by interpolation). In the above results we can thus replace $L^\infty(0, T^*; H^r(\mathbb{R}^d))$ by $\mathscr{C}(0, T^*; H^r(\mathbb{R}^d))$.

(2) As in addition $u \in L^\infty(0, T^*; H^s(\mathbb{R}^d))$, we deduce that in fact, u is continuous with respect to the time, with values in H^s equipped with its weak topology. Showing finally, by the same type of estimates as those already used, that $t \mapsto [u(t)]_s$ is continuous, we deduce the result stated, that is to say that $u \in \mathscr{C}(0, T^*; H^s(\mathbb{R}^d))$ (use the fact that a weakly convergent sequence in a Hilbert space, of which the limit of the norms is equal to the norm of the limit, converges strongly).

(3) The smooth solution is in fact unique, as we can convince ourselves by recalling the inequality (3.42) either for two approximate solutions or for two smooth solutions of the same Cauchy problem; then we have $\beta = 0$ and so $y \le \delta y$ which gives $y = 0$, that is to say that the difference between the two solutions is null in $L^\infty(0, T^*; L^2(\mathbb{R}^d))$, and therefore null.

More generally, this calculation can be carried out with two solutions u and v corresponding to two distinct initial conditions. If $z := v - u$, we obtain a Gronwall inequality

$$\frac{\mathrm{d}}{\mathrm{d}t}\int_{\mathbb{R}^d} zA_0(u)z\,\mathrm{d}x \le C(R)[z]^2,$$

which produces

$$\sup_{0\le t\le T^*} |v-u|_2 \le C\mathrm{e}^{Ct}|v_0-u_0|_2,$$

where the constants C and $T^* > 0$ depend only on $\|u_0\|_s$ and $\|v_0\|_s$.

(4) An important question concerns the manner in which the solution ceases to be smooth, when that is the case. The answer, necessarily partial, is furnished by the inequality (3.35). Let T be the maximal time of existence of the solution of class H^s with $s > 1+\frac{1}{2}d$, and suppose T to be finite. Then the Gronwall inequality shows us that

$$\int_0^T c_1(|u|_\infty, |\mathrm{d}_x u|_\infty)\,\mathrm{d}t = +\infty.$$

We deduce easily (again using the estimate of the local time of existence of the smooth solution) that

$$\lim_{t\to T} \max(|u(t)|_\infty, |\mathrm{d}_x u(t)|_\infty) = +\infty. \tag{3.44}$$

If there is a blow-up in finite time, it thus is produced in the same manner as in the scalar case, by the blow-up of the first derivatives, unless of course u itself becomes unbounded. Returning to a domain $\mathscr{U}$ of admissible states, this should signify that $u(x,t)$ approaches the boundary of $\mathscr{U}$ at a certain point when $t\to T$.

(5) However, and contrary to what occurs in dimension $d = 1$, it is possible that the solution remains smooth for all time even for very non-linear systems provided that the dimension is high enough and that u_0 is sufficiently small and with compact support. The first observation in this direction seems to be due, for a non-linear wave equation, to Klainerman [53, 54] and a comprehensive study of the subject is to be found in the work of Li Ta-Tsien [64]. In this will be found numerous references to the works of other authors, among them L. Hörmander and F. John.

In the special case of the full gas dynamics with the perfect gas law ($p = (\gamma-1)\rho e$), the author and Magali Grassin have recently obtained global existence theorems for non-small data. These are chosen with a small density, an entropy close to a constant, and an initial velocity field which makes the particles spread.

3.7 The wave equation

The wave equation, through second order in the time and in space, comes into the category studied in §3.1 if we put $u = (\partial_t v,\ -\nabla_x v)^{\mathrm{T}}$. In fact,

$$\partial_t^2 v = \Delta v, \qquad v(x,0) = v_0(x), \qquad \partial_t v(x,0) = v_1(x)$$

is equivalent to

$$\partial_t u + \begin{pmatrix} 0 & \mathrm{div} \\ \nabla & 0_d \end{pmatrix} u = 0, \qquad u(x,0) = (v_1(x),\ -\nabla v_0(x)).$$

It is clearly a linear first order system with $n = d + 1$ and

$$A(\xi) = \begin{pmatrix} 0 & \xi^{\mathrm{T}} \\ \xi & 0_d \end{pmatrix}.$$

This system is symmetric and therefore hyperbolic. Its propagation velocities are clearly ± 1 and 0 in each direction ξ but as we are interested only in solutions satisfying $\partial_\alpha u_{\beta+1} = \partial_\beta u_{\alpha+1}$, only ± 1 remain.

The wave equation arises in many physical problems, specially in relativity and electromagnetism, often coupled with other equations. For that reason, it is important to have qualitative results concerning the behavior of solutions. These are based on an explicit formula, much easier to use than that obtained by Fourier analysis. It is extracted from three facts:

the invariance of the equation under the action of isometries of $\mathbb{R}^d$,
the easy solvability in the case $d = 1$,
the possibility of passing from a dimension d to a dimension $d + 2$.

Huygens' principle

We introduce the *spherical mean,* for $x \in \mathbb{R}^d$, $t, r > 0$:

$$I(x,t,r) := \frac{1}{\omega_{d-1} r^{d-1}} \int_{S(x,r)} v(y,t)\, \mathrm{d}s(y),$$

where $S(x,r)$ denotes the sphere with centre x and radius r in $\mathbb{R}^d$, ω_{d-1} is the $(d-1)$-dimensional measure of the unit sphere S^{d-1} of $\mathbb{R}^d$ and $\mathrm{d}s(y)$ is the usual measure on $S(x,r)$. If $x \in \mathbb{R}^d$ and if v is a solution, we verify that $(z,t) \mapsto I(x,t,\|z\|)$ is also a solution of the wave equation[3] which is written

$$\partial_t^2 I = \partial_r^2 I + \frac{d-1}{r} \partial_r I. \tag{3.45}$$

[3] As far as here, the method is valid for every linear partial differential equation invariant under the action of isometries of $\mathbb{R}^d$; for example the heat equation $\partial_t v = \Delta v$.

This is called the *Euler–Poisson–Darboux equation.* We easily show I is a solution if and only if $J := r\partial_r I + (d-2)I$ is a solution of the analogous equation where d is replaced by[4] $d-2$ .When d is odd, that allows us to reduce easily to the trivial case $d=1$. For example, if $d=3$, $K := rI$ is a solution of $\partial_t^2 K = \partial_r^2 K$. We thus have $K(x,t,r) = p(x,t+r) + q(x,r-t)$. Here $p(x,\cdot)$ is defined on $\mathbb{R}^+$, while $q(x,\cdot)$ is defined on the whole of $\mathbb{R}$.

We determine p and q with the help of the initial conditions and of the limits on K:

$$K(x,0,r) = rI_0(x,r), \qquad \partial_t K(x,0,r) = rI_1(x,r), \qquad K(x,t,0) = 0,$$

where $I_j(x,r)$ denotes the mean of v_j on the sphere $S(x,r)$. We thus have, for $r > 0$,

$$\partial_r p(x,r) = \frac{1}{2}(\partial_r(rI_0) + rI_1), \quad \partial_r q(x,r) = \frac{1}{2}(\partial_r(rI_0) - rI_1).$$

Finally, for $r < 0$, we have $q(x,r) = -p(x,-r)$.

We recover the solution v by the relation

$$\begin{aligned} v(x,t) = I(x,t,0) &= \lim_{r\to 0+} \frac{K(x,t,r)}{r} \\ &= \lim_{r\to 0+} \frac{p(x,r+t) - p(x,t-r)}{r} \\ &= 2\partial_r p(x,t) = \partial_t(tI_0(x,t)) + tI_1(x,t). \end{aligned}$$

Finally, the solution of the Cauchy problem in dimension $d=3$ is given by the formula

$$v(x,t) = \frac{\partial}{\partial t}\left(\frac{1}{4\pi t}\int_{S(x;t)} v_0(y)\,\mathrm{d}s(y)\right) + \frac{1}{4\pi t}\int_{S(x,t)} v_1(y)\,\mathrm{d}s(y).$$

Of course, the above calculation proceeds by necessary conditions only, but as we end up with an explicit formula, it is easy to verify that this actually defines a solution of the Cauchy problem, for example, when v_0 and v_1 are sufficiently smooth.

The case of the dimension $d=2$, which is likewise interesting, is solved by associating an extra spatial variable. If the functions $v_j(x_1,x_2)$ are the Cauchy data and $v(x_1,x_2,t)$ is the solution then $V(x_1,x_2,x_3,t) := v(x_1,x_2,t)$ is the solution of the wave equation in dimension 3 for the data $V_j(x_1,x_2,x_3) := v_j(x_1,x_2)$. We

[4] On the other hand, there is no simple way to pass from d to $d-1$; we shall see that that has important consequences.

thus have

$$v(x,t) = \frac{\partial}{\partial t}\frac{1}{4\pi t}\left(\int_{S^2(x,0;t)} v_0(y_1,y_2)\,\mathrm{d}s(y)\right) + \frac{1}{4\pi t}\int_{S^2(x,0;t)} v_1(y_1,y_2)\,\mathrm{d}s(y).$$

Parametrising each hemisphere of $S^2(x,t)$ by (y_1, y_2) we obtain the formula

$$v(x,t) = \frac{\partial}{\partial t}\frac{1}{2\pi}\left(\int_{D(x;t)} \frac{v_0(y)}{\sqrt{t^2 - \|x-y\|^2}}\,\mathrm{d}y\right) + \frac{1}{2\pi}\int_{D(x;t)} \frac{v_1(y)}{\sqrt{t^2 - \|x-y\|^2}}\,\mathrm{d}y,$$

less elegant than that in dimension 3 ($\mathrm{D}(x;r)$ denotes the disk with center x and radius r).

The two formulae derived above show a very different qualitative behaviour according to the dimension: if $d = 3$, the value of v at (x,t) depends only on the Cauchy data by the restriction of v_1, of v_0 and of ∇v_0 on the light cone $\{y \in \mathbb{R}^3;\, \|y - x\| = t\}$ (*Huygens' principle*). On the other hand, if $d = 2$, the value of v at (x,t) depends on the restriction of the data v_j to the whole disk $D(x;t)$.

Conservation and decay

As the system associated with the wave equation is symmetrical, the energy, which is the norm of $u(t)$ in $(L^2(\mathbb{R}^d))^{d+1}$, remains constant in the course of the time. It is not necessary to show this my making use of the Fourier transformation. It is enough to integrate the conservation law

$$\frac{1}{2}\partial_t(|\partial_t u|^2 + |\nabla u|^2) = \operatorname{div}(\partial_t u \nabla u)$$

over the frustum $\{(x,t); 0 < t < T,\ \|x\| + t \le R\}$. By Green's formula this becomes

$$\frac{1}{2}\int_{B(x;R-T)} (|\partial_t v|^2 + |\nabla v|^2)(x,T)\,\mathrm{d}x$$
$$= \frac{1}{2}\int_{B(x,R)} (|v_1|^2 + |\nabla v_0|^2)\,\mathrm{d}x + \int_\Gamma \left(\partial_t v \partial_r v - \frac{1}{2}(|\partial_t v|^2 + |\nabla v|^2)\right)\mathrm{d}s$$

where Γ denotes the lateral boundary of the domain, $\mathrm{d}s$ a conveniently normalised measure of area and ∂_r the radial derivative. The boundary integral is negative by the Cauchy–Schwarz inequality, with the result that

$$\frac{1}{2}\int_{B(x;R-T)} (|\partial_t v|^2 + |\nabla v|^2)(x,T)\,\mathrm{d}x \le E_0 := \frac{1}{2}\int_{\mathbb{R}^d} (|v_1|^2 + |\nabla v_0|^2)\,\mathrm{d}x.$$

Thus, the energy $E(T) := \int_{\mathbb{R}^d} (|\partial_t v|^2 + |\nabla v|^2)(x,T)\,\mathrm{d}x$ is bounded above by E_0. But reversing the direction of the time, we therefore have $E_0 \le E(T)$ and finally $E(T) \equiv E_0$.

However, this result is mediocre when compared with what can be obtained by making use of other conservation laws. These are consequences of Emmy Noether's theorem and the invariance of the wave equation under the action of the Lorentz group. The most important conservation law is

$$\partial_t e_3 = \operatorname{div} q_3,$$

where[5]

$$e_3(x,t) := (r^2+t^2)\left(|\partial_t v|^2+\sum_{j=1}^{d}\lambda_j^2\right)+4t\,\partial_t v\sum_{j=1}^{d}x_j\lambda_j+(d-1)(d-3)\frac{r^2+t^2}{r^2}|v|^2,$$

with

$$\lambda_j := \frac{\partial v}{\partial x_j}+\frac{d-1}{2}\frac{x_j}{r^2}v.$$

The same method as was used for the energy shows that $E_3(T) := \int_{\mathbb{R}^d} e_3(x,T)\,\mathrm{d}x$ is constant if $E_3(0)$ is finite, that is if

$$\int_{\mathbb{R}^d}\left((r^2+t^2)\left(|v_1|^2+\sum_{j=1}^{d}\Lambda_j^2\right)\right.$$
$$\left.+4tv_1\sum_{j=1}^{d}x_j\Lambda_j+(d-1)(d-3)(1+t^2/r^2)|v_0|^2\right)\mathrm{d}x,$$

is finite, where we have written $\Lambda_j := \partial_j v_0 + (d-1)x_j v_0/(2r^2)$. We remark that the integrand Q is a positive semi-definite quadratic form of (v_1, Λ, v_0) provided that d is different from 2:

$$Q = (r^2+t^2)|\Lambda_{\mathrm{T}}|^2+\frac{1}{2}(r-t)^2(v_1-\Lambda_{\mathrm{R}})^2+\frac{1}{2}(r+t)^2(v_1+\Lambda_{\mathrm{R}})^2$$
$$+(d-1)(d-3)(1+t^2/r^2)|v_0|^2,$$

where we have decomposed Λ into its radial and tangential components

$$\Lambda_{\mathrm{R}} := \frac{x}{t};\Lambda = \partial_r v_0+\frac{1}{2}(d-1)v_0/r,\quad \Lambda_{\mathrm{T}} := \Lambda-\Lambda_{\mathrm{R}}\frac{x}{r} = (\nabla v_0)_{\mathrm{T}}.$$

For $t > 0$, we have

$$Q \geq t^2\left(|\Lambda_{\mathrm{T}}|^2+\frac{1}{2}(v_1+\Lambda_{\mathrm{R}})^2\right).$$

[5] Exercise: calculate the expression for q_3.

We deduce

$$\int_{\mathbb{R}^d} |\nabla_T v|^2 \mathrm{d}x + \frac{1}{2}\int_{\mathbb{R}^d} \left(\partial_t v + \partial_r v + \frac{d-1}{2r} v\right)^2 \mathrm{d}x \le (2/t^2) E_3(0). \quad (3.46)$$

Similarly, if we restrict ourselves to the complement of a conical neighbourhood of the light cone we have

$$\int_{|r-t|>\theta t} \left(\partial_t v - \partial_r v - \frac{d-1}{2r} v\right)^2 \mathrm{d}x \le 4/(\theta t)^2 E_3(0). \quad (3.47)$$

The upper bound (3.46) shows that the solution behaves asymptotically, for $t \to +\infty$, as a function V which satisfies

$$\nabla_T V = 0, \quad \partial_t V + \partial_r V + \frac{d-1}{2r} V = 0,$$

that is to say $V = V(t, r)$ and $V = r^{\frac{1}{2}(1-d)} W(t-r)$. Finally (3.47) reduces to saying that

$$\int_{|s|>\eta} (W'(s))^2 \,\mathrm{d}s \le \frac{\text{const}}{\eta^2}.$$

The reader wishing to go further on the dispersion properties of the wave equation, for example in the presence of a potential or of an obstacle, should consult the memoir of Cathleen Morawetz [79].

4

Dimension $d = 1$, the Riemann problem

4.1 Generalities on the Riemann problem

We work in a space of dimension $d = 1$. The systems studied have the conservative form

$$u_t + f(u)_x = 0, \tag{4.1}$$

f being a smooth vector field, defined on a convex set $\mathcal{U}$ of $\mathbb{R}^n$, with a non-empty interior. We assume that $A(u) = \mathrm{d}f(u)$ is diagonalisable over $\mathbb{R}$ with eigenvalues of constant multiplicities, which to fix ideas we arrange in increasing order: $\lambda_1(u) < \lambda_2(u) < \cdots < \lambda_p(u)$.

As in the scalar case, the study of the Riemann problem is essential, both for numerical methods and for the understanding of the Cauchy problem for (4.1). It allows us, for example, to define numerical schemes which are sufficiently precise. With one among them Glimm has been able to prove [32] the sole global existence theorem in time of weak solutions which is of some significance (see Chapter 5).

The Riemann problem consists of solving the Cauchy problem for (4.1) when the initial condition takes the form

$$u_0(x) = \begin{Bmatrix} u_\mathrm{L}, & x < 0, \\ u_\mathrm{R}, & x > 0. \end{Bmatrix} \tag{4.2}$$

If u is a weak solution (respectively an entropy solution in the case where (4.1) has a strictly convex entropy) of (4.1), (4.2) and if $a > 0$, then $u^a(x, t) := u(ax, at)$ defines another solution. We hope that the solution, at least one that makes sense physically, is unique, without which the system is worthless. If such is the case, we have $u^a \equiv u$ for all $a > 0$, which reduces to saying that u depends only on the variable $\xi := x/t$. We shall denote by $v\colon \mathbb{R} \to \mathcal{U}$ the function $u(\cdot, 1)$, with the result that $u(x, t) = v(x/t)$. The Riemann problem thus reduces to the ordinary differential equation

$$(f(v))'(\xi) = \xi v'(\xi), \quad \xi \in \mathbb{R}, \quad \text{with } g' = \mathrm{d}g/\mathrm{d}\xi, \tag{4.3}$$

$$v(-\infty) = u_{\mathrm{L}}, \tag{4.4}$$

$$v(+\infty) = u_{\mathrm{R}}. \tag{4.5}$$

As in the scalar case the solution of the Riemann problem will be a juxtaposition of constant states, of rarefaction waves and of discontinuities. These last could be shock waves or contact discontinuities. The case of the (semi-)characteristic shocks will be ignored *a priori* although it presents an interest for applications. We have preferred to put the stress on the most fundamental questions.

An essential difference from the scalar case occurs as long as $f(u) = Au$ with A diagonalisable on $\mathbb{R}$ (the linear case). In this case, let us decompose the vector $u_{\mathrm{R}} - u_{\mathrm{L}}$ into a series of eigenvectors. We have $u_{\mathrm{R}} - u_{\mathrm{L}} = \sum_{1 \le j \le p} v_j$ with $Av_j = \lambda_j v_j$. The solution of the Riemann problem is given by

$$u(x,t) = u_{\mathrm{L}} + \sum_{j: x > \lambda_j t} v_j.$$

In the non-linear case, the solution will be equally composed of p waves, clearly differentiated by their physical meanings, separated by $p+1$ constant states $u^0 = u_{\mathrm{L}}, u^1, \ldots, u^p = u_{\mathrm{R}}$.

The great variety of the class of strictly hyperbolic conservative systems hinders a truly general study of the Riemann problem, with the notable exception of the case where the initial data satisfy $|u_{\mathrm{R}} - u_{\mathrm{L}}| \ll 1$, which is the object of Theorem 4.6.1 below. In particular, we shall be led to make a hypothesis of a geometrical nature which ensures that each of the p waves mentioned above is *simple*, that is to say that it consists of a shock, a contact discontinuity or a rarefaction wave, but not of several of these waves. In two examples, the p-system and gas dynamics, we shall give the complete solution of the Riemann problem without a hypothesis concerning the smallness of $u_{\mathrm{R}} - u_{\mathrm{L}}$.

4.2 The Hugoniot locus

Local description of the Hugoniot locus

We begin by describing the possible discontinuities (a, b, σ) with respect to the Rankine–Hugoniot condition, which is written here:

$$f(b) - f(a) = \sigma(b - a). \tag{4.6}$$

In the first instance, we are interested in the possible pairs (a, b), reducing thus by projection the trival triplets to the single point (a, a). Fixing the left state (or the right as for the moment we have perfect symmetry), $a \in \mathcal{U}$, we define the *Hugoniot locus of a* by

$$H(a) := \{b \in \mathcal{U} \colon \exists \sigma \in \mathbb{R},\ f(b) - f(a) = \sigma(b - a)\}.$$

The theorem below describes the structure of $H(a)$ in the neighbourhood of a.

Theorem 4.2.1 *We suppose that the eigenvalues of* $\mathrm{d}f$ *are simple (and hence* $p = n$*). In the neighbourhood of* a*, the Hugoniot locus of* a *is the union of* n *smooth curves* $H_k(a)$*,* $1 \leq k \leq n$*. The* k*-th curve is tangent at* a *to the eigenvector* $r_k(a)$ *of* $\mathrm{d}f(a)$*; it is in fact second order tangent at* a *to the integral curve of the eigenfield* r_k*.*

Exercises

4.1 In the case of eigenvalues λ_j with constant multiplicities n_j, $1 \leq j \leq p$, show that $H(a)$ is locally the union of p sub-manifolds $H_j(a)$ of respective dimensions n_j, the jth being tangent to the eigenspace $E_j(a) := \ker(A(a) - \lambda_j I_n)$ (we still suppose that A is diagonalisable in $\mathbb{R}$). If $n_j \geq 2$ show that $H_j(a)$ is in fact an integral manifold of the associated eigenvector field, that is that $H_j(a)$ is tangent to $E_j(b)$ at each of its points b (*Hint*: make use of Theorem 3.3.3.)

4.2 Describe $H(a)$ in the linear case.

4.3 Describe $H(a)$ for a system of two decoupled equations

$$v_t + g(v)_x = 0,$$
$$w_t + h(w)_x = 0.$$

4.4 Describe $H(a)$ for the p-system.

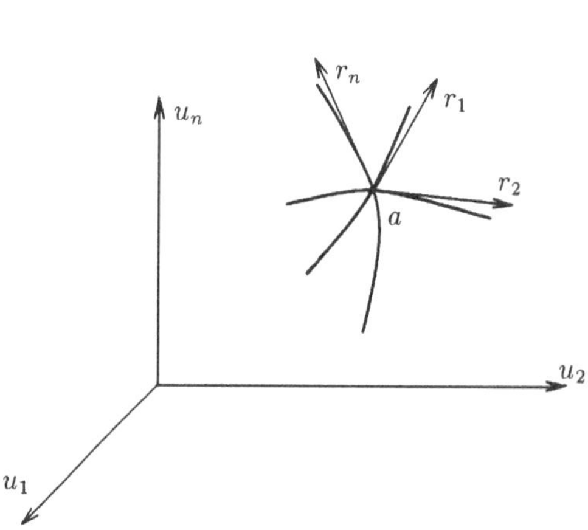

Fig. 4.1: The Hugoniot locus $H(a)$. For simplicity, we have supposed that the eigenvalues are simple.

4.5 Let $s \mapsto u(s)$ be the solution of the differential equation

$$\frac{\mathrm{d}u}{\mathrm{d}s} = r_j(u), \qquad u(0) = a.$$

Let $s \mapsto v(s)$ be a parametrisation of $H_j(a)$ (which is not necessarily the same as that previously introduced, in the case where $H_j(a)$ was the integral curve of r_j).

(1) Let $G(s) := (f(u(s)) - f(a)) \wedge (u(s) - a)$ which is an element of $\Lambda^2(\mathbb{R}^n)$ (an element of degree 2 in the exterior algebra of $\mathbb{R}^n$). Calculate G', G'', G''' and verify that $G(0) = G'(0) = G''(0) = G'''(0) = 0$.
(2) Show (without calculating G^{iv} completely) that

$$G^{iv}(0) = 2(\mathrm{d}\lambda_j \cdot r_j)(\mathrm{d}r_j \cdot r_j) \wedge r_j.$$

(3) Without making use of the preceding calculations, show that if $u(s) - v(s) = O(s^4)$, then $G(s) = O(s^5)$.
(4) We suppose that at every point $a \in \mathcal{U}$, $H_j(a)$ is tangent of the third order to the integral curve of r_j. Show that either the jth characteristic field is linearly degenerate, or the integral curves of r_j are straight lines in $\mathcal{U}$ (see Temple [103]).
(5) In both cases, show that $H_j(a)$ is the integral curve of r_j passing through a.

Some symmetric functions

Since a and b play symmetric rôles in the Rankine–Hugoniot condition, it is convenient to use symmetrical functions in a and b which generalise objects already defined on $\mathcal{U}$. For example, writing

$$A(u, v) := \int_0^1 \mathrm{d}f((1-t)u + tv)\,\mathrm{d}t,$$

we have $A(u, v) = A(v, u)$ and Taylor's formula gives $f(v) - f(u) = A(u, v)(v - u)$. The Rankine–Hugoniot condition is thus written

$$(A(a, b) - \sigma I_n)(b - a) = 0. \tag{4.7}$$

When $v - u$ is small, a symmetric function in u and v possesses a precise equivalent to the second order:

Lemma 4.2.2 *Let $(u, v) \mapsto M(u, v)$ be a symmetric function of class $\mathcal{C}^2$ defined in $\mathcal{U}$ and let $m(u) = M(u, u)$. Then, when $v \to u$, we have*

$$M(u, v) = m\left(\frac{u + v}{2}\right) + O(|v - u|^2).$$

Proof From Taylor's formula,

$$M(u, v) - m\left(\frac{u+v}{2}\right) = (\mathrm{d}_v M - \mathrm{d}_u M)\left(\frac{u+v}{2}, \frac{u+v}{2}\right) \cdot \left(\frac{v-u}{2}\right) + \mathrm{O}(|v-u|^2).$$

But the symmetry implies (differentiate the equality $M(a, b) = M(b, a)$ with respect to one of the vectors, then put $a = b = u$)

$$\mathrm{d}_v M(u, u) = \mathrm{d}_u M(u, u) = \frac{1}{2}\mathrm{d}m(u). \tag{4.8}$$

■

Since the eigenvalues of $A(u)$ are real and simple, every real matrix close to $A(u)$ has its eigenvalues real and simple. This is the case of $A(a, b)$ with b a neighbour of a since $A(a, a) = A(a)$. We shall denote by $\mu_j(a, b)$ these eigenvalues, and by $R_j(a, b)$ some associated eigenvector fields, chosen in a smooth manner, that is of class $\mathscr{C}^p$ as functions of a and b if f is of class $\mathscr{C}^{p+1}$. These functions are symmetric and we have $\mu_j(a, a) = \lambda_j(a)$.

Proof of Theorem 4.2.1

The formulation (4.7) of the Rankine–Hugoniot condition shows that if $b \in H(a)$ is in the neighbourhood of a but is distinct from it, then there exist an integer j, $1 \le j \le n$, and a small real number $s \neq 0$ such that

$$\sigma = \mu_j(a, b), \quad b = a + sR_j(a, b).$$

To the integer j corresponds the sub-set $H_j(a)$ of the Hugoniot curve of a. The discontinuities (a, b, σ) where $\sigma = \mu_j(a, b)$ are called the *j-discontinuities*.

Let us define, j being fixed, a smooth function N from $\mathbb{R} \times U$ into $\mathbb{R}^n$ by

$$N(s, u) := u - a - sR_j(a, u).$$

As $N(0, a) = 0$ and $\mathrm{d}_u N(0, a) = I_n$ is invertible, the implicit function theorem shows that $H_j(a)$ is, in the neighbourhood of a, a smooth curve parametrised by $s \in (-s_0, s_0)$, $s \mapsto \varphi_j(s; a)$, of which we are going to study a Taylor expansion at the origin.

To the first order, since $\varphi_j(0; a) = a$,

$$\begin{aligned}
\varphi_j(s; a) &= a + sR_j(a, a + sr_j(a) + \mathrm{O}(s^2)) \\
&= a + sr_j(a) + s^2 \mathrm{d}_v R_j(a, a) \cdot r_j(a) + \mathrm{O}(s^3) \\
&= a + sr(a) + \frac{1}{2}s^2(\mathrm{d}r_j \cdot r_j)(a) + \mathrm{O}(s^3)
\end{aligned}$$

which shows that $H_j(a)$ is second order tangent to the integral curve of the vector field r_j and the theorem is proved. ■

We notice that these two curves are not in general third order tangents (see Exercise 4.5 above).

4.3 Shock waves

Entropy balance

Theorem 4.2.1 does not indicate, among the discontinuities, those which have a physical sense. Let us suppose that (4.1) is of a physical nature, its entropy, strictly convex, being denoted by E (with $\mathrm{D}^2 E > 0$) and its flux by F. Lax's entropy condition expresses that the rate of dissipation of $[F] - \sigma[E]$ is non-positive. The following theorem provides an equivalent rate when b is close to a.

Theorem 4.3.1 *Let E be an entropy of class $\mathscr{C}^3$ of the system (4.1) and F its flux. Then, for $b \in H_j(a)$ a neighbour of a ($b = \varphi_j(s, a)$, that is $b - a \sim s r_j(a)$), we have*

$$[F] - \sigma[E] = \frac{s^3}{12}(\mathrm{d}\lambda_j \cdot r_j)\mathrm{D}^2 E(r_j, r_j) + \mathrm{O}(s^4),$$

the values of $\mathrm{d}\lambda_j$, r_j, *and* $\mathrm{D}^2 E$ *being calculated at* a.

Proof Since $\mathrm{d}F = \mathrm{d}E\,\mathrm{d}f$, we have

$$\begin{aligned}[F] - \sigma[E] &= \int_0^1 (\mathrm{d}F - \sigma \mathrm{d}E)(a + t(b-a)) \cdot (b-a)\,\mathrm{d}t \\ &= \int_0^1 \mathrm{d}E(a + t(b-a))(\mathrm{d}f(a + t(b-a)) - \sigma) \cdot (b-a)\,\mathrm{d}t. \qquad (4.9)\end{aligned}$$

Similarly and with the Rankine–Hugoniot condition we have

$$0 = [f] - \sigma[u] = \int_0^1 (\mathrm{d}f(a + t(b-a)) - \sigma) \cdot (b-a)\,\mathrm{d}t. \qquad (4.10)$$

Taking the scalar product of (4.10) by $\mathrm{d}E((a+b)/2)$ and subtracting from (4.9) the result is

$$\begin{aligned}&[F] - \sigma[E] \\ &= \int_0^1 \left(\mathrm{d}E(a + t(b-a)) - \mathrm{d}E\left(\frac{a+b}{2}\right)\right)(A(a + t(b-a)) - \sigma) \cdot (b-a)\,\mathrm{d}t.\end{aligned}$$

But with (4.7)

$$[F] - \sigma[E] = \int_0^1 \left(\mathrm{d}E(a + t(b-a)) - \mathrm{d}E\left(\frac{a+b}{2}\right)\right)(A(a+t(b-a)) \\ - A(a,b)) \cdot (b-a)\, \mathrm{d}t.$$

We now make use of Lemma 4.2.2:

$$A(a+t(b-a)) - A(a,b) = A(a+t(b-a)) - A\left(\frac{a+b}{2}\right) + \mathrm{O}(|b-a|^2) \\ = \left(t - \frac{1}{2}\right)\mathrm{D}^2 f(a) \cdot (b-a) + \mathrm{O}(|b-a|^2).$$

Similarly

$$\mathrm{d}E(a+t(b-a)) - \mathrm{d}E\left(\frac{a+b}{2}\right) = \left(t - \frac{1}{2}\right)\mathrm{D}^2 E(a) \cdot (b-a) + \mathrm{O}(|b-a|^2).$$

Thus,

$$[F] - \sigma[E] = C\,\mathrm{D}^2 E(b-a, \mathrm{D}^2 f(b-a, b-a)) + \mathrm{O}(|b-a|^4) \\ = C s^3 \mathrm{D}^2 E(r_j, \mathrm{D}^2 f(r_j, r_j)) + \mathrm{O}(|b-a|^4)$$

where

$$C := \int_0^1 \left(t - \frac{1}{2}\right)^2 \mathrm{d}t = \frac{1}{12}.$$

The theorem thus results from the following two important lemmas.

Lemma 4.3.2 *For $1 \le j \le n$, we have*

$$\mathrm{D}^2 f(r_j, r_j) = (\mathrm{d}\lambda_j \cdot r_j) r_j + \sum_{k \neq j} c_{jk} r_k,$$

where the c_{jk} depend on the normalisation of the eigenbasis.

Lemma 4.3.3 *The basis $(r_j)_{1 \le j \le n}$ is orthogonal for the symmetric bilinear form $\mathrm{D}^2 E$.*

Actually, we have from these lemmas

$$\mathrm{D}^2 E(r_j, \mathrm{D}^2 f(r_j, r_j)) = (\mathrm{d}\lambda_j \cdot r_j)\mathrm{D}^2 E(r_j, r_j) + \sum_{k \neq j} c_{jk} \mathrm{D}^2 E(r_j, r_k)$$

where the terms of the last sum are zero by orthogonality.

Proof of Lemmas 4.3.2 and 4.3.3

Let us begin with Lemma 4.3.3.

Proof Differentiating the equality $\mathrm{d}F \cdot h = \mathrm{d}E \cdot (\mathrm{d}f \cdot h)$, we have

$$\mathrm{D}^2 F(h, k) = \mathrm{D}^2 E(\mathrm{d}f \cdot h, k) + \mathrm{d}E \cdot (\mathrm{D}^2 f(h, k)).$$

By symmetry we derive

$$\mathrm{D}^2 E(\mathrm{d}f \cdot h, k) = \mathrm{D}^2 E(\mathrm{d}f \cdot k, h).$$

Putting $h = r_j$ and $k = r_k$ in the above equality we obtain

$$(\lambda_j - \lambda_k)\mathrm{D}^2 E(r_j, r_k) = 0,$$

which proves the lemma. ∎

Now, let us look at Lemma 4.3.2.

Proof We differentiate the relation $(\mathrm{d}f - \lambda_j)r_j = 0$ in the direction h:

$$\mathrm{D}^2 f(r_j, h) + (\mathrm{d}f - \lambda_j)(\mathrm{d}r_j \cdot h) = (\mathrm{d}\lambda_j \cdot h)r_j.$$

We put $h = r_j$ in this formula, then we remark that $\mathrm{Im}(\mathrm{d}f - \lambda_j)$ is spanned by the vectors r_k for $k \neq j$ since $\mathrm{d}f$ is diagonalisable. ∎

Genuinely non-linear characteristic fields

Of course, when $\mathrm{D}^2 E > 0$, that is for what concerns us, we have $\mathrm{D}^2 E(r_j, r_j) > 0$ and the sign of $[F] - \sigma[E]$ is uniquely determined by those of s and of $\mathrm{d}\lambda_j \cdot r_j$ provided that this latter number is not zero. This justifies the following definition.

Definition 4.3.4 We say that the jth characteristic field is *genuinely non-linear at* a if $\mathrm{d}\lambda_j \cdot r_j$ is non-zero at a. We say that it is *genuinely non-linear* if it is genuinely non-linear at every point of $\mathscr{U}$.

The notion of a genuinely non-linear field means that λ_j is monotonic along the integral curves of r_j and thus also in the neighbourhood of a along $H_j(a)$. This is the antithesis of a linearly degenerate field, which does not mean a field is one or the other: the rate of variation $d\lambda_j \cdot r_j$ of the eigenvalue along the eigenfield can be zero on a closed set of $\mathscr{U}$ with empty interior, for example a hypersurface transverse to r_j. In this case, λ_j is not monotonic along the integral curves of r_j, or along the curves of the Hugoniot locus H_j.

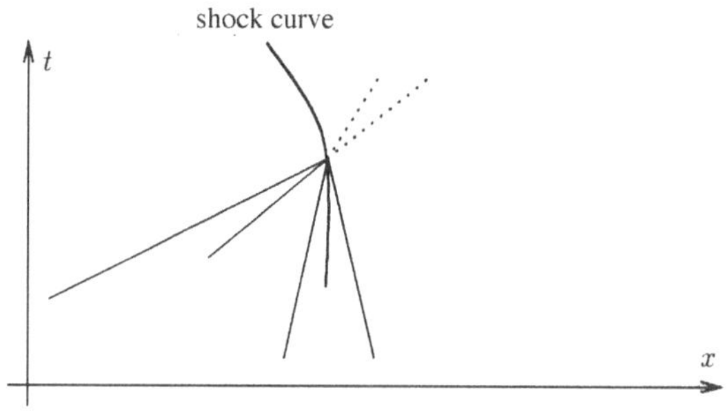

Fig. 4.2: Lax shocks. Here $n = 3$: in full line (resp. in dotted line) the characteristics incoming (resp. outgoing).

For a genuinely non-linear field, there is canonical choice of right or left eigenfields (note that because of Theorem 3.3.3, a genuinely non-linear field corresponds to a simple eigenvalue) by the normalisation

$$\mathrm{d}\lambda_j \cdot r_j \equiv 1, \quad l_j \cdot r_j \equiv 1.$$

We shall take care not to confuse the differential forms $\mathrm{d}\lambda_j$ and l_j, as $l_j \cdot r_k \equiv 0$ for $k \neq j$, while this is not the case in general for $\mathrm{d}\lambda_j$ (this causes the coupling of the equations of the system).

We now have

Proposition 4.3.5 *We suppose that the j-th characteristic field is genuinely non-linear and that we have adopted the above normalisation.*

If $b \in H_j(a)$ is in the neighbourhood of a, the discontinuity $(a, b, \sigma = \mu_j(a, b))$ satisfies Lax's entropy condition if and only if $s \leq 0$.

We have seen in the scalar case an inequality comparing the speed of the discontinuity with those of the waves to the right and to the left of a shock. Lax [59] has introduced for systems the following definition.

Definition 4.3.6 We say that the discontinuity (a, b, σ) is a *j-shock* in the sense of Lax if it satisfies the inequalities

$$\lambda_j(b) \leq \sigma \leq \lambda_j(a), \qquad \lambda_{j-1}(a) < \sigma < \lambda_{j+1}(b).$$

Lax's shock condition is one of numerous conditions of admissibility of discontinuities, in fact the simplest. It has the great merit of having a geometrical interpretation in terms of the stability of a discontinuity subject to a perburbation of small amplitude [64]. It expresses that at a point of discontinuity there are $n + 1$ incoming characteristics, of which the speeds are the eigenvalues $\lambda_1(b), \ldots, \lambda_j(b)$, $\lambda_j(a), \ldots, \lambda_n(a)$, leading to $n + 1$ scalar data (instead of n at a point of continuity), which shows up the fact that the speed of the discontinuity is itself an unknown (the shock curve is a *free boundary*).

The major inconvenience of Lax's shock condition is that it is unable to be expressed for a weak solution, but only for a piecewise smooth solution, contrary to the entropy condition. On the other hand, it keeps its meaning for piecewise smooth solutions of (4.1) even when the system (4.1) does not possess a non-trivial convex entropy. Finally these two entropy conditions are equivalent for discontinuities of small amplitude.

Theorem 4.3.7 *We suppose that the j-th characteristic field is genuinely non-linear. If $b \in H_j(a)$ in the neighbourhood of a, the discontinuity $(a, b, \sigma = \mu_j(a, b))$ satisfies the Lax entropy condition if and only if it is a j-shock.*

In fact, $\sigma \neq \lambda_j(a), \lambda_j(b)$ for $b \neq a$, the two inequalities $\sigma < \lambda_j(a)$ and $\lambda_j(b) < \sigma$ are equivalent while $\lambda_{j-1}(a) < \sigma < \lambda_{j+1}(b)$ is trivial.

Proof From Lemma 4.2.2,

$$\sigma = \lambda_j\left(\frac{a+b}{2}\right) + \mathrm{O}(|b-a|^2) = \lambda_j(a) + \frac{1}{2} s \mathrm{d}\lambda_j \cdot r_j + \mathrm{O}(s^2),$$

with the result that $\sigma \neq \lambda_j(a)$ if $s \neq 0$, that is, if $b \neq a$. Similarly $\sigma \neq \lambda_j(b)$. In fact $\sigma - \lambda_j(a)$ is of the same sign as s, as is $\lambda_j(b) - \sigma$ because

$$\sigma = \lambda_j(b) - \frac{1}{2} s \mathrm{d}\lambda_j \cdot r_j + \mathrm{O}(s^2).$$

Finally, $\lambda_{j-1}(a) < \lambda_j(a) \sim \sigma$ and $\sigma \sim \lambda_j(b) < \lambda_{j+1}(b)$ complete the proof. ■

Exercise

4.6 We consider the p-system

$$\left.\begin{aligned} u_t + v_x &= 0, \\ v_t + p(u)_x &= 0, \end{aligned}\right\}$$

where $p' > 0$.

(1) Calculate the eigenvalues $\lambda_1 < \lambda_2$ of the system and the associated vectors. Show that each field is genuinely non-linear in (u, v) if and only if $p''(u) \neq 0$.

(2) Let $(a, b) \in \mathbb{R}^2$ and $1 \leq i \leq 2$. Describe $H_i(a, b)$ as a curve parametrised by $v = b + \varepsilon\varphi(u, a)$ where $\varepsilon = (-1)^i$.

(3) Show that $E(u, v) := \frac{1}{2}v^2 + e(u)$ where $e' = p$ is a strictly convex entropy. Calculate its flux.

(4) Let $(u, v) \in H(a, b)$. Calculate the rate $[F] - \sigma[E]$ of production of entropy as a function of u and a only. Show that its sign is equal to that of $\varepsilon[u]$ (respectively of $-\varepsilon[u]$) if p is convex (respectively concave) between a and u.

(5) We suppose that $(u - a)p''(u) > 0$ for $u \neq a$. Show that $((a, b), (u, v), \sigma)$ with $(u, v) \in H_1(a, b)$ satisfies Lax's entropy condition, and Lax's shock condition, but that those for which $(u, v) \in H_2(a, b)$ satisfy neither the one nor the other. Compare with Proposition 4.3.5.

4.4 Contact discontinuities

If the jth characteristic curve is linearly degenerate, Theorem 4.3.1 does not allow us to determine from among the j-discontinuities those which satisfy the entropy condition. In fact all satisfy it, for these are contact discontinuities. They are thus reversible: (a, b, σ) and (b, a, σ) are admissible. More precisely:

Theorem 4.4.1 *We suppose that the j-th characteristic field is linearly degenerate with one simple eigenvalue.*

Then $H_j(a)$ coincides with the integral curve of r_j and the rate of production of entropy is zero for every j-discontinuity. Finally $\lambda_j(a) = \sigma = \lambda_j(b)$.

Proof First of all, it suffices to show that if b is on the integral curve $\gamma_j(a)$ of r_j through a, then $b \in H_j(a)$. Let us notice, first of all, that, since $\mathrm{d}\lambda_j \cdot r_j \equiv 0$, λ_j is constant on $\gamma_j(a)$. Thus

$$\begin{aligned} f(b) - f(a) - \lambda_j|_{\gamma_j(a)}(b - a) &= \int_0^s \frac{\mathrm{d}}{\mathrm{d}t}(f(u) - f(a) - \lambda_j|_{\gamma_j(a)}(u - a))\,\mathrm{d}t \\ &= \int_0^s (\mathrm{d}f(u) - \lambda_j(u))r_j(u)\,\mathrm{d}t \\ &= 0. \end{aligned}$$

Hence $\gamma_j(a) \subset H_j(a)$, that is to say that these curves are identical. Then

$$\begin{aligned} &F(b) - F(a) - \lambda_j|_{\gamma_j(a)}(E(b) - E(a)) \\ &= \int_0^s \frac{\mathrm{d}}{\mathrm{d}t}(F(u) - F(a) - \lambda_j|_{\gamma_j(a)}(E(u) - E(a)))\,\mathrm{d}t \\ &= \int_0^s \mathrm{d}E(u)(\mathrm{d}f(u) - \lambda_j)r_j\,\mathrm{d}t \\ &= 0. \end{aligned}$$

∎

When λ_j is of constant multiplicity $m > 1$ (and therefore is linearly degenerate) the algebraic properties described above are still valid for the integral manifold Γ_j of $\ker(\mathrm{d}f - \lambda_j)$: if $b \in \Gamma_j(a)$, then $b \in H(a)$ and the discontinuities $(a, b, \lambda_j(a))$ and $(b, a, \lambda_j(a))$ are both admissible. We shall again therefore denote by $H_j(a)$ this integral manifold.

Most physical systems, for which $n \geq 2$, possess simultaneously genuinely non-linear fields and one (or several) linearly degenerate fields. The two concepts are therefore equally important. The presence of a linearly degenerate field is often associated with an invariance group of the system, for example a rotation group [24]. It would be erroneous to believe that the linearly degenerate fields are simpler in their structure, easier to understand, or to treat, under the pretext that the linear systems are less complicated. It is rather the contrary that occurs. For example, on account of their dissipative aspect, the genuinely non-linear fields lead to stable structures (the shocks) which are less perturbed, even by the addition of a parabolic term (a viscosity) into the system [70] (see Chapter 7). On the other hand the contact discontinuities have a marginal stability in dimension $d = 1$ and can even be plainly unstable in higher spatial dimensions (Kelvin–Helmholtz or Richtmyer–Meshkov instability for gas dynamics). Even in dimension 1, their behaviour with respect to a parabolic perturbation of the system is extremely complex and cannot in general be described with the help of conservative integrals. Finally, linear degenerate fields can lead to solutions which display large amplitude oscillations even if of high frequency, for example sequences of entropy solutions of the form $u^\varepsilon(x, t) = v(\varepsilon^{-1}\varphi(x - ct))$, (see Chapter 10). The persistence of structures of large amplitude and arbitrarily high frequency renders null and void the linearisation by which we have justified hyperbolicity as a geometrical condition allowing the Cauchy problem to be well-posed. In Chapter 10, we shall see therefore an extension of the notion of hyperbolicity which takes into account the large amplitudes all over a linearly degenerate field and which reduces to the actual notion in the case without linearly degenerate fields and in the case of linear fields.

Riemann invariants

The integral curves of a vector field can be described as level sets of a list of $n - 1$ independent functions defined on $\mathcal{U}$. Let us consider a simple eigenvalue λ_j of $\mathrm{d}f$ and its field of eigenvectors r_j. Let us choose arbitrarily a hypersurface Σ transverse to r_j and a regular function $v_0\colon \Sigma \to \mathbb{R}$. Under sufficiently general hypotheses, Σ meets each integral curve in one point and one only. The Cauchy problem

$$\left.\begin{aligned} \mathrm{d}v \cdot r_j &= 0, & u &\in \mathcal{U}, \\ v(u) &= v_0(u), & u &\in \Sigma, \end{aligned}\right\} \tag{4.11}$$

has a unique global solution. Let us choose on Σ independent functions v_0^α, $1 \leq \alpha \leq n-1$, that is to say such that the differential form $\omega_0 := \mathrm{d}v_0^1 \wedge \cdots \wedge \mathrm{d}v_0^{n-1}$ does not vanish on the space tangent to Σ. We verify easily that for the corresponding solutions v^α of (4.11), the form $\omega := \mathrm{d}v^1 \wedge \cdots \wedge \mathrm{d}v^{n-1}$ satisfies a differential equation of the form $\mathrm{d}\omega \cdot r_j = L(\omega, \mathrm{d}r_j)$ where L is bilinear. From the Cauchy–Lipschitz theorem, ω is zero at a point of $\mathcal{U}$ only if it is identically zero on the integral curve passing through that point. By hypothesis, ω is thus not zero on any part: the functions v^α remain independent at every point. In particular, the level sets of $(v^1, \ldots, v^{n-1})$ are smooth curves, which are the integral curves of r_j since each v^α is constant along these.

A method of describing the integral curves of r_j is thus to find $n-1$ independent solutions of the linear differential equation $\mathrm{d}v \cdot r_j = 0$. Each non-trivial solution (that is to say of which the differential does not vanish) is called a *weak Riemann invariant*. There is no general method of solving that equation as that comes down to knowing how to integrate all the differential equations. But in most applications, separation of variables, homogeneity properties or considerations of symmetry enable us to set up an explicit list. In the following sections, we shall see how to proceed for the p-system and for gas dynamics, where we shall solve globally the Riemann problem.

If the field is linearly degenerate of multiplicity m_j, (4.11) must be replaced by

$$\mathrm{d}v|_{\ker(\mathrm{d}f-\lambda_j)} = 0, \qquad u \in \mathcal{U}. \tag{4.12}$$

The initial data are specified over a sub-manifold of codimension m_j, transverse to $\ker(\mathrm{d}f - \lambda_j)$. From Theorem 3.3.3, the system (4.12) has $n - m_j$ independent solutions v^α. The level sets of $(v^1, \ldots, v^{n-m_j})$ are again integral manifolds of the *eigenfield*. We notice that if $\mathrm{d}\lambda_j$ does not vanish, the condition of linear degeneracy makes λ_j be a Riemann invariant.

Exercises

4.7 The following theorem, due to B. Sévennec [93], is a difficult geometrical problem. Its physical interpretation is still not clear, at the moment of publication of this work.

Theorem 4.4.2 *Let λ be an eigenvalue of constant multiplicity m of a linearly degenerate field for a physical system. let Γ be an integral manifold of the field of the corresponding eigenspaces (see Theorem 3.3.3). Using the Legendre transformation $q := \mathrm{d}_u E$, the set*

$$\Gamma^* := \{(q, E^*(q)) \colon u \in \Gamma\}$$

is included in an affine subspace of dimension $m + 1$.

Verify this statement for the following examples (in each case, first of all identify the genuine non-linearity or the linear degeneracy of the fields).

(1) The system of Keyfitz and Kranzer $u_t + (\varphi(r)u)_x = 0,\ r := \|u\|$.
(2) Gas dynamics in eulerian variables:

$$\left.\begin{aligned} &\rho_t + (\rho v)_x = 0,\\ &(\rho v)_t + (\rho v^2 + p(\rho, e))_x = 0,\\ &\left(\frac{1}{2}\rho v^2 + \rho e\right)_t + \left(\left(\frac{1}{2}\rho v^2 + \rho e + p\right)v\right)_x = 0. \end{aligned}\right\}$$

(3) Gas dynamics in lagrangian variables:

$$\left.\begin{aligned} &v_t = z_x,\\ &z_t + q(v, e)_x = 0,\\ &\left(e + \frac{1}{2}z^2\right)_t + (qz)_x = 0. \end{aligned}\right\}$$

(4) The dynamics of an elastic string:

$$\left.\begin{aligned} &v_t = w_x,\\ &w_t = (T(r)v/r)_x, \quad r = \|v\|. \end{aligned}\right\}$$

4.8 Let $u_t + f(u)_x = 0$ be a strictly hyperbolic system, of which we choose a characteristic field, of multiplicity p. We also choose independent Riemann invariants $w_1, \dots, w_{n-p}$ for this field.

(1) Show that there exists a mapping $u \mapsto B(u)$ with values in $M_{n-p}(\mathbb{R})$ such that, for every classical solution of the system, we have

$$w(u)_t + B(u)w(u)_x = 0,$$

with $w = (w_1, \dots, w_{n-p})^{\mathrm{T}}$.
(2) What are the eigenvalues of $B(u)$?
(3) How is B transformed when we change the choice of Riemann invariants?

4.5 Rarefaction waves. Wave curves

Rarefaction waves are, as in the scalar case, the solutions of (4.3) which are of class $\mathscr{C}^1$ in an interval (ξ_1, ξ_2), with $v \neq 0$. Expanding the differential equation, we arrive at $(\mathrm{d}f(v) - \xi)v' = 0$ which shows that $v'(\xi)$ is an eigenvector of $\mathrm{d}f(v(\xi))$, associated with the eigenvalue ξ. Thus there exists an integer j, $1 \le j \le n$, such that

$$\xi = \lambda_j(v(\xi)), \qquad \xi \in (\xi_1, \xi_2), \tag{4.13}$$

$$v'(\xi) \parallel r_j(v(\xi)), \qquad \xi \in (\xi_1, \xi_2). \tag{4.14}$$

Differentiating (4.13) with respect to ξ we deduce $1 = \mathrm{d}\lambda_j(v(\xi)) \cdot v'(\xi)$, which by (4.14) implies that $\mathrm{d}\lambda_j \cdot r_j \neq 0$. Hence the jth characteristic field is genuinely non-linear. Conversely, if a characteristic field, let us say the jth, is genuinely non-linear in an open set $\mathscr{V}$ of $\mathscr{U}$, let us consider a curve γ_j, connected in $\mathscr{V}$, being an integral curve of the vector field r_j. Then λ_j is monotonic increasing along γ_j, varying from ξ_- to ξ_+. If $\xi_- < \xi_1 < \xi < \xi_2 < \xi_+$, then the equality (4.13) determines a unique point $v(\xi)$ of γ_j, and the mapping $\xi \mapsto v(\xi)$ clearly defines a rarefaction, which we call a *j-rarefaction*.

Let us now consider the possibility of having a j-wave, that is a j-rarefaction, a j-shock or a j-contact-discontinuity, which passes from a fixed state a to a neighbouring state b in going from left to right, that is, in the direction of ξ increasing. We suppose that the jth field is linearly degenerate or else that it is genuinely non-linear at a.

If it is linearly degenerate, then we go from a to b by a j-contact-discontinuity whenever $b \in \gamma_j(a)$, the integral curve of r_j passing through a (to be replaced by the integral manifold of the eigenfield if λ_j is of multiplicity $m \geq 2$).

If the field is genuinely non-linear at a there are two possibilities. Either $b \in \gamma_j(a)$, but then $b - a \sim sr_j(a)$ with $s > 0$ and the wave is a rarefaction (we have normalised r_j by $\mathrm{d}\lambda_j \cdot r_j(a) = 1$), or $b \in H_j(a)$, but now $b - a \sim sr_j(a)$ with $s < 0$, and the wave is a shock. In each of these two situations, b describes a curve parametrised by s and of which a is one extremity. The union of these two curves and of a forms a curve of class $\mathscr{C}^2$ (from Theorem 4.2.1), tangent at a to the vector $r_j(a)$, the j-wave *curve* originating at a and which we denote by $\mathscr{O}_j(a)$. In the linearly degenerate case, the curve (or the variety of dimension m) of the j-wave is $\gamma_j(a)$, which we still denote by $\mathscr{O}_j(a)$.

Finally, to each field, genuinely non-linear at a or linearly degenerate, there corresponds a manifold of j-wave indexed by a, defined in the neighbourhood of a and denoted by $\mathscr{O}_j(a)$, such that for $b = \mathscr{O}_j(a)$, there exists a jth simple wave passing from a to b. To be perfectly clear, it is even necessary to speak of the direct wave curve $\mathscr{O}_j^{\mathrm{d}}(a)$ (we fix the left state of the wave) in contrast to the reverse wave curve $\mathscr{O}_j^{\mathrm{r}}(a)$ (where we fix the right state of the wave). The relation between the two families of curve is given by the equivalence

$$b \in \mathscr{O}_j^{\mathrm{d}}(a) = \mathscr{O}_j(a) \Leftrightarrow a \in \mathscr{O}_j^{\mathrm{r}}(b).$$

The extension of the wave curves $\mathscr{O}_j(a)$ beyond a neighbourhood of a is an important question when we have in mind realistic problems where the variation of the solution is not small. The procedure is clear in the case of a linearly degenerate field: as long as the field is linearly degenerate extend $\mathscr{O}_j(a)$ by an integral curve of r_j, that is as a Hugoniot curve. For genuinely non-linear field at a, we can also extend $\mathscr{O}_j(a)$ on the side of $s > 0$ by an integral curve of r_j, as long as λ_j is strictly increasing along it. On the other hand, on the side of $s < 0$ (shocks), the monotonicity of λ_j is neither a necessary condition (as can be seen in the scalar case) nor a sufficient condition, since it does not imply in an obvious way Lax's entropy condition or Lax's shock condition. The extension must follow the Hugoniot curve (in so far as it is a curve) until a suitable entropy condition leads to the exclusion of certain discontinuities. When a field is genuinely non-linear, except on a hypersurface, transverse to r_j, there will correspond to it composite waves, in which (semi-)characteristic shocks are combined with expansion waves, as in the scalar case. The description of these curves is much more complicated than any we have seen up until now and can only be made by taking a well-defined example and treating it thoroughly.

The most satisfactory entropy condition for a characteristic field of which the expression $d\lambda_j \cdot r_j$ changes sign is that of Liu [66, 67] which generalises to systems Oleĭnik's criterion. First of all, let us denote by $\sigma(a, b)$ the speed of the discontinuity between a and b, when $f(b) - f(a)$ is parallel to $b - a$ and $a \neq b$.

Definition 4.5.1 Let $(u_L, u_R; \sigma(u_L, u_R))$ be a discontinuity of the system $u_t + f(u)_x = 0$. We say that it is *admissible (in the sense of Liu)* if the following conditions are fulfilled.

(1) There exists an index j such that λ_j is simple, such that $H_j(u_L)$ extends to a curve of class $\mathscr{C}^1$ as far as u_R and that $H_j(u_R)$ extends to a curve of class $\mathscr{C}^1$ as far as u_L.

(2) For all $u \in H_j(u_L)$, located between u_L and u_R, we have

$$\sigma(u_L, u_R) \leq \sigma(u_L, u).$$

(3) For all $u \in H_j(u_R)$, located between u_L and u_R, we have

$$\sigma(u, u_R) \leq \sigma(u_L, u_R).$$

Let (u_L, u_R, s) be a discontinuity satisfying Liu's criterion. When u tends to u_L while remaining in $H_j(u_L)$, $\sigma(u_L, u)$ tends to $\lambda_j(u_L)$. We thus have $\sigma(u_L, u_R) \leq \lambda_j(u_L)$. Similarly, letting u tend to u_R along $H_j(u_R)$, we obtain $\lambda_j(u_R) \leq \sigma(u_L, u_R)$. Liu's criterion is therefore more precise than Lax's criterion.

Exercise: Prove that properties (2) and (3) in Definition 4.5.1 are equivalent to each other.

Parametrisation of wave curves

We have seen that a canonical choice of r_j is possible for a non-linear field. Similarly, there exists a canonical parametrisation of the wave curve $\mathscr{O}_j(a)$, by $b = \varphi_j(s; a)$ where $s := \lambda_j(b) - \lambda_j(a)$. We recall the inequality $s > 0$ on the side of the rarefactions and the inequality $s < 0$ on the side of the shocks. We notice that this parametrisation is of class $\mathscr{C}^2$ but not any better in general. On the side of the rarefactions, s is exactly the time it takes to pass from a to b in solving the differential equation $u' = r_j(u)$ where r_j is normalised by $\mathrm{d}\lambda_j \cdot r_j = 1$.

On the other hand, there is not one favoured parametrisation of a wave curve of a linearly degenerate field and anyone must find that which is most suitable for the calculations of this or that example.

Finally, let us note, what will serve in the proof of Theorem 4.6.1, that each function φ_j is of class $\mathscr{C}^2$ with respect to its arguments (ε, a). In fact, the integral curves of a vector field, being the solutions of a differential equation, depend in a $\mathscr{C}^\infty$ way (if the field itself is $\mathscr{C}^\infty$) on the 'time' ε and the initial point a. Similarly, the Hugoniot curves are projections on $\mathbb{R}^n$ of a manifold (that of the pairs (a, b) for which $(f(b) - f(a)) \wedge (b - a) = 0$, which is of class $\mathscr{C}^\infty$ if f is itself $\mathscr{C}^\infty$) and if this projection is made transversely to the tangent space at (a, a); these curves are thus regular. Finally, one glues together the relevant pieces of the Hugoniot and integral curves in a $\mathscr{C}^2$ way with the following Taylor expansion at a point of coincidence:

$$\begin{aligned}\varphi_j(\varepsilon, b) - \varphi_j(0, a) = b - a + \varepsilon r_j(a) + \varepsilon \mathrm{d}r_j(a) \cdot (b - a) \\ + \frac{1}{2}\varepsilon^2(\mathrm{d}r_j \cdot r_j) + \mathrm{O}(\varepsilon^3 + \|b - a\|^3).\end{aligned} \tag{4.15}$$

4.6 Lax's theorem

The form of the solution of the Riemann problem

If $i < j$, an i-wave which joins a to b and a j-wave that joins b to c have different velocities, that is to say that, if they are centred at the origin, with the view of solving a Riemann problem, the zones where they do vary can be disjoint. More precisely, let $x/t \in [s_1, s_2]$ and $x/t \in [s_3, s_4]$ be these zones. The i-wave has value b for $x > s_2 t$ whereas the j-wave has value b for $x < s_3 t$. We can construct a self-similar solution, which will be a solution of the Riemann problem between a and c, by gluing these two waves provided that $s_2 < s_3$. We shall see that this condition is realised except perhaps if the two waves are shock waves of sufficiently large amplitudes.

If one of the waves is a shock wave, for example the i-wave, then $s_1 = s_2 =: \sigma_i$, the velocity of the shock wave, and this satisfies the Lax inequalities $\lambda_i(b) \le \sigma_i \le \lambda_i(a)$ and $\lambda_{i-1}(a) < \sigma_i < \lambda_{i+1}(b)$. If the i-wave is a rarefaction wave or a contact discontinuity, then $\lambda_i(a) = s_1$ and $\lambda_i(b) = s_2$, with $s_1 = s_2$ in the latter case.

If both waves are rarefaction waves or contact discontinuity, then $s_2 = \lambda_i(b) < \lambda_j(b) = s_3$.

If the i-wave is a rarefaction wave or a contact discontinuity and the j-wave is a shock wave, then $s_2 = \lambda_i(b) \le \lambda_{j-1}(b) < \sigma_i = s_3$.

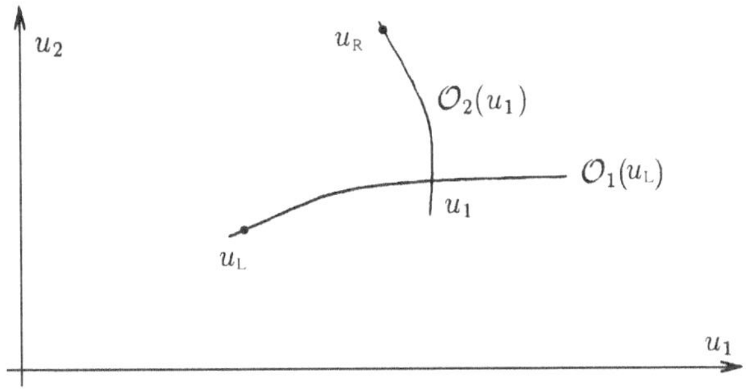

Fig. 4.3: Solution of the Riemann problem. Here $n = 2$.

The case of an i-shock-wave and a j-rarefaction-wave or contact-discontinuity is symmetric with the preceding case.

If both waves are shock waves and $j = i + 1$, we are unable to deduce the order of s_2 and s_3 from the inequalities $s_2 < \lambda_{i+1}(b)$, $s_2 \leq \lambda_i(a)$, $s_3 > \lambda_{j-1}(b)$ and $s_3 \geq \lambda_j(c)$, unless at least one of the shock waves is weak. For example if the i-shock-wave is of small amplitude, then $s_2 \sim \lambda_i(b) \leq \lambda_{j-1}(b) < s_3$, from which we deduce that $s_2 < s_3$.

Of course, the case of shock waves of large amplitude remains to be examined case by case. For a great number of physical systems, the gluing of an i-shock-wave and a j-shock-wave is always possible for $i < j$.

Generalising the idea developed above, we therefore seek the solution of the Riemann problem between two states u_L and u_R under a succession of constant states $u_0 = u_L, u_1, \ldots, u_{n-1}, u_n = u_R$, separated by simple waves. For $1 \leq p \leq n$ a p-wave localised in a sector of the form $x/t \in [s_{2p-1}, s_{2p}]$ takes u_{p-1} to u_p and the sequence $(s_k)_{1 \leq k \leq n}$ is increasing. We have $u_{p+1} \in \mathcal{O}_p(u_p)$, that is $u_{p+1} = \varphi_p(\varepsilon_p; u_p)$ where $\varepsilon_p \in \mathbb{R}$ has the value $\lambda_p(u_{p+1}) - \lambda_p(u_p)$ if the p-th field is genuinely non-linear.

Now, let us define a mapping $\Phi(\cdot, a)$, from a neighbourhood of the origin in $\mathbb{R}^n$ into a neighbourhood of $a \in U$, by

$$\Phi(\varepsilon; a) := \varphi_n(\varepsilon_n; \varphi_{n-1}(\varepsilon_{n-1}; \ldots; \varphi_1(\varepsilon_1; a) \ldots)).$$

The preceding construction relies on the solution of the equation

$$\Phi(\varepsilon; u_L) = u_R, \tag{4.16}$$

where $\varepsilon \in \mathbb{R}^n$ is the unknown vector. In fact, we have seen that a solution of the Riemann problem obtained by the gluing of a simple wave of each family (n waves in all) corresponds to a solution of (4.16). Conversely if ε is a solution of (4.16), if we define $u_0 = u_L$, then $u_{p+1} = \varphi_p(\varepsilon; u_p)$ by induction on p, we have $u_n = u_R$ and we can join u_p to u_{p+1} by a p-wave, these waves gluing except perhaps in the case where a p-wave and a $(p+1)$-wave are shock waves of large amplitude, that is if $\varepsilon_p < 0$ and $\varepsilon_{p+1} < 0$ are both large. Let us note that the solution of (4.16), if it exists, can have one (or several) vanishing component(s); for example $\varepsilon_p = 0$ means that there is not a p-wave, that is that $u_{p+1} = u_p$. In this case, because of Lax's inequalities, the $(p-1)$-wave and the $(p+1)$-wave are always gluable.

Local existence of the solution of the Riemann problem

The main theorem, due to Lax [59], is the following.

Theorem 4.6.1 (Lax) *Let $u_t + f(u)_x = 0$ be a strictly hyperbolic system of conservation laws in an open set of $\mathcal{U}$. We suppose that each characteristic field is either genuinely non-linear or linearly degenerate in the neighbourhood of a.*

For every neighbourhood $W \subset \mathcal{U}$ of a, there exists a neighbourhood $V \subset W$ of a such that, for $u_L, u_R \in V$, the Riemann problem has a solution of the form described above and with values in W. In addition such a solution is unique.

Because we do not know of a uniqueness theorem suitable for the Cauchy problem in the case of systems, we cannot guarantee that the solution constructed in this way is the only one, although it is difficult to imagine a solution which does not have the structure imposed here: self-similar with simple waves separating $n+1$ constant states. For a physical system, Heibig [39, 40] shows that a self-similar solution of the Riemann problem which satisfies the entropy inequality necessarily possesses this structure. This shows that if the non-linearity of the fields is well-defined, a sole strictly convex entropy might be sufficient to characterise the mathematically reasonable solutions.

Proof Let us carry out the proof in the case of simple eigenvalues.

Since each function φ_j is of class $\mathcal{C}^2$, Φ is $\mathcal{C}^2$ with respect to (ε, u) throughout a neighbourhood of $(0, a)$ in $\mathbb{R}^n \times U$ and similarly for the partial functions Φ_k defined by

$$\Phi_k(\varepsilon_1, \ldots, \varepsilon_k; a) := \varphi_k(\varepsilon_k; \varphi_{k-1}(\ldots; \varphi_2(\varepsilon_2; \varphi_1(\varepsilon_1; u))\ldots)).$$

Let us calculate the differential of Φ with respect to ε at $\varepsilon = 0$, by induction on k.

We have $\varphi_1(\varepsilon_1; a) = a + \varepsilon_1 r_1(a) + O(\varepsilon_1^2)$. If

$$\Phi_k(\varepsilon_1, \dots, \varepsilon_k; a) = a + \sum_{1 \le j \le k} \varepsilon_j r_j(a) + O(\|\varepsilon\|^2),$$

then

$$\begin{aligned}\Phi_{k+1}(\varepsilon_1, \dots, \varepsilon_{k+1}; a) &= \Phi_k(\varepsilon_1, \dots, \varepsilon_k; a) + \varepsilon_{k+1} r_{k+1}(\Phi_k(\varepsilon_1, \dots, \varepsilon_k; a)) \\ &\quad + O(\|\varepsilon\|^2) \\ &= a + \sum_{1 \le j \le k} \varepsilon_j r_j(a) + \varepsilon_{k+1}(r_{k+1}(a) + O(\|\varepsilon\|)) + O(\|\varepsilon\|^2) \\ &= a + \sum_{1 \le j \le k+1} \varepsilon_j r_j(a) + O(\|\varepsilon\|^2),\end{aligned}$$

which justifies the induction hypothesis. For $k = n$, we find that $d_\varepsilon \Phi(0; a)$ is the matrix whose column vectors are the eigenvectors r_j of $df(a)$. These forming a basis in $\mathbb{R}^n$, this matrix is invertible. We shall now make use of the implicit function theorem in the following quantitative form.

Let G be a function of class $\mathscr{C}^2$ defined on a ball $B(x_0; \rho)$ of $\mathbb{R}^n$, with values in $\mathbb{R}^n$. We suppose that $dG(x_0)$ is invertible. Then there exist two numbers $\eta > 0$ and $L > 0$, which depend only on ρ, on $\|dG(x_0))\|$ and on $\sup_{B(x_0, \rho)} \|D^2 G(x)\|$, such that

G is injective on $B(x_0; \eta)$,
the image under G of the ball $B(x_0; \alpha)$ contains the ball $B(G(x_0); \alpha/L)$ for all $\alpha < \eta$.

Now, let K be a compact neighbourhood of a, contained in W, and $\rho > 0$ such that $B(a; 2\rho)$ is contained in K. The implicit function theorem gives two constants $L > 0$ and $\eta > 0$ corresponding to ρ, to $\inf_{U \in K} \|d_\varepsilon \Phi(0; u)\|$ and to $\sup_{(\varepsilon, u) \in B_1 \times K} \|D^2 \Phi(\varepsilon; u)\|$. If $u_L \in B(a; \rho)$ then $B(u_L; \rho) \subset K$ and the preceding argument applies: for all $u_R \in U$ satisfying $\|u_R - u_L\| < \eta/L$, there exists one and only one $\varepsilon \in \mathbb{R}^n$ such that $\|\varepsilon\| < L\eta$ and $u_R \in \Phi(\varepsilon; u_L)$. In particular, the Riemann problem clearly has a solution when $u_L, u_R \in B(a; \frac{1}{2}\eta/L)$, and we have $\|\varepsilon\| < L\|u_R - u_L\|$. Making use of the fact that each Φ_k is locally Lipschitz, we see that all the values taken by the solution of the Riemann problem, even those which are to be found in a k-rarefaction-wave (corresponding to $\Phi_k(\varepsilon_1, \dots, \varepsilon_{k-1}, \alpha; u)$ with $0 < \alpha < \varepsilon_k$), are in W provided that $\max(\|u_L - a\|, \|u_R - a\|)$ is sufficiently small. ∎

The proof gives an equivalent of ε and of the constant intermediate states when u_{L} and u_{R} are close. In fact $0 = \Phi(\varepsilon; u_{\mathrm{L}}) - u_{\mathrm{R}} = -[u] + \sum_{1\le j\le n} \varepsilon_j r_j(u_{\mathrm{L}}) + \mathrm{O}(\|\varepsilon\|^2)$. We thus have $\varepsilon_j \sim l_j(u_{\mathrm{L}}) \cdot [u]$ when $[u] = u_{\mathrm{R}} - u_{\mathrm{L}}$ is small, with the usual normalisation $l_j \cdot r_j = 1$, l_j being a left eigenvector associated with the eigenvalue λ_j. In particular, the intermediate states are given by

$$u_k = u_{\mathrm{L}} + \sum_{1\le j\le k} (l_j(u_{\mathrm{L}}) \cdot [u]) r_j(u_{\mathrm{L}}) + \mathrm{O}(\|[u]\|^2).$$

In the strictly hyperbolic case where the eigenvalues $\lambda_1, \dots, \lambda_p$ are not necessarily simple, the solution of the Riemann problem contains only p distinct waves, but each set $H_j(a)$, of dimension m_j equal to the multiplicity of λ_j, can be parametrised by a vector ε_j running over a neighbourhood of the origin in $\mathbb{R}^{m_j}$, this parametrisation being smooth. The preceding proof carries over without change because the tangent spaces to $H_j(u_{\mathrm{L}})$ are linearly independent and their direct sum is $\mathbb{R}^n$. The calculation of the intermediate states up to $\|[u]\|^2$ is still easy: we decompose $[u]$ into a sum of eigenvectors of $\mathrm{d}f(u_{\mathrm{L}})$,

$$[u] = \sum_{1\le j\le p} v_j.$$

We then have

$$u_k = u_{\mathrm{L}} + \sum_{1\le j\le p} v_j + \mathrm{O}(\|[u]\|^2).$$

Comments B. Riemann, in his memoir to the Royal Academy of Sciences of Göttingen (1860), introduced most of the essential ideas for 2×2 systems in restricting himself to the study of the system of isentropic gas dynamics. He calculates the expressions r and s which we today call the Riemann invariants. He shows that (x, t), considered as a function of r and s, satisfies a linear hyperbolic system with variable coefficients: this is the hodograph method. For such a system, written in the form of a single equation of the second order, he describes the method of duality (which bears his name) which reduces the solution of the non-characteristic Cauchy problem to that of Goursat problems for the adjoint equation. Noting that the velocity of propagation depends on the state, he shows that the first derivatives r_x and s_x blow up in a finite time for general data. Riemann deduces the appearance of shock waves, writes the Rankine–Hugoniot condition (of which he seems to have no previous knowledge). In the fifth section there appears explicitly 'Lax's shock condition'. Finally, Riemann solves the 'Riemann problem' for isothermal gas dynamics ($p = \rho$), thus avoiding a discussion of the possible vacuum.

4.7 The solution of the Riemann problem for the p-system

Hypotheses

Let us consider the p-system, which is equivalent to the non-linear wave equation $y_{tt} = (p(y_x))_x$:

$$\left.\begin{aligned} u_t + v_x &= 0, \\ v_t + p(u)_x &= 0. \end{aligned}\right\} \tag{4.17}$$

Since $f(u, v) = (v, p(u))^{\mathrm{T}}$ the matrix $\mathrm{d}f$ has the value

$$\begin{pmatrix} 0 & 1 \\ p'(u) & 0 \end{pmatrix}.$$

Its eigenvalues are the roots of $x^2 = p'(u)$. If $p'(u) = 0$, the matrix $\mathrm{d}f$ is not diagonalisable. The system is hyperbolic if and only if $p'(u) > 0$, which we suppose from now. The eigenvalues are $\lambda_1 = -\sqrt{p'(u)}$ and $\lambda_2 = \sqrt{p'(u)}$. The corresponding eigenvectors are

$$r_1 = \begin{pmatrix} -1 \\ \sqrt{p'(u)} \end{pmatrix}, \qquad r_2 = \begin{pmatrix} 1 \\ \sqrt{p'(u)} \end{pmatrix}.$$

We therefore have $\mathrm{d}\lambda_j \cdot r_j = p''(u)/2\sqrt{p'(u)}$: the characteristic fields are simultaneously genuinely non-linear or linearly degenerate. To solve the Riemann problem we assume the genuinely non-linear case. At the expense of a change of variables $(x, u) = (-x, -u)$, we can suppose that $p''(u) > 0$, that is that p is strictly convex. This hypothesis ensures that $\lim_{u\to\infty} \sqrt{p'(u)} > 0$, but not that $\lim_{u\to-\infty} \sqrt{p'(u)} > 0$, that is that the system might not be uniformly hyperbolic when $u \to -\infty$. We are therefore driven by subsequent needs to make a slightly stronger hypothesis concerning the hyperbolicity: we suppose that $\int_{-\infty} \sqrt{p'(u)}\,\mathrm{d}u = +\infty$.

Rarefaction waves

A 1-Riemann-invariant is a non-trivial solution of the equation $dw \cdot r_j = 0$, that is of

$$\frac{\mathrm{d}w}{\mathrm{d}u} = \sqrt{p'(u)}\,\frac{\mathrm{d}w}{\mathrm{d}v}.$$

The simplest solution of this equation is $w := v + g(u)$ where $g(u) := \int_0^u \sqrt{p'(s)}\,\mathrm{d}s$. The integral curves r_1 are thus parametrised by u and have the form

$$v = v_0 - g(u) + g(u_0) = v_0 - \int_{u_0}^{u} \sqrt{p'(s)}\,\mathrm{d}s.$$

Similarly, the parametrisation by u of the integral curves of r_2 is

$$v = v_0 + g(u) - g(u_0) = v_0 + \int_{u_0}^{u} \sqrt{p'(s)}\,\mathrm{d}s.$$

Two points (u_-, v_-) and (u_+, v_+) of the same integral curve of r_j are linked in this order by a j-rarefaction-wave if and only if λ_j is strictly increasing along the length of this curve from (u_-, v_-) to (u_+, v_+). As λ_2 increasing with u, a 2-rarefaction-wave is characterised by

$$v_+ = v_- + \int_{u_-}^{u_+} \sqrt{p'(s)}\,\mathrm{d}s, \quad u_- < u_+. \tag{4.18}$$

Similarly, a 1-rarefaction-wave is characterised by

$$v_+ = v_- - \int_{u_-}^{u_+} \sqrt{p'(s)}\,\mathrm{d}s, \quad u_- > u_+. \tag{4.19}$$

We notice that in both cases, $v_- < v_+$. In addition, if instead of ordering the states following the xs increasing we order them following the times increasing, as this is possible since the waves have non-zero speeds, then we see that, in every expansion wave, u decays with time.

Shocks

The Rankine–Hugoniot condition between two states $[u_-, v_-]$ and $[u_+, v_+]$ is written

$$[v] = \sigma[u], \qquad [p(u)] = \sigma[v].$$

We derive $[p(u)] = \sigma^2[u]$, which gives $\sigma = \pm S$ where $S = \sqrt{[p(u)]/[u]}$. Since p is strictly convex, we have $(S - \lambda_2^+)(S - \lambda_2^-) < 0$ and $(S + \lambda_1^+)(S + \lambda_1^-) < 0$, where we have written $\lambda_j^+ = \lambda_j(u_+, v_+)$, etc. In the 1-shock-wave, the condition $\sigma < \lambda_2^+$ thus implies that $\sigma = -S$. Lax's inequality is then written $\sqrt{p'(u_-)} < \sqrt{[p(u)]/[u]} < \sqrt{p'(u_+)}$, which because of the convexity of p is equivalent to $u_- < u_+$. Finally, a 1-shock-wave is characterised by

$$v_+ = v_- - [u]\sqrt{[p(u)/[u]} = v_- - \sqrt{[p(u)][u]}, \quad u_- < u_+. \tag{4.20}$$

Similarly, in a 2-shock-wave, $\sigma = S$ while the Lax inequality $\sqrt{p'(u_+)} < \sqrt{[p(u)]/[u]} < \sqrt{p'(u_-)}$ is equivalent to $u_+ < u_-$. A 2-shock is thus characterised by

$$v_+ = v_- + [u]\sqrt{[p(u)/[u]} = v_- - \sqrt{[p(u)][u]}, \quad u_- > u_+. \tag{4.21}$$

Now, let us show that the shock condition which we are about to treat is here equivalent to Lax's entropy condition, with the result that our analysis does not depend on the admissibility criterion adopted.

The natural entropy for this problem is a total energy, sum of the kinetic energy and of a potential energy $e(u)$ defined within a constant $e' = p$:

$$E(u, v) = \frac{1}{2}v^2 + e(u), \qquad F(u, v) = vp(u).$$

The rate of entropy production is

$$\begin{aligned}
[F] - \sigma[E] &= [vp(u)] - \sigma\left[\frac{1}{2}v^2 + e(u)\right] \\
&= [v]p_- + v_+[p] - \frac{\sigma}{2}[v](v_- + v_+) - \sigma[e] \\
&= \sigma[u]p_- + \frac{1}{2}[p][v] - \sigma[e] \\
&= \sigma\left(\frac{1}{2}[u](p_- + p_+) - [e]\right),
\end{aligned}$$

where we have used the Rankine–Hugoniot condition to eliminate v_+. In addition

$$[u]\frac{p_- + p_+}{2} - [e] = (u_+ - u_-)\int_0^1 (sp_+ + (1-s)p_- - p(su_+ + (1-s)u_-))\,\mathrm{d}s$$

is of the same sign as $[u]$ since p is strictly convex. Thus $[F] - \sigma[E]$ has the same sign as $\sigma[u]$. For $\sigma < 0$, as discontinuity is thus entropic if and only if $u_- < u_+$, while for $\sigma > 0$, it is so if and only if $u_+ < u_-$. This confirms the criterion obtained via Lax's shock inequalities.

Wave curves

To resume the two preceding sections, each wave curve is parametrised by u. A single function of two variables is enough to make this point obvious. Let us put

$$\theta(u, u_1) = \begin{cases} \displaystyle\int_u^{u_1} \sqrt{p'(s)}\,\mathrm{d}s, & u < u_1, \\ -\sqrt{(p(u) - p(u_1))(u - u_1)}, & u > u_1. \end{cases}$$

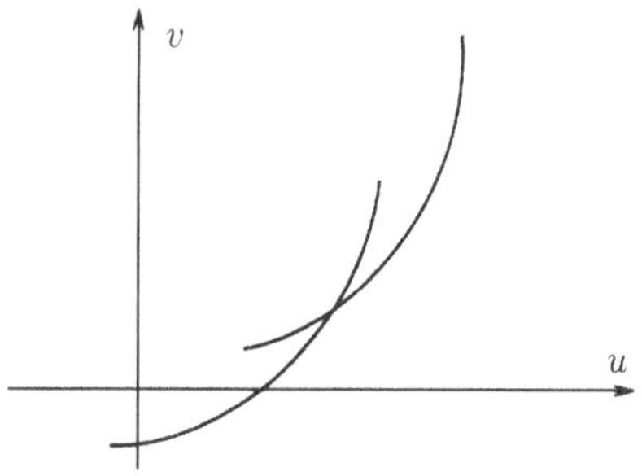

Fig. 4.4: Two 2-wave curves for the p-system. The relation $P \in \mathcal{O}_2(Q)$ is not transitive; the curves do not permit the definition of a coordinate system.

Being given two states (a, b) and (u, v), we have

$$(u, v) \in \mathcal{O}_1(a, b) \Leftrightarrow v = b + \theta(u, a),$$

just as

$$(u, v) \in \mathcal{O}_2(a, b) \Leftrightarrow v = b + \theta(a, u).$$

In the plane $\mathbb{R}^2$, the 1-wave curves are strictly decreasing and we infer one from the other by vertical translation. The 2-wave curves are strictly increasing and we infer one from the other by vertical translation. These curves are unbounded in the vertical direction. In fact we have made the hypothesis that $\theta(-\infty, u_1) = +\infty$. On the other hand, if $u \to +\infty$, then $\theta(u, a) < -\sqrt{(p(a+1) - p(a))(u - a)}$ gives $\theta(+\infty, a) = -\infty$, and so on.

The solution of the Riemann problem

Let (u_L, v_L) and (u_R, v_R) be the two initial states of the Riemann problem. The solution is *a priori* made up of a 1-wave and a 2-wave which separate two initial states and an intermediate state (u_0, v_0). Making use of the parametrising of the wave curves, the solution of the Riemann problem comes down to finding the solution (u_0, v_0) of the system

$$\left.\begin{aligned} v_0 &= v_L + \theta(u_0, u_L), \\ v_R &= v_0 + \theta(u_0, u_R) \end{aligned}\right\} \tag{4.22}$$

and verifying that the two waves are gluable, that is if they are two shock waves, then $\sigma_1 < \sigma_2$. This last point is trivial since we always have $\sigma_1 < 0 < \sigma_2$.

Eliminating v_0 from the equations (4.22), we are led to the scalar equation in the single unknown u_0,

$$G(u_0)=0, \tag{4.23}$$

where $G(u) := \theta(u, u_L)+\theta(u, u_R)-[v]$. The function G is continuous (it is in fact of class $\mathscr{C}^2$) and satisfies $G(\pm\infty) = \mp\infty$. Also $G' < 0$ as for $u \le a$, $\theta_u(u, a) = -\sqrt{p'(u)} < 0$ and on the other hand, for $u > a$, $2\theta\theta_u(u, a) = p'(u)(u - a) + p(u) - p(a) > 0$ (as the sum of two positive terms), while $\theta \ge 0$.

The equation (4.23) thus has one and only one solution u_0. The pair $(u_0, v_L + \theta(u_0, u_L))$ is then the unique solution of (4.22). Finally, the Riemann problem for the p-system has one and only one solution. We can even make precise the nature of the waves produced as a function of the values of u_L, u_R and $[v]$ by considering the signs of $G(u_L) = \theta(u_L, u_R) - [v]$ and of $G(u_R) = \theta(u_R, u_L) - [v]$, since G is decreasing and $G(u_0) = 0$.

Case $u_L \le u_R$. Then we have $G(u_R) \le G(u_L)$.

If $[v] < \theta(u_R, u_L)$, then $u_0 > u_R$. There are two shock waves.
If $\theta(u_R, u_L) < [v] < \theta(u_L, u_R)$, then $u_L < u_0 < u_R$. There are one 1-shock-wave and one 2-rarefaction wave.
If $\theta(u_L, u_R) < [v]$, then $u_L > u_0$. There are two rarefaction waves.

Case $u_R \le u_L$. Then $G(u_L) \le G(u_R)$.

If $[v] < \theta(u_L, u_R)$, then $u_0 > u_L$. There are two shocks.
If $\theta(u_L, u_R) < [v] < \theta(u_R, u_L)$, then $u_R < u_0 < u_L$. There are one 1-rarefaction and one 2-shock.
If $\theta(u_R, u_L) < [v]$, then $u_0 < u_R$. There are two rarefactions.

Comments (1) We note that in these criteria, the two values of $\theta(u_L, u_R)$ and $\theta(u_R, u_L)$ are of opposite signs. In particular, if $v_R = v_L$ then one of the waves is a shock wave while the other is a rarefaction wave, this remaining true if $v_R - v_L$ is small with respect to $|u_R - u_L|$.

(2) When $[v]$ is equal to one of the values $\theta(u_L, u_R)$ and $\theta(u_R, u_L)$, one of the waves disappears, that is the median state (u_0, v_0) is equal to (u_L, v_L) or to (u_R, v_R).

4.8 The solution of the Riemann problem for gas dynamics

Hypotheses

We consider the system of gas dynamics in one spatial dimension with eulerian coordinates. The choice of lagrangian coordinates would make the solution of the Riemann problem more difficult as the eventual appearance of the vacuum is represented by a Dirac measure for the specific volume. Also, when we approach multi-dimensional configurations only the eulerian coordinates have a practical interest. The system is thus

$$\left.\begin{aligned}
&\rho_t + (\rho v)_x && = 0, \\
&(\rho v)_t + (\rho v^2 + p(\rho, e))_x && = 0, \\
&\left(\rho\left(e + \frac{1}{2}v^2\right)\right)_t + \left(\rho\left(e + \frac{1}{2}v^2\right)v + pv\right)_x && = 0.
\end{aligned}\right\} \tag{4.24}$$

The velocity v takes its values in $\mathbb{R}$ while ρ and e take positive values or zero. Many of the calculations are simpler if we use the state variables (ρ, e) and v, because of the simplicity of the non-conservative form of the system in these variables, when $\rho > 0$:

$$\left.\begin{aligned}
&(\partial_t + v\partial_x)\rho + \rho v_x && = 0, \\
&(\partial_t + v\partial_x)v + \rho^{-1} p_x && = 0, \\
&(\partial_t + v\partial_x)e + \rho^{-1} p v_x && = 0.
\end{aligned}\right\} \tag{4.25}$$

The matrix of this system is

$$A = vI_3 + \begin{pmatrix} 0 & \rho & 0 \\ \rho^{-1} p_\rho & 0 & \rho^{-1} p_e \\ 0 & \rho^{-1} p & 0 \end{pmatrix}.$$

The eigenvalues are the solutions of $(\lambda - v)^3 = (\lambda - v)(p_\rho + \rho^{-2} p p_e)$. In the form (4.25) we see that the system has a singularity all over the plane $\rho = 0$. This corresponds to the fact that, when ρ is identically zero, the conservative variables $\rho, q = \rho v$ and $\varepsilon = \rho(e + \frac{1}{2}v^2))$ are not independent of each other since they are all zero together (with the result that for (4.24) the singularity reduces to a single point $(0, 0, 0)$). The density being zero on an interval expresses the fact that this interval is free of gas. We cannot exclude this situation in the solution of the Riemann problem, which introduces an indeterminacy in the variables which describe the flow. Indeed, it is clear that the vacuum has zero energy, momentum and pressure, on the other hand the velocity is not defined (this is the quotient q/ρ), which prevents giving a sense to the energy flux $\rho(e + \frac{1}{2}v^2)v + pv$. We admit that generally, the speed having to be a bounded variable, the energy flux is also identically zero in the vacuum.

The system is therefore hyperbolic for $\rho > 0$ if, and only if $p_\rho + \rho^{-2} p p_e > 0$, which we shall henceforth assume to be the case (the matrix is not diagonalisable if $p_\rho + \rho^{-2} p p_e = 0$).

We express the eigenvectors and the eigenvalues of A as a function $c := (p_\rho + \rho^{-2} p p_e)^{1/2}$, the speed of sound in the gas:

$$\lambda_1 = v - c, \qquad \lambda_2 = v, \qquad \lambda_3 = v + c,$$

$$r_1 = \begin{pmatrix} -\rho \\ c \\ -\rho^{-1} p \end{pmatrix}, \qquad r_2 = \begin{pmatrix} p_e \\ 0 \\ -p_\rho \end{pmatrix}, \qquad r_3 = \begin{pmatrix} \rho \\ c \\ \rho^{-1} p \end{pmatrix}.$$

We have $d\lambda_j \cdot r_j = (\partial_\rho + \rho^{-2} p \partial_e)(\rho c)$ for $j = 1, 3$. The first and the third fields are of the same nature, in general genuinely non-linear, as for a perfect gas (state law $p = (\gamma - 1)\rho e$ where $\gamma > 1$ is a constant, $c^2 = \gamma(\gamma - 1)e = \gamma p/\rho$ and $d\lambda_j \cdot r_j = \frac{1}{2}(1 + \gamma)c > 0$). On the other hand, the second field is always linearly degenerate, independent of the state law chosen.

Let us note finally that the same uncertainty as formerly occurs for the speed of sound in the vacuum. For example for a perfect gas, $c^2 = \gamma(\gamma - 1)e$ which does not make sense.

The rarefaction waves

Let us examine the 1-rarefaction-waves by first calculating the corresponding Riemann invariants. A 1-Riemann-invariant is a solution w of $dw \cdot r_1 = 0$, that is of $\rho w_\rho - c w_v + \rho^{-1} p w_e = 0$. The simplest solutions are $v + g(\rho, e)$ and S where g is a solution of the linear first order equation.

$$\rho g_\rho + \rho^{-1} p g_e = c,$$

and S, the specific entropy considered by the physicists (but which is not an entropy in the mathematical sense) satisfies the homogeneous equation $\rho S_\rho + \rho^{-1} p S_e = 0$. Let us remark that the non-linearity condition of the first and third fields can be interpreted by saying that ρc and S are two independent functions.

In a symmetric way, the Riemann invariants for the third field are S and $v - g(\rho, e)$. A translation parallel to the velocity axis thus preserves the family of integral curves of r_j for $j = 1, 3$. Also, the symmetry $(\rho, v, e) \mapsto (\rho, -v, e)$ exchanges the two families of curves. More precisely, if there exists a 1-rarefaction to pass from a state $u_- = (\rho_-, v_-, e_-)$ to a state $u_+ = (\rho_+, v_+, e_+)$, then $v - c(\rho, e)$ is monotonic increasing in going from u_- to u_+ following the integral curve of r_1. It follows that $v + c(\rho, e)$ is monotonic decreasing from $U_- = (\rho_-, -v_-, e_-)$ to $U_+ = (\rho_+, -v_+, e_+)$ along the integral curve of r_3, that is we can pass from U_- to

U_+ by a 3-rarefaction-wave. The symmetry thus as a matter of fact exchanges the curves of the rarefaction waves relative to the first and third characteristic fields.

The curve of the 1-rarefaction-wave issuing from u_- can be parametrised by the pressure p_+. Indeed, $\mathrm{d}p \cdot r_1 = -\rho c^2 < 0$ shows that p is strictly monotonic along the integral curve of r_1. If we suppose in fact that $\mathrm{d}\lambda_1 \cdot r_1 > 0$ (which is true for known gases and in particular for perfect gases, knowing that $\gamma > 1$), then the pressure decreases along the 1-rarefactions (this is the origin of the name *rarefaction wave*). Writing that S and $v + g(\rho, e)$ are conserved, we obtain a parametrisation of the form

$$\left.\begin{aligned} \rho_+ &= \varphi(p_+; \rho_-, p_-), \\ v_+ &= v_- - \theta(p_+; \rho_-, p_-) \end{aligned}\right\} (p_+ \leq p_-).$$

By symmetry, the 3-rarefaction-waves have the parametrised form

$$\left.\begin{aligned} \rho_- &= \varphi(p_-; \rho_+, p_+), \\ v_- &= v_+ + \theta(p_-; \rho_+, p_+) \end{aligned}\right\} (p_- \leq p_+).$$

Since the first and the third components of r_1 are negative, ρ and e also decrease along the 1-rarefaction-wave: φ is increasing with respect to its first argument. Additionally, $\theta(p; \rho_-, p_-) = g(p, S_-) - g(p_-, S_-)$ where

$$\rho c = (\rho^2 \partial_\rho + p\partial_e) g = g_p(\rho^2\partial_\rho + p\partial_e)p + g_S(\rho^2\partial_\rho + p\partial_e)S = \rho^2 c^2 g_p.$$

Hence, θ is strictly increasingly with respect to p. As p diminishes, the velocity grows in a 1-rarefaction-wave. We can write

$$v_+ = v_- + \int_{p_-}^{p_+} \frac{\mathrm{d}p}{\rho c}$$

where the integral is evaluated along the rarefaction curve. By symmetry, we obtain the following result.

Proposition 4.8.1 *With respect to x, the variations of ρ, p, e and v across a rarefaction wave are as follows, provided* $\mathrm{d}\lambda_1 \cdot r_1 > 0$.

In a 1-rarefaction-wave, the density, the pressure and the specific energy diminish while the speed grows.

In a 3-rarefaction-wave, the density, the pressure, the specific energy and the speed grow.

On the other hand, when we follow the particles paths, the density, the pressure and the specific energy decrease across a rarefaction wave.

Remark As we are now going to see, in the 1-shock-waves and the 3-shock-waves, the variations of the pressure and of the speed are opposite to those stated above.

The shocks

Let us write the Rankine–Hugoniot condition for a discontinuity of velocity s:

$$\begin{aligned}[\rho v] &= s[\rho],\\ [\rho v^2 + p] &= s[\rho v],\\ \left[\left(\frac{1}{2}\rho v^2 + \rho e + p\right)v\right] &= s\left[\frac{1}{2}\rho v^2 + \rho e\right].\end{aligned}$$

By defining $z := v - s$, we obtain the reduced conditions which amount to saying that $(\rho, v - s, e)_\pm$ satisfy the Rankine–Hugoniot condition for a zero velocity:

$$\begin{aligned}[\rho z] &= 0,\\ [\rho z^2 + p] &= 0,\\ \left[\left(\frac{1}{2}\rho z^2 + \rho e + p\right)z\right] &= 0.\end{aligned}$$

Then, let us denote by m the common value of $\rho_- z_-$ and $\rho_+ z_+$. There follow

$$\begin{aligned}m[z] + [p] &= 0,\\ m\left[\frac{1}{2}z^2 + e\right] + [pz] &= 0.\end{aligned}$$

By combining these two equalities, we obtain

$$\begin{aligned}m\left[\frac{1}{2}z^2 + e\right] &= -[p]z_+ - p_-[z] = (mz_+ - p_-)[z] = (mz_- - p_+)[z]\\ &= \frac{1}{2}(m(z_- + z_+) - (p_- + p_+))[z]\\ &= m\left[\frac{1}{2}z^2\right] - \frac{1}{2}(p_- + p_+)[z];\end{aligned}$$

this is to say

$$m[e] = -\frac{1}{2}(p_- + p_+)[z] = -\frac{1}{2}m(p_- + p_+)[\rho^{-1}].$$

There are now two cases, according as m is zero or not. We shall see the case $m = 0$ later since it corresponds to contact discontinuities. We therefore assume that $m \neq 0$ which implies the fundamental relation across the discontinuities from which the velocities of the gas and of the discontinuity itself have been eliminated:

$$[e] + \frac{1}{2}(p_+ + p_-)\left[\frac{1}{\rho}\right] = 0. \tag{4.26}$$

This relation, which appears to be a necessary condition to satisfy the Rankine–Hugoniot condition, is merely sufficient: if ρ_-, ρ_+, e_-, e_+ satisfy it we choose for m one of the roots, which we hope are real, of $m^2[\rho^{-1}] + [p] = 0$, and then we define $z_+ = m/\rho_+$. The (arbitrary) choice of s brings to an end the construction of $v_\pm$ with the definitions $v_\pm = z_\pm + s$.

Now, let us look at the admissibility of the shock waves. The convex entropy is of the form $E = \rho h(S)$ where h is a suitable numerical function. Its flux is $F = vE$, whence the rate of entropy production is $P = m[h(S)]$. Its sign depends on the one hand on that of $z_\pm$ and on the other on that of $[S]$, provided that h is strictly monotonic. Again, the symmetry $(x, \rho, v, e) \mapsto (-x, \rho, -v, e)$ exchanges the admissible discontinuities of one family $(s < v)$ with the admissible discontinuities of the other family $(s > v)$: if (U_-, U_+, s) is admissible then $(U_-, U_+; -s)$ is also, in the notation adopted above.

Lemma 4.8.2 *Let $S = S(\rho, e)$ be an entropy density in the sense of thermodynamics, that is to say a solution of $\rho^2 S_\rho + pS_e = 0$ satisfying $S_e > 0$ (the inverse $T = 1/S_e$ is called the absolute temperature).*

For every numerical function h, the function $E := \rho h \circ S$ an entropy in the mathematical sense, of flux $F = vE$. If $(\rho, q, \varepsilon) := (\rho, \rho v, \frac{1}{2}\rho v^2 + \rho e) \mapsto E$ is convex, then h is decreasing.

Proof Because of (4.25), we verify immediately that $(\partial_t + v\partial_x)S = 0$ for the smooth solutions: the specific entropy is constant along the trajectories of the gas particles (this is no longer true when a particle goes through a shock wave). It is the same for $h(S)$. Combining with the conservation law for mass, it becomes $E_t + (vE)_x = 0$: E is an entropy of flux vE.

If E is convex, then the mapping $a \mapsto k(a) := E(\rho, a, \varepsilon + v(a - q))$ is convex for every choice of (ρ, q, ε), that is, writing $S^1 := h \circ S$,

$$0 \le k''(a) = \rho^{-1}\big(-S_e^1 + (v - \rho^{-1}a)^2 S_{ee}^1\big).$$

In particular,

$$0 \le k''(q) = -\rho^{-1} S_e^1.$$

We thus find that $S_e^1 \le 0$, that is, $h' \le 0$. If the entropy is not trivial, h is strictly decreasing. ∎

Let us note finally that if $m \neq 0$, a part of the Lax inequalities is satisfied. For example, if $m > 0$, then $s < \min(\lambda_2^+, \lambda_2^-) = \min(v_+, v_-)$ which leads us to associate the discontinuity with the first characteristic field: it will be admissible if and only if this is a 1-shock-wave. Similarly, if $m < 0$, it will be admissible if and only if it is a 3-shock-wave.

The 1-shock-waves

Taking advantage of all the admissibility conditions introduced already, we require that the shock waves satisfy simultaneously the entropy condition, that is,

$m[S] \geq 0$ (since h is decreasing), and the Lax shock condition which for a 1-shock is written

$$v_+ - c_+ < s < \min(v_- - c_-, v_+).$$

We thus have $m > 0$ and the entropy condition reduces to $[S] > 0$. Observing that the gas particles traverse the 1-shock-wave from the left towards the right, after $m > 0$, we reformulate this condition by saying that the entropy density, which is constant along the trajectories in the absence of a shock, grows along these across a shock wave.

The 3-shock-waves

In a symmetrical manner, a 3-shock-wave satisfies the Lax inequalities

$$\max(v_-, v_+ + c_+) < s < v_- + c_-.$$

The entropy condition becomes $[S] < 0$. Since $m < 0$, the trajectories traverse the 3-shock-waves from right to left, with the result that the entropy again grows along the trajectories. Finally:

Theorem 4.8.3 *Let $(u_-, u_+; s)$ be a discontinuity which is not of contact (i.e. $m \neq 0$). Then it satisfies Lax's entropy criterion if and only if the entropy density S grows across the discontinuity when we follow the motion of the particles.*

This statement is nothing but the second law of thermodynamics, due to Carnot. An important consequence of this criterion is the maximum principle for S.

Corollary 4.8.4 *For a bounded and piecewise smooth entropy solution of the Cauchy problem, we have the maximum principle*

$$\inf_{x \in \mathbb{R}} S(x, t) \geq \inf_{x \in \mathbb{R}} S(x, 0).$$

Proof For $t > 0$ and $x \in \mathbb{R}$, we consider the particle path which has arrived at x at the time t. It clearly started out at the initial instant $t_0 = 0$ since the trajectories only traverse the 1-shock-waves and 3-shock-waves and never meet with the 2-waves which are contact discontinuities of the same speed as the flow ($s = \lambda_2^- = \lambda_2^+$). The value of S along the length of the trajectory only varies when crossing a shock wave and it increases in doing so. We therefore have

$$S(x, t) \geq S(x_0, 0) \geq \inf_{y \in \mathbb{R}} S(y, 0),$$

which proves the corollary. ■

Parametrisation of shock curves

Let us consider for example the 1-shock-waves, because of the symmetry evoked above with the 3-shock-waves. Because of the omnipresence of the operator $L := \partial_\rho + \rho^{-2} p \partial_e$ and the condition of hyperbolicity, it is more convenient to use the variables (p, S) than (ρ, e). In fact, $Lp = c^2 > 0$ and $LS = 0$ show that $(\rho, e) \mapsto (p, S)$ is a change of variables. We then have the formulae

$$c^2 \partial_p = L,$$
$$c^2 \partial_S = T(p_\rho \partial_e - p_e \partial_\rho).$$

The genuinely non-linear hypothesis (adapted from the case of perfect gases) is $\partial_p(\rho c) > 0$.

The function g used to define the Riemann invariants of the 1- and 3-characteristic-fields is a solution of the equation $\partial_p g = \rho^{-1} c^{-1}$. Finally we have the formulae

$$\rho^2 c^2 \partial_p e = p > 0,$$
$$\rho^2 c^2 \partial_p \frac{1}{\rho} = -1 < 0,$$
$$\rho^2 c^2 \partial_S e = \rho^2 T p_\rho,$$
$$\rho^2 c^2 \partial_S \frac{1}{\rho} = T p_e.$$

Since S is a Riemann invariant associated with the 1- and 3-waves, we know that, for a weak shock, the jump $[S]$ is of the order of the cube of the amplitude of the shock wave ($\mathscr{C}^2$ matching of shock and rarefaction curves). As p is not a Riemann invariant for these fields (we have seen that $\mathrm{d}p \cdot r_1 < 0$), we can measure this amplitude by the jump $[p]$ of p. Hence, we have $[S] = \mathrm{O}([p]^3)$. Now, using the relation (4.26), we write, with the now classical notation ($\langle k \rangle := \frac{1}{2}(k_+ + k_-)$),

$$\begin{aligned}
0 &= \int_{A_-}^{A_+} \left(\mathrm{d}e + \langle p \rangle \, \mathrm{d}\frac{1}{\rho} \right) \\
&= \int_{A_-}^{A_+} \left(T \, \mathrm{d}S + (\langle p \rangle - p) \, \mathrm{d}\frac{1}{\rho} \right) \\
&= \langle T \rangle [S] + \int_{A_-}^{A_+} \left((T - \langle T \rangle) \, \mathrm{d}S + (\langle p \rangle - p) \, \mathrm{d}\frac{1}{\rho} \right) \\
&= \mathrm{O}([p^4] + \langle T \rangle [S] + \int_{A_-}^{A_+} (\langle p \rangle - p) \, \mathrm{d}\frac{1}{\rho} \\
&= \mathrm{O}([p]^4) + \langle T \rangle [S] + \int_{A_-}^{A_+} (\langle p \rangle - p) \left(\left(\partial_p \frac{1}{\rho} \right) \mathrm{d}p + \left(\partial_S \frac{1}{\rho} \right) \mathrm{d}S \right) \\
&= \mathrm{O}([p]^4) + \langle T \rangle [S] + \int_{p_-}^{p_+} (p - \langle p \rangle) \rho^{-2} c^{-2} \mathrm{d}p.
\end{aligned}$$

Now,

$$\rho^{-2}c^{-2} = \langle \rho^{-2}c^{-2} \rangle - a(p - \langle p \rangle) + \mathrm{O}([p]^3),$$

where $a = -\partial_p(\rho^{-2}c^{-2}) > 0$ from the non-linearity hypothesis we have made. Hence, we have

$$\begin{aligned} 0 &= \mathrm{O}([p]^4) + \langle T \rangle [S] - a \int_{p_-}^{p_+} (p - \langle p \rangle)^2 \mathrm{d}p \\ &= \mathrm{O}([p]^4) + \langle T \rangle [S] - \frac{a}{12}[p]^3. \end{aligned}$$

Finally,

$$[S] \sim \frac{a}{12\langle T \rangle}[p]^3.$$

The sign of variation of S is thus the same as that of p across a discontinuity. From the argument developed above, p thus grows across the 1-shock-waves and decays across the 3-shock-waves with respect to x, which is the opposite situation to that of the rarefaction waves (just as we have already observed for the p-system). Of course, if instead of making x vary while keeping t constant, we follow the particles, we find that p grows along the trajectories, that is to say, that the shock waves are *compression waves*.

In the favourable cases (those of the perfect gases for example as we shall see later) equation (4.26) defines globally a curve $p_+ \mapsto (p_+, S_+)$ where $S_+ = \Sigma(p_+; p_-, S_-)$. On the other hand from the first two Rankine–Hugoniot relations we derive the formula

$$[v]^2 = \frac{[p]}{\langle p \rangle}[e] = -[p]\left[\frac{1}{\rho}\right]. \tag{4.27}$$

Also, we have seen that $m[z] + [p] = 0$, that is $m[v] + [p] = 0$. For all the shock waves (at least for weak shocks, but we shall admit, this being true for a perfect gas, that it is so for shock waves of arbitrary amplitude), we have $[v] < 0$, then either $m < 0$ and $[p] < 0$ or $m > 0$ and $[p] > 0$. Again, this is the opposite to the case of rarefaction waves. The curves of the 1-shock-waves are thus of the form

$$\begin{aligned} S_+ &= \Sigma(p_+; p_-, S_-), \\ v_+ &= v_- - \psi(p_+; p_-, S_-), \\ p_+ &> p_-, \end{aligned}$$

where $\psi := \sqrt{[p]/\langle p \rangle [e]}$ is a function defined for $p_+ > p_-$. To avoid having to extend ψ to other values of the variables, we make use of the symmetry

$(x, t, p, v, S) \mapsto (-x, t, p, -v, S)$ to write the 3-shock-wave in the form

$$S_- = \Sigma(p_-; p_+, S_+),$$
$$v_- = v_+ + \psi(p_-; p_+, S_+),$$
$$p_- > p_+.$$

Wave curves

Finally, the 1-wave curves are described in the form

$$S_+ = \sigma(p_+; p_-, S_-),$$
$$v_+ = v_- - \tau(p_+; p_-, S_-),$$

by defining on the one hand $\tau := \theta$ when $p_+ \leq p_-$, and $\tau := \psi$ when $p_+ > p_-$, and on the other hand $\sigma := \Sigma$ when $p_+ > p_-$ and $\sigma := S_-$ when $p_+ \leq p_-$. By symmetry, the 3-wave curves are described by

$$S_- = \sigma(p_-; p_+, S_+),$$
$$v_- = v_+ + \tau(p_-; p_+, S_+).$$

Concerning the 2-waves, which are contact discontinuities because the second field is linearly degenerate, their curves are the integrals of the vector field r_2, which are given by

$$v_+ = v_-,$$
$$p_+ = p_-.$$

The solution of the Riemann problem

Being given a state to the left (p_L, v_L, S_L) and a state to the right (p_R, v_R, S_R) the usual procedure is to seek two intermediate states, of indices 1 and 2, and three waves linking these four states. The central wave being a contact discontinuity, we have $p_1 = p_2$ and $v_1 = v_2$. We shall denote these common values by p and v. There are thus a 1-wave from (p_L, v_L, S_L) to (p, v, S_1) and a 3-wave from (p, v, S_2) to (p_R, v_R, S_R). This results in the equations

$$S_1 = \sigma(p; p_L, S_L), \tag{4.28}$$

$$v = v_L - \tau(p; p_L, S_L), \tag{4.29}$$

$$S_2 = \sigma(p; p_R, S_R), \tag{4.30}$$

$$v = v_R + \tau(p; p_R, S_R). \tag{4.31}$$

The cancellation of v from equations (4.29) and (4.31) reduces the Riemann problem to that of solving a single scalar equation in the unknown p:

$$\tau(p; p_L, S_L) + \tau(p; p_R, S_R) = v_L - v_R. \tag{4.32}$$

Once this equation has been solved, (4.28) and (4.30) yield the values of S_1 and S_2. Finally v is given by (4.29). However, the equation (4.32) does not always have a solution. We shall see for example that for a perfect gas, τ is an increasing function of p (this is true generally at least as long as $p < p_-$, since then $\tau = g(p, S_-) - g(p_-, S_-)$ and $\partial_p g > 0$ is our hypothesis of non-linearity). If we admit this property, then the left-hand side of (4.32) is minimal for $p = 0$, taking a value $V(p_L, S_L, p_R, S_R) = \tau(0; p_L, S_L) + \tau(0; p_R, S_R)$, finite in general. If $v_L - v_R < V(p_L, S_L, p_R, S_R)$, equation (4.32) does not have a solution and we do not find a solution of the Riemann problem by the classical method.

This difficulty is removed by observing that a vacuum can occur when the pressure is zero. In this case, there is no contact discontinuity and the vacuum is found between the straight lines of slopes v_1 and v_2. We have necessarily $v_1 < v_2$ and $p_1 = p_2 = 0$. Finally there are two cases:

Either $v_L - v_R \geq V(p_L, S_L, p_R, S_R)$, and then the solution of the Riemann problem is made up of a wave of each family and there is no vacuum. In this case the fact that the 1- and 3-waves are rarefaction waves or shock waves can be determined by considering the position of $v_L - v_R$ with respect to $\tau(p_R; p_L, S_L)$ and $\tau(p_L; p_R, S_R)$, as for the p-system.

or $v_L - v_R < V(p_L, S_L, p_R, S_R)$, and then the solution of the Riemann problem is made up of a 1-rarefaction-wave (leading to a state of zero pressure), followed by a vacuum, followed by a 3-rarefaction-wave starting from zero pressure. We verify clearly that in this case $v_1 < v_2$, since $v_2 - v_1 = v_R - v_L + V(p_L, S_L, p_R, S_R)$.

The case of a perfect gas

We now take up the case where $p := (\gamma - 1)\rho e$, where $\gamma > 1$ is a constant. The first calculations are immediate and give

$$\begin{aligned}
c &= \sqrt{\gamma(\gamma - 1)e},\\
g &= 2\sqrt{\frac{\gamma e}{\gamma - 1}},\\
S &= (1 - \gamma)\log \rho + \log e,\\
T &= e.
\end{aligned}$$

The functions of the form $\rho h \circ S$ are mathematical entropies of flux $\rho v h \circ S$. We have seen in Exercise (3.7) that the mapping $(1/\rho, v, e + \frac{1}{2}v^2) \mapsto S$ is concave. It is written then as the infimum of a family A of affine functions:

$$S = \inf\left\{l\left(\frac{1}{\rho}, v, e + \frac{1}{2}v^2\right) : l \in A\right\}.$$

To every affine function $l(\tau, v, E) = \alpha_0 + \alpha_1 \tau + \alpha_2 v + \alpha_3 E$, we make a corresponding affine function $m(\rho, q, e) := \alpha_0 \rho + \alpha_1 + \alpha_2 q + \alpha_3 \varepsilon$. This correspondence sends the set A onto the set B. Then

$$\rho S = \inf\left\{m\left(\rho, \rho v, \rho\left(\frac{1}{2}v^2 + e\right)\right) : m \in B\right\},$$

which shows that ρS is a *concave* entropy of the conservative variables

$$u := \left(\rho, \rho v, \rho\left(\frac{1}{2}v^2 + e\right)\right).$$

Let us look at the shock curves, in the light of equations (4.27) and (4.26). This latter is written

$$(\gamma - 1)(\rho_+ p_+ - \rho_- p_-) + (\gamma + 1)(\rho_+ p_- - \rho_- p_+) = 0$$

which reduces to

$$[v]^2 = \frac{(p_+ - p_-)^2}{\rho_-\left(\frac{\gamma+1}{2} p_+ + \frac{\gamma-1}{2} p_-\right)}.$$

In a 1-shock-wave, $[p] > 0$, and hence we have (since $[v] < 0$)

$$v_+ - v_- = -\frac{p_+ - p_-}{\sqrt{\rho_-\left(\frac{\gamma+1}{2} p_+ + \frac{\gamma-1}{2} p_-\right)}}.$$

The parametrisation by p of the wave curves is done by means of the function $\tau(p; p_-, S_-)$ defined here by

$$\tau(p; p_-, S_-) = \begin{cases} (p - p_-)\sqrt{\rho_-\left(\dfrac{\gamma+1}{2}p + \dfrac{\gamma-1}{2}p_-\right)}, & p > p_-, \\ 2c_-\left(\left(\dfrac{p}{p_-}\right)^{(\gamma-1)/2\gamma} - 1\right)/(\gamma - 1), & p \le p_-. \end{cases}$$

The second line of the above formula is obtained by writing that, in a 1-rarefaction-wave, on the one hand $\tau = g_- - g_+$, on the other hand $S_+ = S_-$, that is to say $p_+^{1-\gamma} e_+^{\gamma} = p_-^{1-\gamma} e_-^{\gamma}$. We find that the minimal value $V(p_L, \rho_L, p_R, \rho_R)$ below which the value of $v_L - v_R$ leads to the creation (if we venture to so express it) of a vacuum,

$$V = -\frac{2}{\gamma - 1}(c_L + c_R).$$

Finally, noticing that $\tau(+\infty; p_-, S_-) = +\infty$, we discover that there is no upper limit imposed on $v_L - v_R$. From the intermediate value theorem (τ is certainly continuous) equation (4.32) possesses at least one solution as long as $v_L - v_R \geq V$. But as $\tau(\cdot; p_-, S_-)$ is obviously increasing, this solution is unique.

Theorem 4.8.5 *The Riemann problem for the dynamics of perfect gases has a unique solution. That is of classical form (a 1-wave followed by a contact discontinuity followed by a 3-wave) if $v_L - v_R \geq V(p_L, S_L, p_R, S_R)$. Otherwise, it is made up of a 1-rarefaction-wave and of a 3-rarefaction-wave which join respectively the states (p_L, v_L, S_L) and (p_R, v_R, S_R) to a vacuum.*

4.9 Exercises

4.9 For gas dynamics, we consider one of the Hugoniot curves $H_j(u_-)$ with $j = 1, 3$.

(1) Express the differential of the restriction of p to $H_j(u_-)$ as a function of the differential of that of $1/\rho$.

(2) Calculate the differential of the restriction of S to $H_j(u_-)$. Deduce that S is monotonic along $H_j(u_-)$ so long as $\rho^2c^2[1/\rho] + [p]$ is not zero.

4.10 We consider only classical solutions of gas dynamics.

(1) Let $s(x, t)$ be a quantity satisfying the transport equation $(\partial_t + v\partial_x)s = 0$. Show that $\rho^{-1}\partial_x s$ does too.

(2) Let g be a numerical function. Show that $E := \rho g(\rho^{-1}S_x)$ is an 'entropy' with 'flux' $F = vE$.

(3) We construct by induction $S_0 := S, \dots, S_n := \rho^{-1}\partial_x S_{n-1}$. Let g be a real function of $m + 1$ real variables. Show that $\rho g(S_0, \dots, S_m)$ is an 'entropy' of 'flux' $F = vE$.

4.11 We choose to express every thermodynamic quantity as a function of the entropy–pressure pair (p, S).

(1) Show that

$$d\rho = c^{-2}(dp - p_e T\, dS),$$
$$de = c^{-2}(\rho^{-2} p dp + p_\rho T\, dS).$$

(2) We consider the (unrealistic) case where all the fields are linearly degenerate. Show that there exists a numerical function h such that $\rho^2c^2h(S) \equiv 1$.

(3) Deduce that e and ρ are expressed in the form

$$e = \frac{1}{2}h(S)p^2 + k(S), \qquad \frac{1}{\rho} = -h(S)p + l(S).$$

(4) Show that $[S] = 0$ implies the relation (4.26). Deduce that in the neighbourhood of u_-, $S \equiv S_-$ along the curves $H_j(u_-)$ for $j = 1, 3$.

4.12 We consider the interaction of two shock waves of the same family, $(u_-, u_0; s_-)$, $(u_0, u_+; s_+)$. More precisely, the initial condition is

$$u(x, 0) = \begin{cases} u_-, & x < -1, \\ u_0, & -1 < x < 1, \\ u_+, & x > 1. \end{cases}$$

We denote by $z := 1/\rho$ the specific volume

(1) Calculate the solution of the Cauchy problem up to the time T of interaction. Show that T is finite.

(2) We suppose that the by-product of the interaction, that is the solution of the Riemann problem with $u_L = u_-$, and $u_R = u_+$, reduces to a single shock wave. Derive the formula

$$p_-(z_+ - z_0) + p_+(z_0 - z_-) + p_0(z_- - z_+) = 0.$$

(3) In a shock wave with velocity s, show that

$$\frac{[p]}{[z]} = -(\rho(v - s))^2,$$

(ρ, v) taking either of the values, downstream or upstream.

(4) Deduce from what has come before that the result of the interaction of two shock waves of the same family cannot reduce to a single shock wave (show that $s_- = s_+$ before finishing).

4.13 Solve the Riemann problem for the dynamics of barotropic gases:

$$\left.\begin{aligned} \rho_t + (\rho v)_x &= 0, \\ (\rho v)_t + (\rho v^2 + p(\rho))_x &= 0, \end{aligned}\right\} \tag{4.33}$$

where $\rho(\geq 0) \mapsto p(\rho) \geq 0$ is the equation of state of the gas. We shall make the appropriate hypotheses on p for the system to be hyperbolic and to have two genuinely non-linear characteristic fields (for the sense of variation of λ_j along the integral curve of r_j, we shall adopt the same assumptions as for a polytropic gas). We shall take account of the fact that, as for a non-isothermal gas (i.e. a general gas), it must sometimes admit a vacuum. Following the position of $v_R - v_L$ with respect to the appropriate values of $H(\rho_L, \rho_R)$ and $H(\rho_R, \rho_L)$ we shall determine the nature (shock or rarefaction) of the waves.

4.14 (See [5], [77], [78], [52], [94])

We consider the system which describes the dynamics of an elastic string in $\mathbb{R}^2$. We have $v = (v^1, v^2)$ and $w = (w^1, w^2)$:

$$v_t = w_x,$$
$$w_t = (T(r)v/r)_x, \quad r := \|v\|.$$

We write also $q := v/r$, which is a unit vector.

We suppose that $T(1) = 0$ (in the reference frame, the string is at rest), $T' > 0$ and $T'' > 0$.

(1) Show that the system is hyperbolic if and only if $r > 1$. We shall denote this zone by $\mathscr{H}$, which describes the string in extension.
(2) Show that the 1- and 4-characteristic-fields are genuinely non-linear, while the 2- and 3-fields are linearly degenerate. Establish a symmetry relation between the j-waves and the $(5 - j)$-waves.
(3) Show that in parametric form each wave curve can be written in the form

$$w_+ = w_- + \varphi_j(v_+, v_-).$$

What can we deduce about the φ_j, from the symmetry relations?
(4) We wish to solve the Riemann problem in the usual form. We suppose that $r_L, r_R > 1$ and we denote the intermediate states by (u_j, v_j) for $1 \le j \le 3$. Show that $q_1 = q_L$, $q_2 = q_R$, and $r_1 = r_2 = r_3$. We denote this common value by r.
(5) Eliminating the vectors w_j (and using (3) above) reduce the solution of the Riemann problem to that of a single vector equation with values in $\mathbb{R}^2$, whose unknowns are q_2 and r.
(6) Writing that q_2 is a unit vector leads to the solution of a scalar equation of the form

$$J(r; w_R - w_L, v_L, v_R) = 0.$$

(7) Study the variations of J with respect to its first variable on the interval $(1, \infty)$. Deduce that the Riemann problem has one and only one solution with values in $\mathscr{H}$.
(8) Characterise the nature of the 1-wave and of the 4-wave as a function of the values, which should be calculated explicitly, of $J(r_L; w_R - w_L, v_L, v_R)$ and of $J(r_R; w_R - w_L, v_L, v_R)$.

5

The Glimm scheme

It is not the purpose of this work to introduce the schemes for the numerical simulation of the systems of conservation laws, which is very well done in other works [62, 34]. But it is impossible to study the theory of systems without describing Glimm's scheme, which gives the sole result of any generality concerning Cauchy's problem. Curiously, this scheme, the only one for which we have at our disposal a convergence theorem for systems in one space dimension, is rarely used, no doubt by reason of its random aspect (which prevents it attaining a high precision) and also because its extension to several space dimensions is disappointing.

We therefore restrict ourselves to systems of n conservation laws in one space dimension,

$$u_t + f(u)_x = 0, \quad x \in \mathbb{R}, t > 0, \tag{5.1}$$

$$u(x, 0) = \bar{u}(x), \quad x \in \mathbb{R}, \tag{5.2}$$

for which we know *a priori* the Riemann problem has a solution. For example, since the principal result concerns an initial datum u near to a constant, we can suppose that the system is strictly hyperbolic and that each of its characteristic fields is either genuinely non-linear or linearly degenerate, as we can use Lax's theorem for the local solution of the Riemann problem.

5.1 Functions of bounded variation

Glimm's theory makes use of the class of functions of bounded total variation on $\mathbb{R}$. We recall that if E is a normed vector space and if I is an interval of $\mathbb{R}$ (which can be either open or closed), the *total variation* of a function $v: \mathbb{R} \to E$ on I, denoted by $\mathrm{TV}(v; I)$, is the upper bound of

$$\sum_{j=1}^{r} \|v(x_j) - v(x_{j-1})\|,$$

when $x = (x_0, \ldots, x_r)$ runs through the set of finite increasing sequences with values in I. We say that v is of bounded total variation on I if $\mathrm{TV}(v; I) < +\infty$. The set $\mathrm{BV}(I; E)$ of these functions is a Banach space when we equip it with the norm $\mathrm{TV}(v; I) + \|v(y)\|$, y being a point chosen in I. The main property of the space BV is Helly's theorem.

Theorem 5.1.1 *We suppose that E is of finite dimension and that I is bounded. Then the canonical mapping of* $\mathrm{BV}(I; E)$ *into* $L^1(I; E)$ *is compact.*

We take heed of the fact that this mapping is not injective since the functions in L^1 are defined only almost everywhere. We could make it injective by replacing BV by its quotent modulo the null functions almost everywhere. The norm of a class of functions v would then be the lower bound of the norms of its elements. This theorem can be seen as a variant of the Rellich–Kondrachov theorem for Sobolev spaces since the image of BV is also the space of locally integrable functions whose distributional derivative is a bounded measure. This space is particularly appropriate (at least in one space dimension) in the study of weak solutions involving shock waves and contact discontinuities and which are smooth elsewhere.

> In several space dimensions, the situation is clearly less favourable since we know [4, 85] that the space BV is not suitable for the linear systems $u_t + \sum_{1\le j\le d} A_j u_{x_j} = 0$, except when the matrices A_j commute pairwise. Although it is not excluded that the non-linearity of certain characteristic fields contributes to partially regularise the solution, the physical systems also have linearly degenerate fields and it is improbable that functions of bounded variation settle the question. However, no other satisfactory function space has been suggested until now to study weak solutions.

Now, let us introduce some rules of calculation. When a function $v: I \to E$ is piecewise continuous, discontinuous only at the points $a_1, \ldots, a_s$ and piecewise $\mathscr{C}^1$ between these, the total variation of v is calculated simply by the formula

$$\mathrm{TV}(v; I) = \sum_{j=1}^{s} (\|v(a_j+0) - v(a_j)\| + \|v(a_j) - v(a_j-0)\|) + \int_{I\setminus\{a_1,\ldots,a_s\}} \left\|\frac{\mathrm{d}v}{\mathrm{d}x}\right\| \mathrm{d}x.$$

If $I = (a, b)$ and if $c \in (a, b)$, then

$$\mathrm{TV}(v; I) = \mathrm{TV}(v; (a, c)) + \mathrm{TV}(v; (c, b)) + |v(c) - v(c-0)| + |v(c) - v(c+0)|.$$

The following inequality is a consequence of the definition.

Proposition 5.1.2 *If* $v \in \mathrm{BV}(\mathbb{R})$ *and if* $h > 0$, *we have*

$$\int_{\mathbb{R}} |v(x+h) - v(x)|\, \mathrm{d}x \leq h\, \mathrm{TV}(v; \mathbb{R}).$$

In fact the number of jumps of a function of bounded variation is at most denumerable and, free to modify v at these points with the result that $v(x)$ is on the segment of extremities $v(x-0)$ and $v(x+0)$, we have

$$\mathrm{TV}(v; \mathbb{R}) = \lim_{h \to 0} \frac{1}{h} \int_{\mathbb{R}} |v(x+h) - v(x)|\, \mathrm{d}x.$$

Proof This uses only the triangle inequality and the relation of Chasles:

$$\begin{aligned}
\int_{\mathbb{R}} |v(x+h) - v(x)|\, \mathrm{d}x &= \sum_{k \in \mathbb{Z}} \int_{kh}^{(k+1)h} |v(x+h) - v(x)|\, \mathrm{d}x \\
&= \sum_{k \in \mathbb{Z}} \int_0^h |v(y + (k+1)h) - v(y + kh)|\, \mathrm{d}y \\
&= \int_0^h \sum_{k \in \mathbb{Z}} |v(y + (k+1)h) - v(y + kh)|\, \mathrm{d}y \\
&\leq \int_0^h \mathrm{TV}(v; \mathbb{R})\, \mathrm{d}y = h\, \mathrm{TV}(v; \mathbb{R}).
\end{aligned}$$

It is easy now to prove Helly's theorem. In fact if $\mathscr{F}$ is a bounded family of real functions defined on a bounded interval I we go on to consider the family F of their extensions by 0 to the whole of $\mathbb{R}$. This family is again bounded in $\mathrm{BV}(\mathbb{R})$ since each extension satisfies $\mathrm{TV}(\tilde{f}; \mathbb{R}) \leq \mathrm{TV}(f; I) + 2\|f\|_\infty \leq 3\|f\|_{\mathrm{BV}(I)}$. From the above proposition, F is uniformly equi-integrable. In addition, the elements of F are with support in $\bar{I}$, which is a fixed compact set, and are uniformly bounded on $\mathbb{R}$, since $\|f\|_\infty \leq \|f\|_{\mathrm{BV}(I)}$. Thus, the set F satisfies the hypotheses of the relative compactness theorem of Kolmogorov in $L^1(\mathbb{R})$ and we conclude that $\mathscr{F}$ is relatively compact in $L^1(I)$. ∎

The last result which we need concerns functions which also depend on a time variable.

Theorem 5.1.3 *We suppose that E is finite-dimensional and we consider a sequence* $(v_m)_{m \in \mathbb{N}}$ *of functions defined on* $[0, T] \times \mathbb{R}$, *with values in E satisfying the following three hypotheses:*

(1) *there exists a positive real number M such that* $\mathrm{TV}(v_m(t); \mathbb{R}) \leq M$ *for all* $m \in \mathbb{N}$,

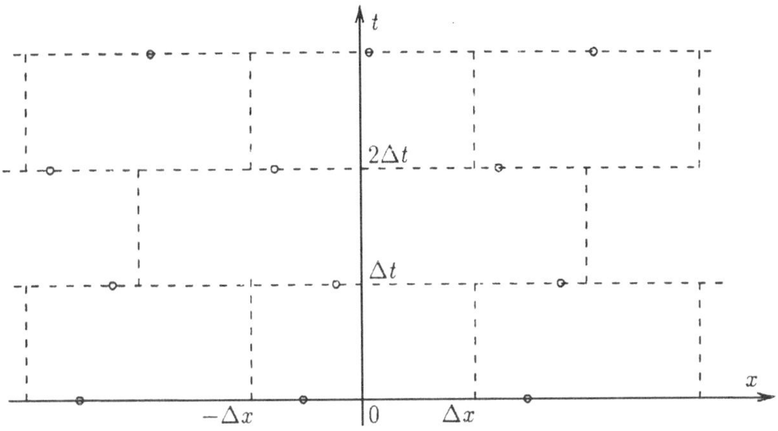

Fig. 5.1: The grid with sampling points.

(2) $\|v_m(t, 0)\| \leq M$ *for all* $m \in \mathbb{N}$,

(3) *there exists a sequence* $(\varepsilon_m)_{m \in \mathbb{N}}$, *which converges to* $0+$, *such that*

$$\int_{\mathbb{R}} \|v_m(t, x) - v_m(s, x)\| \, \mathrm{d}x \leq \varepsilon_m + M|t - s|$$

for all $m \in \mathbb{N}$ *and all* $s, t \in [0, T]$.

Then this sequence is relatively compact in $L^1_{\text{loc}}((0, T) \times \mathbb{R})$: *we can extract a subsequence whose restriction to every bounded open set* $\Omega \subset (0, T) \times \mathbb{R}$ *converges in* $L^1(\Omega)$. *Also the limit* v *belongs to* $\mathscr{C}([0, T]; L^1(\mathbb{R}))$.

In fact the limit also satisfies

$$\int_{\mathbb{R}} \|v(t, x) - v(s, x)\| \, \mathrm{d}x \leq M|t - s|,$$

and $v(t, \cdot)$ (which is well-defined as an element of $L^1(\mathbb{R})$ for all values of t) admits for $t \in (0, T)$ a representation with bounded variation satisfying $\mathrm{TV}(v(t, \cdot); \mathbb{R}) \leq M$ and $\|v(t, 0)\| \leq M$.

The proof of this theorem will be given in §5.8.

5.2 Description of the scheme

The approximate solution of the Cauchy problem given by Glimm's scheme depends on the choice of the space step Δx and of the time step Δt and on that of a sequence $a = (a_0, a_1, \ldots)$ of which each term is an element of $(-1, 1)$. In general, the ratio $\rho := \Delta x / \Delta t$ is chosen *a priori* to ensure a sufficient speed of propagation (the Courant–Friedrichs–Lewy condition) and remains constant when we make $h = \Delta x$

tend to zero. We then denote by u_a^h the approximate solution, although it depends also on ρ, as we only vary h and a in the sequel.

This is defined by induction on $n \in \mathbb{N}$ in each strip $[n\Delta t, (n+1)\Delta t] \times \mathbb{R}$. Possibly this induction may fail at a stage N and the solution will be defined only on $(0, N\Delta t) \times \mathbb{R}$. At the instants of the form $n\Delta t$, u_a^h is piecewise constant, constant on the meshes $I_j := ((j-1)\Delta x, (j+1)\Delta x)$ for $j \in n + 2\mathbb{Z}$. The meshes are thus arranged in alternate rows. At the initial instant, we define $u_a^h(0, I_j) = \bar{u}((j + a_0)\Delta x)$, which is a sampling of the initial condition. Similarly, for $n \geq 1$, we pass from $n\Delta t - 0$ to $n\Delta t + 0$ through a sampling value

$$u_a^h(n\Delta t, I_j) := u_a^h(n\Delta t - 0, (j + a_n)\Delta x - 0), \quad j \in n + 2\mathbb{Z}.$$

This, the choice of the value to the left, is conventional and is intended to remove the ambiguity when we must sample a discontinuity in the approximate solution (which happens only exceptionally).

It remains to define u_a^h in the strip $(n\Delta t, (n+1)\Delta t) \times \mathbb{R}$ as *the exact solution of the Cauchy problem* in this strip, of which the initial condition is the piecewise constant function $u_a^h(n\Delta t, \cdot)$. This exact solution is known to us during a certain time interval δt_n: it is the gluing of solutions of the Riemann problem. Let us write $u_{n,j} := u_a^h(n\Delta t, I_j)$. The Riemann problem centred in $t = n\Delta t$ and $x = k\Delta x$ (for $k \in n + 1 + 2\mathbb{Z}$), of which the left and right states are respectively $u_{n,k-1}$ and $u_{n,k+1}$, admits a solution $v_{n,k}$ by hypothesis. The function $v(\cdot, \cdot)$ defined for $t \geq n\Delta t$ by

$$v_n(t, x) = v_{n,k}(t, x), \qquad \forall (x, t) : x \in I_k,$$

is a weak entropy solution of the Cauchy problem considered while v_n is continuous at the interfaces between the meshes, that is to say for $x = j\Delta x$, $j \in n + 2\mathbb{Z}$. It is enough for this that the waves of the Riemann problem issuing from the node $k\Delta x$ do not reach the boundaries of the mesh I_k, this for all k. Denoting by V_n the upper bound of the speeds of the waves of all the Riemann problems solved at the instant $n\Delta t$, we observe that v_n is the exact solution sought in the time interval $\delta t_n = V_n^{-1}\Delta x$. The calculation of u_a^h is thus effective until the following instant $(n+1)\Delta t$ provided the Courant–Friedrichs–Lewy (CFL) condition is satisfied:

$$\rho V_n \leq 1. \tag{5.3}$$

We note that V_n depends only on the (ordered) list of the states $u_{n,j}$, $j \in n + 2\mathbb{Z}$. In particular, suppose that $\bar{u}$ takes its values in a domain $K \subset \mathcal{U}$, compact and invariant for the Riemann problem, that is to say satisfying the following property:

> For all $v_L, v_R \in K$, the solution of the Riemann problem between v_L and v_R has its values in K.

Under this hypothesis, it is immediate that u_a^h takes its value in K while the approximate solution is defined, that is, while the condition (5.3) is satisfied. But as K is compact, there exists a bound V^K of the speed of the waves in the Riemann problems whose initial data are in K. And as $V_n \leq V^K$, it is sufficient to choose *a priori* $\rho = (V^K)^{-1}$ for (5.3) to hold and for the approximate solution to be defined for all time. This remark is due to D. Hoff [42].

In the general case, *grosso modo* for systems of at least three equations, there does not exist an invariant compact domain for the Riemann problem (see Chapter 8). We therefore define the approximate solution by induction on the strips $(t_n, t_{n+1}) \times \mathbb{R}$ with $t_{n+1} - t_n = V_n^{-1}\Delta x$, hoping that $\sum_n V_n^{-1}$ diverges. However, Glimm's theorem, which we are going to prove, states that for a small enough initial datum the approximate solution remains in a fixed compact set, independent of ρ, with the result that there again exists a value of ρ which satisfies the CFL condition at all stages of the calculation.

Theorem 5.2.1 **(Glimm)** *Let $\hat{u}$ be an interior point of $\mathcal{U}$. We suppose that each characteristic field is either genuinely non-linear or linearly degenerate in the neighbourhood of $\hat{u}$. Then, there exist two numbers $\delta > 0$ and $C > 0$ such that for every given initial function satisfying the hypothesis*

$$\|\bar{u} - \hat{u}\|_{L^\infty} + \mathrm{TV}(\bar{u}) \leq 0, \tag{5.4}$$

the Cauchy problem (5.1), (5.2) possesses a weak solution in $\mathbb{R}_t^+ \times \mathbb{R}_x$ satisfying in addition

$\|u(t,\cdot) - \hat{u}\|_{L^\infty} \leq C(\|\bar{u} - \hat{u}\|_{L^\infty} + \mathrm{TV}(\bar{u}))$,

$\mathrm{TV}(u(t,\cdot)) \leq C\mathrm{TV}(\bar{u})$,

$\|u(t,\cdot) - u(s,\cdot)\|_{L^1} \leq C|t-s|\mathrm{TV}(\bar{u})$,

u satisfies Lax's entropy inequalities $E_t + F_x \leq 0$ for every convex entropy E of flux F:

$$\iint_{\mathbb{R}^+\times\mathbb{R}} (E(u)\varphi_t + F(u)\varphi_x)\,\mathrm{d}x\,\mathrm{d}t + \int_{\mathbb{R}} E(\bar{u})\varphi(x,0)\,\mathrm{d}x \geq 0, \quad \forall \varphi \in \mathcal{D}^+(\mathbb{R}^2).$$

Of course, this theorem is tied to Glimm's scheme by a statement of convergence. However, as we shall see in the following section, the approximate solution, though it converges *in general*, does not converge to a weak solution of the Cauchy problem when we choose certain sequences a. The convergence is linked to a random property of the sequence a, which explains that we have recourse to a parameter as complex as a generic element of $A = (-1, 1)^{\mathbb{N}}$.

First of all, let us provide A with the uniform probability measured $\mathrm{d}\nu$, product of the uniform measures $\mathrm{d}m_j = \frac{1}{2}\mathrm{d}a_j$ on each factor $(-1, 1)$. This is the unique measure defined on the class of Borel sets of A which satisfies the identities

$$\int_A g(a)\,\mathrm{d}\nu(a) = 2^{-1-N} \int_{-1}^{1} \cdots \int_{-1}^{1} G(a_0, \ldots, a_N)\,\mathrm{d}a_0 \ldots \mathrm{d}a_N$$

when $g(a) := G(a_0, \ldots, a_N)$ has only a finite number of arguments. We see from the Stone–Weierstrass theorem that these functions form a dense sub-space of $\mathscr{C}(\bar{A})$ so the above formula allows us to define in a unique manner the integral of a function, continuous on $\bar{A}$, with respect to $\mathrm{d}\nu$.

Here is the convergence result.

Theorem 5.2.2 *Under the hypotheses of Theorem 5.2.1, there exist two numbers $h_0 > 0$ and $\rho_0 > 0$ such that, for all $a \in A$ and every step $h = \Delta x$, $0 < h < h_0$, $\Delta t = \rho \Delta x$ $(0 < \rho < \rho_0)$, the approximate solution is defined for all time and satisfies at each instant*

$$\left\| u_a^h - \hat{u} \right\|_{L^\infty} \leq C(\|\bar{u} - \hat{u}\|_{L^\infty} + \mathrm{TV}(\bar{u})),$$
$$\mathrm{TV}\left(u_a^h\right) \leq C\,\mathrm{TV}(\bar{u}).$$

In addition, there exists a subset $\mathscr{N}$ of measure zero in A such that, for all $a \in A \backslash \mathscr{N}$, the sequence $(u_a^h)_{h \to 0+}$ is relatively compact in $L^1_{\mathrm{loc}}(\mathbb{R}^+ \times \mathbb{R})$, its limits being the weak solutions of the Cauchy problem such as are described in Theorem 5.2.1.

Of course, the uniqueness of the entropy solution in the class where we show the existence not being known,[1] it is not possible to write a simpler statement. The scalar case is the most favourable because of Theorem 2.3.5 and in this case it is the whole sequence $(u_a^h)_h$ which converges, this for almost all a. Also, the proof of the stability of the scheme (obtaining *a priori* estimates) is much more simple and general in the scalar case (we have for example $C = 1$): we no longer suppose that the given initial function is small. It is the same for the systems said to be of B. Temple (see Chapter 13) which extend in a natural manner the scalar conservation laws. This remark is also of value for a class of systems which we shall study in §5.6 and which contains the isothermal model of gas dynamics. In the general case, we can ask if the hypothesis of a small datum is essential for the existence of a weak solution to the Cauchy problem, since the real world does not consist of such data. We do not have the means to answer this question, but it has been observed that the estimate $\mathrm{TV}(u(t, \cdot)) \leq C\,\mathrm{TV}(\bar{u})$ cannot be true if the constant C depends solely on $\|\bar{u} - \hat{u}\|_\infty$ for rather general systems of at least three conservation laws [47]. The

[1] Actually, a recent result of A. Bressan states that the limit is unique whenever Glimm's estimate and consistency hold.

number 3 is here optimal since in the case of 2×2 systems ($n = 2$ equations), only the norm of $\bar{u} - \hat{u}$ in L^∞ must be supposed small (see §5.9).

As often in numerical analysis, the stability of the scheme implies its convergence (Theorem 5.4.1). It is the possibility of showing the stability which distinguishes the general case from particular cases. In fact, only the convergence is the object of a probabilistic analysis, while that of the stability, which leads to the estimate (5.4), proceeds from physical ideas and makes use of the notion of *interaction potential.*

Before dealing with the proof of Theorem 5.2.2 we shall explain in the next section why certain sequences $a \in A$ are not suitable.

5.3 Consistency

For most of the systems of physical interest, one at least of the characteristic fields is genuinely non-linear and the work of Lax shows that there exist pairs of states $(u_{\mathrm{L}}, u_{\mathrm{R}})$ linked by a shock wave. We denote by c the speed of this shock wave and we calculate the approximate solution provided by Glimm's scheme when the given initial condition is

$$\bar{u}(x) = \begin{cases} u_{\mathrm{L}}, & x < 0, \\ u_{\mathrm{R}}, & x > 0. \end{cases}$$

Since the solution of the Riemann problem between two equal states is constant and since the scheme proceeds by sampling and by solutions of Riemann problems, we see immediately that the states $u_{n,j}$ are all with values in $\{u_{\mathrm{L}}, u_{\mathrm{R}}\}$. More precisely, there exists a number $j_n \in n + 1 + 2\mathbb{Z}$ such that $u_{n,j} = u_{\mathrm{L}}$ if $j < j_n$ and $u_{n,j} = u_{\mathrm{R}}$ if $j > j_n$. The approximate solution is thus completely known if we realise the recurrence $j_n \mapsto j_{n+1}$. In the strip $(t_n, t_{n+1}) \times \mathbb{R}$, the approximate solution takes the values

$$u_a^h(t, x) = \begin{cases} u_{\mathrm{L}}, & x - j_n h < c(t - t_n), \\ u_{\mathrm{R}}, & x - j_n h > c(t - t_n). \end{cases}$$

The CFL condition is written $\rho|c| \leq 1$, and this does not depend on n. The approximate solution is thus defined for all time. Also, $u_a^h(t_{n+1} - 0, x)$ takes the value u_{L} on I_j for all $j < j_n$ and the value u_{R} for $j > j_n$. Thus, $u_{n+1,j}$ takes the value u_{L} for $j < j_n$ and the value u_{R} for $j > j_n$. As $j_{n+1} - j_n$ is odd, we thus have $j_{n+1} = j_n \pm 1$ and in fact

$$j_{n+1} = j_n + \begin{cases} 1, & a_{n+1} < \rho c, \\ -1, & a_{n+1} \geq \rho c. \end{cases}$$

Finally

$$j_n = n + 1 - 2\,\mathrm{card}\{m \in \mathbb{N}\colon 0 \le m \le n, a_m \ge \rho c\}.$$

The approximate solution takes the values u_{L} and u_{R} at one side and the other of the broken line which passes through the nodes $P_n = (n\Delta t, j_n h)$. The slope of the straight line OP_n has the value

$$p(a; h, n\Delta t) = p_n = \frac{j_n h}{n\Delta t} = \frac{j_n}{\rho n} \sim \frac{1}{\rho} - \frac{2}{\rho n}\mathrm{card}\{m \in \mathbb{N}\colon 0 \le m \le n, a_m \ge \rho c\}.$$

If the approximate solution converges to the exact solution of the Cauchy problem, which is the shock wave of speed c, we must have $\lim_{h\to 0} p(a; h, t) = c$, that is to say,

$$\frac{2}{n}\mathrm{card}\{m \in \mathbb{N}\colon 0 \le m \le n, a_m \ge \rho c\} \sim 1 - \rho c. \tag{5.5}$$

If we wish to maintain that this convergence takes place for all the possible systems, that is for shock waves of arbitrary speed and numbers ρ compatible with the CFL condition, it is necessary that

$$\frac{2}{n}\mathrm{card}\{m \in \mathbb{N}\colon 0 \le m \le n, a_m \ge \theta\} \sim 1 - \theta, \tag{5.6}$$

when $\theta \in (-1, 1)$ and $n \to \infty$. A sequence a which satisfies this property is called *equi-distributed* in $(-1, 1)$. An equivalent condition is the convergence of quadrature formulae:

$$\lim_{n\to+\infty} \frac{1}{n+1}\sum_{k=0}^{n} f(a_k) = \frac{1}{2}\int_{-1}^{1} f(x)\,\mathrm{d}x, \quad \forall f \in \mathscr{C}([0, 1]). \tag{5.7}$$

A classic example of an equi-distributed sequence is $a_n = n\xi \pmod 2$ where ξ is an irrational number. But to put Glimm's scheme into operation, Collela [10] has proposed the sequence of Van der Corput [38] in which the filling of the interval $(-1, 1)$ is mades at an optimal speed (this refers to the notion of discrepancy of a sequence which goes beyond the objective of this section). This sequence is obtained by writing each integer n as a number to the base 2, $n = \alpha_r \cdots \alpha_1$ with $\alpha_j = 0$ or 1 and $\alpha_r = 1$, then calculating the number b_n written in binary mode as $b_n = \alpha_1, \cdots \alpha_r$. Finally we put $b_n = 1 - a_n$. The first elements of this sequence are

$$\begin{array}{cccccccc} -1 & & & & 0 & & & \\ & & -1/2 & & & & 1/2 & \\ & -3/4 & & -1/4 & & 1/4 & & 3/4 \\ & & \vdots & & & \vdots & & \end{array}$$

Thus, only the equi-distributed sequences let us hope for convergence of u_a^h to a weak solution of the Cauchy problem. Happily we have

Proposition 5.3.1 *Almost every (with respect to* $\mathrm{d}\nu$*) sequence* $a \in A$ *is equi-distributed.*

> This proposition explains that the convergence of the scheme can take place for almost every sequence a except for those that are badly distributed. In fact, T.-P. Liu has improved Glimm's theorem in proving that convergence takes place for every equi-distributed sequence [69].

Proof Let F be a dense denumerable subset of $\mathscr{C}([-1,1])$. For $f \in \mathscr{C}([-1,1])$ and $a \in A$, we write

$$I(f) = \frac{1}{2}\int_{-1}^{1} f(x)\,\mathrm{d}x\,, \qquad I_n(a;f) = \frac{1}{n+1}\sum_{k=0}^{n} f(a_k).$$

If a sequence a satisfies $\lim_{n\to+\infty} I_n(a;f) = I(f)$ for all f in F, then it is equi-distributed for if $g \in \mathscr{C}([-1,1])$ and if $\varepsilon > 0$, there exists f in F such that $\sup_x |f(x) - g(x)| \leq \varepsilon$ and there exists $N \in \mathbb{N}$ such that, for $n > N$, we have $|I_n(a;f) - I(f)| \leq \varepsilon$. But then, for $n \geq N$, we have

$$|I_n(a;g) - I(g)| \leq |I_n(a;g-f)| + |I_n(a;f) - I(f)| + |I(f-g)| \leq 3\varepsilon.$$

We thus have $\lim_{n\to+\infty} I_n(a;g) = I(g)$ for every continuous function g and this is equivalent to the equi-distribution of a. The set of sequences badly distributed is thus the (denumerable) union of sets N_f defined by

$$J(a;f) := \limsup_{m\to+\infty} |I_m(a;f) - I(f)|^2 > 0.$$

Now making use of Fatou's lemma,

$$\int_A J(a;f)\,\mathrm{d}\nu(a) \leq \limsup_{m\to+\infty} \int_A |I_m(a;f) - I(f)|^2\,\mathrm{d}\nu(a),$$

where the integral on the right-hand side has value

$$\frac{1}{(m+1)^2}\sum_{j,k=0}^{m}\int_A f(a_j)f(a_k)\,\mathrm{d}\nu(a) - \frac{2}{m+1}I(f)\sum_{k=0}^{m}\int_A f(a_k)\,\mathrm{d}\nu(a) + I(f)^2$$
$$= I(f)^2\left(\frac{m}{m+1} - 2 + 1\right) + \frac{1}{m+1}I(f^2)$$
$$= \frac{1}{m+1}(I(f^2) - I(f)^2) \overset{m\to+\infty}{\longrightarrow} 0.$$

Thus $\int_A J(a; f)\,d\nu(a)$ is null, with the result that N_f is of null measure. Finally, the (denumerable) union of the N_f for f ranging over F is of null measure. ■

5.4 Convergence

We show in this section that the stability of Glimm's scheme in BV implies the convergence to a weak entropy solution for almost every choice of the sequence a.

Theorem 5.4.1 *Let $\bar{u} \in \mathrm{BV}(\mathbb{R})$ and $\rho > 0$. We suppose that for $h \in (0, h_0)$ (with $h_0 > 0$ suitably chosen) and for all $a \in A$, Glimm's scheme defines a global approximate solution u_a^h and that there exists a constant $M > 0$ such that we have for all time*

$$\left\|u_a^h - \hat{u}\right\|_{L^\infty} + \mathrm{TV}\left(u_a^h\right) \leq M.$$

Then the sequences $(u_a^h)_{h>0}$ are relatively compact in $L^1_{\mathrm{loc}}(\mathbb{R}^+ \times \mathbb{R})$ and their limit are, for $d\nu$-almost all $a \in A$, weak solutions of the Cauchy problem (5.1), (5.2).

Let us note that we do not assume that the given initial condition is small in $\mathrm{BV}(\mathbb{R})$, with the result that, if we know how to prove the stability of the scheme for arbitrarily large given initial data, we immediately deduce an existence theorem for such data.

Compactness

The compactness of the sequences $(u_a^h)_{h>0}$ will follow from Theorem 5.1.3, once we have verified the third hypothesis. Let us first suppose that $s = n\Delta t + 0$ and $n\Delta t < t < (n+1)\Delta t - 0$. In these calculations let us write $u = u_a^h$. Then $u(s, x)$ has value $u_{n,j}$ on I_j ($j+n$ even), in the same manner as $u(t, jh)$. We therefore have

$$\begin{aligned}\int_{\mathbb{R}} |u(t) - u(s)|\,dx &= \sum_{j\in n+2\mathbb{Z}} \int_{I_j} |u(t) - u(s)|\,dx \\ &= \sum_{j\in n+2\mathbb{Z}} \int_{I_j} |u(t, x) - u(t, jh)|\,dx \\ &\leq \sum_{j\in n+2\mathbb{Z}} 2h\,\mathrm{TV}(u(t); I_j) = 2h\,\mathrm{TV}(u(t)) \leq 2Mh.\end{aligned}$$

Then, if $s = n\Delta t - 0$, and $t = n\Delta t + 0$,

$$\begin{aligned}\int_{\mathbb{R}} |u(t) - u(s)|\,dx &= \sum_{j\in n+2\mathbb{Z}} \int_{I_j} |u(s, (j + a_n)h) - u(s, x)|\,dx \\ &\leq \sum_{j\in n+2\mathbb{Z}} 2h\,\mathrm{TV}(u(s); I_j) \leq 2Mh.\end{aligned}$$

Combining these two inequalities we bound the integral for any s and t by $2(N+1)Mh$ where N is the number of integers n such that $s \le n\Delta t \le t$. Finally,

$$\int_{\mathbb{R}} |u(t) - u(s)| \, \mathrm{d}x \le 2M(|t-s| + h).$$

We can therefore apply Theorem 5.1.3.

We note that we have implicity assumed that the CFL condition is satisfied, by supposing that u_a^h is globally defined. We make use of this when we write $u(t, jh) = u_{n,j}$ for $n\Delta t < t < (n+1)\Delta t$.

Estimate of the error

The error due to the scheme is the value of the expression which should be null for a weak solution in the variational formulation of the Cauchy problem, namely

$$e(a;\varphi,h) := \iint_{\mathbb{R}^+\times\mathbb{R}} \left(u_a^h \cdot \varphi_t + f\left(u_a^h\right) \cdot \varphi_x\right) \mathrm{d}x \, \mathrm{d}t + \int_{\mathbb{R}} \bar{u}(x) \cdot \varphi(x,0) \, \mathrm{d}x.$$

We prove here the following lemma.

Lemma 5.4.2 *Under the hypotheses of Theorem 5.4.1 we have*

$$\lim_{h\to 0+} \int_A |e(a;\varphi,h)|^2 \, \mathrm{d}\nu(a) = 0$$

for all $\varphi \in \mathscr{D}(\mathbb{R}^2)^n$.

Proof Let $\varphi \in \mathscr{D}(\mathbb{R}^2)^n$. In each strip $B_n = (n\Delta t, (n+1)\Delta t) \times \mathbb{R}$, u_a^h is a weak solution, piecewise smooth, of a Cauchy problem; that is we have

$$\iint_{B_n} \left(u_a^h \cdot \varphi_t + f\left(u_a^h\right) \cdot \varphi_x\right) \mathrm{d}x \, \mathrm{d}t = \int_{\mathbb{R}} u_a^h((n+1)\Delta t - 0, x) \cdot \varphi((n+1)\Delta t, x) \, \mathrm{d}x - \int_{\mathbb{R}} u_a^h(n\Delta t + 0, x) \cdot \varphi(n\Delta t, x) \, \mathrm{d}x.$$

Summing these equalities (all these integrals except a finite number among them are null) we obtain $e(a;\varphi,h) = \sum_{n\ge 0} e_n(a;\varphi,h)$ where

$$e_n(a;\varphi,h) := \int_{\mathbb{R}} \left(u_a^h(n\Delta t - 0, x) - u_a^h(n\Delta t + 0, x)\right) \cdot \varphi(n\Delta t, x) \, \mathrm{d}x,$$

these quantities being all null but a finite number. It is easy to see that each $e_n(a;\varphi,h)$ is $O(h)$, for by writing $w_n := u_a^h(n\Delta t - 0)$ we have

$$\begin{aligned} e_n(a;\varphi,h) &= \sum_{j\in n+2\mathbb{Z}} \int_{I_j} (w_n(x) - w_n((j+a_n)h)) \cdot \varphi(n\Delta t, x)\,\mathrm{d}x, \\ |e_n(a;\varphi,h)| &\le \sum_{j\in n+2\mathbb{Z}} 2h\,\mathrm{TV}(w_n; I_j)\|\varphi\|_\infty \qquad (5.8) \\ &\le 2h\,\mathrm{TV}(w_n)\|\varphi\|_\infty \le 2hM\|\varphi\|_\infty. \end{aligned}$$

Besides, $e_n(a;\varphi,h)$ in fact depends only on $(a_0,\ldots,a_n)$ and not on the entire sequence a and behaves *in mean* as $O(h^2)$. In fact

$$\begin{aligned} \frac{1}{2}\int_{-1}^{1} e_n(a;\varphi,h)\,\mathrm{d}a_n &= \sum_{j\in n+2\mathbb{Z}} \frac{1}{2h} \iint_{I_j\times I_j} (w_n(x) - w_n(y)) \cdot \varphi(n\Delta t, x)\,\mathrm{d}x\,\mathrm{d}y \\ &= \frac{1}{4h} \sum_{j\in n+2\mathbb{Z}} \iint_{I_j\times I_j} (w_n(x) - w_n(y)) \cdot (\varphi(n\Delta t, x) \\ &\quad - \varphi(n\Delta t, y))\,\mathrm{d}x\,\mathrm{d}y. \end{aligned}$$

Thus

$$\begin{aligned} \left|\frac{1}{2}\int_{-1}^{1} e_n(a;\varphi,h)\,\mathrm{d}a_n\right| &\le h \sum_{j\in n+2\mathbb{Z}} \mathrm{TV}(w_n; I_j)\mathrm{TV}(\varphi(n\Delta t); I_j) \\ &\le 2h^2 \sum_{j\in n+2\mathbb{Z}} \mathrm{TV}(w_n; I_j)\|\varphi_x\|_\infty \le 2h^2\,\mathrm{TV}(w_n)\|\varphi_x\|_\infty \\ &\le 2h^2 M\|\varphi_x\|_\infty. \qquad (5.9) \end{aligned}$$

We develop now the integral over A.

$$\begin{aligned} \int_A |e(a;\varphi,h)|^2\,\mathrm{d}\nu(a) &= \sum_{n,m\ge 0} \int_A e_n e_m\,\mathrm{d}\nu(a) \\ &= \sum_{n\ge 0} \int_A |e_n|^2\,\mathrm{d}\nu(a) + 2\sum_{0\le m<n} \int_A e_n e_m\,\mathrm{d}\nu(a). \end{aligned}$$

Let N be the number of strips B_n whose intersection with the support of φ is not empty. As this support is contained in a half-plane $t \le T$, we have

$$N \le 1 + \frac{T}{\Delta t} = 1 + \frac{T}{\rho h}.$$

The first sum is bounded above, from (5.8), by $N(2hM\|\varphi_x\|_\infty)^2 = O(h)$. Also, each integral $\int_A e_m e_n \,\mathrm{d}\nu(a)$ is $O(h^3)$, as by (5.9) and (5.8),

$$\begin{aligned}\left|\int_A e_m e_n \,\mathrm{d}\nu(a)\right| &= \left|\int_{-1}^{1}\cdots\int_{-1}^{1} e_m\left(\frac{1}{2}\int_{-1}^{1} e_n \,\mathrm{d}a_n\right)2^{-n}\,\mathrm{d}a_0\cdots\mathrm{d}a_{n-1}\right| \\ &\le \left|\int_{-1}^{1}\cdots\int_{-1}^{1} 2hM\|\varphi\|_\infty \cdot 2h^2M\|\varphi_x\|_\infty 2^{-n}\,\mathrm{d}a_0\cdots\mathrm{d}a_{n-1}\right| \\ &= 4h^3M^2\|\varphi\|_\infty\|\varphi_x\|_\infty.\end{aligned}$$

The second sum is thus bounded above by $N(N-1)4h^3M^2\|\varphi\|_\infty\|\varphi_x\|_\infty = O(h)$. Finally

$$\int_A |e(a;\varphi,h)|^2\,\mathrm{d}\nu(a) = O(h),$$

which proves the lemma. ∎

Conclusion

From the lemma, when $\varphi \in \mathscr{D}(\mathbb{R}^2)^n$, there exists a negligible part N_φ of A such that for $a \in A \setminus N_\varphi$, we have $\lim_{h\to 0} e(a;\varphi,h) = 0$. Let us choose a denumerable dense subset F of $\mathscr{D}(\mathbb{R}^2)^n$. The set $\mathscr{N}$, the union of the N_φ when φ ranges over F, is negligible in A. Being given $a \in A \setminus \mathscr{N}$, let us consider a limiting value of the sequence $(u_a^h)_h$ in $L^1_{\mathrm{loc}}(\mathbb{R}^+ \times \mathbb{R})$ (we have seen that there exist such limits). We know that $u(x,t)$ is the pointwise limit, almost everywhere in $\mathbb{R}^+ \times \mathbb{R}$, of a sub-sequence $(u_a^{h_p})_{p\in\mathbb{N}}$ where $h_p \to 0+$ is a suitable sequence of mesh sizes. From the theorem of dominated convergence and since f is continuous, we have

$$\lim_{p\to+\infty}\iint_{\mathbb{R}^+\times\mathbb{R}} f\left(u_a^{h_p}\right)\cdot\varphi\,\mathrm{d}x\,\mathrm{d}t = \iint_{\mathbb{R}^+\times\mathbb{R}} f(u)\cdot\varphi\,\mathrm{d}x\,\mathrm{d}t,$$

and therefore

$$\lim_{p\to+\infty} e(a;\varphi,h_p) = \iint_{\mathbb{R}^+\times\mathbb{R}} (u\cdot\varphi_t + f(u)\cdot\varphi_x)\,\mathrm{d}x\,\mathrm{d}t + \int_{\mathbb{R}} \bar{u}(x)\cdot\varphi(x,0)\,\mathrm{d}x.$$

Finally, since $a \in A \setminus \mathscr{N}$, we deduce

$$\iint_{\mathbb{R}^+\times\mathbb{R}} (u\cdot\varphi_t + f(u)\cdot\varphi_x)\,\mathrm{d}x\,\mathrm{d}t + \int_{\mathbb{R}} \bar{u}(x)\cdot\varphi(0,x)\,\mathrm{d}x = 0$$

for all $\varphi \in F$ and hence for all $\varphi \in \mathscr{D}(\mathbb{R}^2)^n$ since F is dense and since the left-hand side is a continuous linear form on $\mathscr{D}(\mathbb{R}^2)^n$. This completes the proof of Theorem 5.4.1.

Entropy inequalities

When $E\colon \mathscr{U} \to \mathbb{R}$ is a convex entropy of flux F, we prove the entropy inequality in the same manner provided that the shock waves used in the solution of the Riemann problem satisfy this inequality (this is the least that we can ask of them). The approximate solutions satisfy in each band the inequality, for $\varphi \in \mathscr{D}^+(\mathbb{R}^2)$,

$$\int_{n\Delta t}^{(n+1)\Delta t} \int_{\mathbb{R}} \left(E\left(u_a^h\right) \cdot \varphi_t + F\left(u_a^h\right) \cdot \varphi_x\right) \mathrm{d}x\, \mathrm{d}t + \int_{\mathbb{R}} E\left(u_a^h(n\Delta t + 0)\right) \cdot \varphi(n\Delta t)\, \mathrm{d}x$$
$$\geq \int_{\mathbb{R}} E\left(u_a^h((n+1)\Delta t - 0)\right) \cdot \varphi((n+1)\Delta t)\, \mathrm{d}x.$$

Denoting still the error due to the scheme by $e(a; \varphi, h)$, that is to say

$$e(a; \varphi, h) := \int_{\mathbb{R}^+ \times \mathbb{R}} \left(E\left(u_a^h\right) \cdot \varphi_t + F\left(u_a^h\right) \cdot \varphi_x\right) \mathrm{d}x\, \mathrm{d}t + \int_{\mathbb{R}} E(\bar{u}) \cdot \varphi(x, 0)\, \mathrm{d}x,$$

we thus have $e(a; \varphi, h) \geq \sum_{n \geq 0} e_n(a; \varphi, h)$ where

$$e_n(a; \varphi, h) := \int_{\mathbb{R}} \left(E\left(u_a^h(n\Delta t - 0)\right) - E\left(u_a^h(n\Delta t + 0)\right)\right) \cdot \varphi(n\Delta t)\, \mathrm{d}x.$$

On the compact set $B(\hat{u}; M)$, E is Lipschitz with a constant denoted by K. We then have, as in (5.8),

$$|e_n(a; \varphi, h)| \leq 2hKM\|\varphi\|_\infty$$

and similarly

$$\left|\frac{1}{2}\int_{-1}^{1} e_n(a; \varphi, h)\, \mathrm{d}a_n\right| \leq 2h^2 KM\|\varphi_x\|_\infty.$$

Writing $\tilde{e}(a; \varphi, h) := \sum_{n \geq 0} e_n(a; \varphi, h)$, we deduce again that $\int_A |\tilde{e}(a; \varphi, h)|^2\, \mathrm{d}a_n = \mathrm{O}(h)$ and hence that $\lim_{h \to 0} \tilde{e}(a; \varphi, h) = 0$ for almost all $a \in A$, that is

$$\liminf_{h \to 0} e(a; \varphi, h) \geq 0.$$

The same arguments as were used in the preceding section then show the existence of $\mathscr{N}_E$, a negligible part of A, containing $\mathscr{N}$, such that for $a \in A \setminus \mathscr{N}_E$, the limiting values of $(u_a^h)_{h>0}$ satisfy Lax's inequality

$$\iint_{\mathbb{R}^+ \times \mathbb{R}} (E(u) \cdot \varphi_t + F(u) \cdot \varphi_x)\, \mathrm{d}x\, \mathrm{d}t + \int_{\mathbb{R}} E(\bar{u}) \cdot \varphi(0)\, \mathrm{d}x \geq 0$$

for all $\varphi \in \mathscr{D}^+(\mathbb{R}^2)$.

5.5 Stability

Supplements apropos of the local Riemann problem

Since the characteristic fields are each either genuinely non-linear or linearly degenerate, Theorem 4.6.1 ensures, for every neighbourhood ω of $\hat{u}$, the existence of a neighbourhood $\omega_1 \subset \omega$ such that, for all $(u_L, u_R) \in \omega_1 \times \omega_1$, the Riemann problem between u_L and u_R admits a unique solution with values in ω. This solution is made up of the succession of waves of each family, a contact discontinuity if the field is linearly degenerate, a shock wave or a rarefaction wave otherwise. The k-wave links the constant states u_{k-1} and u_k ($u_0 = u_L$, $u_n = u_R$). Using the parametrisation $s \mapsto \varphi_k(s, v)$ of the wave curve issuing from a point v (with $s = \lambda_k(\varphi_k(s, v)) - \lambda_k(v)$ if the kth field is genuinely non-linear), defined in Chapter 4, we construct the solution of the Riemann problem by solving the equation $\Phi(\varepsilon, u_L) = u_R$ where

$$\Phi(\varepsilon, v) := \varphi_n(\varepsilon_n, \varphi_{n-1}(\varepsilon_{n-1}, \ldots, \varphi_1(\varepsilon_1, v)\cdots)),$$

The mapping Φ is of class $\mathscr{C}^2$ on $\mathscr{V} \times \omega$, $\mathscr{V}$ being a neighbourhood of the origin which depends only on the choice of ω (which we take compact) and satisfies

$$\frac{\partial \Phi}{\partial \varepsilon_k}(0, v) = r_k(v),$$

$r_k(v)$ being an eigenvector of $\mathrm{d}f(v)$ associated with the eigenvalue λ_k and normalised by $\mathrm{d}\lambda_k \cdot r_k \equiv 1$ when the field is genuinely non-linear. We thus have $\varepsilon_k = l_k(u_L) \cdot (u_R - u_L) + \mathrm{O}(\|u_R - u_L\|^2)$ denoting by $(l_k)_{1 \le k \le n}$ the dual basis of $(r_k)_{1 \le k \le n}$. The intermediate constant states are given by induction on k: $u_k = \varphi_k(\varepsilon_k; u_{k-1})$.

The essential idea of the stability analysis made by Glimm [32] concerning his scheme is that of the interaction of successive Riemann problems. If three states u_L, u_m, u_R are given in ω_1, we have $u_m = \Phi(\delta, u_L)$, $u_R = \Phi(\varepsilon; u_m)$ and $u_R = \Phi(\gamma; u_L)$. What can we say about the vector γ when we express it as a function of the parameters $(\delta, \varepsilon, u_m)$? The cases $u_L = u_m$ and $u_R = u_m$ give us the formulae $\gamma(0, \varepsilon, u_m) = \varepsilon$ and $\gamma(\delta, 0, u_m) = \delta$ respectively. As γ is of class $\mathscr{C}^2$ (from the implicit function theorem and since Φ is of class $\mathscr{C}^2$), we deduce that $\gamma(\delta, \varepsilon, u_m) - \delta - \varepsilon = \mathrm{O}(\|\varepsilon\|^2 + \|\delta\|^2)$, which we are now going to improve.

In fact, we shall have $\gamma = \delta + \varepsilon$ when u_m is a value taken by the solution of the Riemann problem between u_L and u_R, that is to say when there exists an index $p \in \{1, \ldots, n\}$ such that

for $k > p$, $\delta_k = 0$,
for $k < p$, $\varepsilon_k = 0$,
the pth field is linearly degenerate, or $\varepsilon_p = 0$, or $\delta_p = 0$ or ($\varepsilon_p > 0$ and $\delta_p > 0$).

The last case cited is that of a p-rarefaction-wave that passes through the value u_{m}. If the pth field is genuinely non-linear, the third condition is thus written $\varepsilon_p^- \delta_p^- = 0$ by writing $z^- = \max(0, -z)$. The set of the above condition is thus written $\Delta(\delta, \varepsilon, u_{\mathrm{m}}) = 0$, where Δ is the *quadratic interaction* term

$$\Delta(\delta, \varepsilon, u_{\mathrm{m}}) = \sum_{1 \leq p < q \leq n} |\delta_q \varepsilon_p| + \sum_{p \in \mathrm{GNL}} \varepsilon_p^- \delta_p^-,$$

the symbol $\sum_{p \in \mathrm{GNL}}$ meaning that the summation is over the indices of the genuinely non-linear fields only. Finally:

Lemma 5.5.1 *The function* $(\delta, \varepsilon, u_{\mathrm{m}}) \mapsto \gamma - \delta - \varepsilon$ *is of class* $\mathscr{C}^2$ *in a neighbourhood of* $(0, 0, \hat{u})$. *It satisfies*

(1) $\gamma - \delta - \varepsilon = \mathrm{O}(\|\varepsilon\|^2 + \|\delta\|^2)$,
(2) $\Delta = 0 \Longrightarrow \gamma - \delta - \varepsilon = 0$.

Owing to a geometrical lemma of which we shall give the statement and the proof in §5.8, we deduce the result which will enable us to establish the stability.

Lemma 5.5.2 *There exist a neighbourhood* Ω *of* $(0, 0, \hat{u})$ *and a real number* $c_0 > 0$ *such that in* Ω *we have*

$$\|\gamma(\delta, \varepsilon, u_{\mathrm{m}}) - \delta - \varepsilon\| \leq c_0 \, \Delta(\delta, \varepsilon, u_{\mathrm{m}}).$$

In the sequel we shall take Ω of the form $\mathscr{O} \times \omega_2$, small enough for us to have $\Phi(\varepsilon, \Phi(\delta, u_{\mathrm{m}})) \in \omega_1$ when $(\delta, \varepsilon, u_{\mathrm{m}}) \in \Omega$.

A linear functional

If $\|\bar{u} - \hat{u}\|_\infty$ is small enough, $\bar{u}$ has values in ω_2 and we are able to start to put Glimm's scheme into operation. The ratio $\rho = \Delta t / \Delta x$ is fixed so that $\rho V^{\omega_1} < 1$ and the scheme stops if the approximate solution leaves ω_2. One of our aims is to show that it does not leave if the given initial state is sufficiently close to $\hat{u}$ in $\mathrm{BV}(\mathbb{R})$. At the first iteration, we still have $u_{1,k} \in \omega_1$ since $u_{0,k} \in \omega_2$. As long as $u_a^h(n\Delta t)$ is defined with values in ω_2, we denote by $\varepsilon(n, k)$, $\delta(n, k)$, $\theta(n, k)$ and $\Gamma(n, k) = \theta(n + 1, k)$ the respective solutions of $\Phi(\varepsilon, u_{n+1,k}) = u_{n,k+1}$, $\Phi(\delta, u_{n,k-1}) = u_{n+1,k}$, $\Phi(\theta, u_{n,k-1}) = u_{n,k+1}$, and $\Phi(\Gamma, u_{n+1,k-1}) = u_{n+1,k+1}$. Since $u_{n+1,k}$ is an intermediate state of the Riemann problem between $u_{n,k-1}$ and $u_{n,k+1}$ (this is the sampling principle) we have $\theta(n, k) = \delta(n, k) + \varepsilon(n, k)$ and likewise

$$\left.\begin{array}{lll}\Delta(\delta(n,k),\varepsilon(n,k))=0, & & \\ |\theta_p| & =|\delta_p|+|\varepsilon_p|, & 1\le p\le n, \\ \theta_p^- & =\delta_p^-+\varepsilon_p^-, & p\in \text{GNL}.\end{array}\right\} \tag{5.10}$$

In addition $\Gamma(n,k)=\gamma(\varepsilon(n,k-1),\delta(n,k+1),u_{n,k})$. In future, we shall omit the last argument which is always easy to identify. We define a functional $L(n)$ of which we shall show the equivalence with the total variation of $u_a^h(n\Delta t)$:

$$L(n):=\sum_{k\in n+1+2\mathbb{Z}}\|\theta(n,k)\|.$$

In fact, when u_{R} and u_{L} are in ω_1 with $u_{\mathrm{R}}=\Phi(\varepsilon;u_{\mathrm{L}})$, we have $\varepsilon\sim P(\hat{u})\cdot(u_{\mathrm{R}}-u_{\mathrm{L}})$ where $P(v)^{-1}$ is the matrix whose columns are the eigenvectors of $\mathrm{d}f(v)$, normalised by $\mathrm{d}\lambda_p\cdot r_p\equiv 1$ if $p\in$ GNL. There thus exists a constant $C>1$ such that

$$C^{-1}\|u_{\mathrm{R}}-u_{\mathrm{L}}\|\le\|\varepsilon\|\le C\|u_{\mathrm{R}}-u_{\mathrm{L}}\|,\qquad \forall u_{\mathrm{R}},u_{\mathrm{L}}\in\omega_1.$$

Since $u_{n,k-1}$ and $u_{n,k+1}$ have values in ω_2, the Riemann problem between these states has a solution with values in ω_1 and the above inequality applies to all the quantities $(\delta,\varepsilon,\theta,\Gamma)(n,k)$. The total variation of $u_a^h(n\Delta t)$, by breaking it up on transverse waves (Chasles' relation), is bounded above by

$$C\sum_{p=1}^{n}|\theta_p(n,k)|=C\|\theta(n,k)\|$$

by choosing the norm l^1 on $\mathbb{R}^2$. We deduce that

$$C^{-1}\mathrm{TV}\big(u_a^h(t)\big)\le L(n)\le C\mathrm{TV}\big(u_a^h(t)\big)$$

for $n\Delta t\le t\le(n+1)\Delta t$, and similarly

$$C^{-1}\mathrm{TV}\big(u_a^h((n+1)\Delta t)\big)\le L(n+1)\le C\mathrm{TV}\big(u_a^h((n+1)\Delta t)\big).$$

In particular, $L(0)\le C\mathrm{TV}(\bar{u})$.

From now on, looking to effect an induction from n to $n+1$, we omit the argument n in $(\varepsilon,\delta,\theta,\Gamma)$ and we suppose the sequence $(u_{n,k})_{k\in n+2\mathbb{Z}}$ takes values in ω_2. We shall bound $L(n+1)$ above.

$$L(n+1) = \sum_{k\in n+2\mathbb{Z}} \|\Gamma(k)\| = \sum_{k\in n+2\mathbb{Z}} \|\gamma(\varepsilon(k-1), \delta(k+1))\|$$
$$\leq \sum_{k\in n+2\mathbb{Z}} (\|\varepsilon(k-1)\| + \|\delta(k+1)\| + c_0\Delta(\varepsilon(k-1), \delta(k+1))).$$

Reordering the terms and with (5.10), we find

$$L(n+1) \leq L(n) + c_0\Lambda(n), \tag{5.11}$$

where we have defined

$$\Lambda(n) := \sum_{k\in n+2\mathbb{Z}} \Delta(\varepsilon(k-1), \delta(k+1)).$$

Obviously, the presence of the positive term $c_0\Lambda(n)$ on the right-hand side does not allow us to come to a conclusion.

A quadratic functional

We now introduce the interaction potential

$$Q(n) := \sum_{\substack{j,k\in n+1+2\mathbb{Z};\\ j<k}} \Delta(\theta(j), \theta(k)).$$

In particular

$$\Lambda(n) \leq Q(n). \tag{5.12}$$

Because of the formulae (5.10) and although Δ is not bilinear we can in any case develop

$$\Delta(\theta(j), \theta(k)) = \Delta(\varepsilon(j), \varepsilon(k)) + \Delta(\varepsilon(j), \delta(k))$$
$$+ \Delta(\delta(j), \varepsilon(k)) + \Delta(\delta(j), \delta(k)). \tag{5.13}$$

Since Δ is sub-additive, we have also

$$\Delta(\Gamma(j), \Gamma(k)) \leq \Delta(\delta(j+1), \delta(k+1)) + \Delta(\delta(j+1), \varepsilon(k-1))$$
$$+ \Delta(\delta(j+1), \rho(k)) + \Delta(\varepsilon(j-1), \delta(k+1))$$
$$+ \Delta(\varepsilon(j-1), \varepsilon(k-1))$$
$$+ \Delta(\varepsilon(j-1), \rho(k)) + \Delta(\rho(j), \Gamma(k)),$$

where we have denoted by $\rho(j)$ the difference $\Gamma(j) - \varepsilon(j-1) - \delta(j+1)$. Reordering the terms, we thus have

$$\begin{aligned}
Q(n+1) \leq & \sum_{\substack{j,k\in n+1+2\mathbb{Z};\\ j<k}} \{\Delta(\delta(j),\delta(k)) + \Delta(\delta(j),\varepsilon(k)) + \Delta(\delta(j),\rho(k-1)) \\
& \qquad + \Delta(\varepsilon(j),\delta(k)) + \Delta(\varepsilon(j),\varepsilon(k)) + \Delta(\varepsilon(j),\rho(k+1)) \\
& \qquad + \Delta(\rho(j+1),\Gamma(k+1))\} \\
& + \sum_{j\in n+1+2\mathbb{Z}} \{\Delta(\delta(j),\varepsilon(j)) - \Delta(\varepsilon(j-2),\delta(j))\} \\
= & \; Q(n) - \Lambda(n) \\
& + \sum_{j<k} \{\Delta(\delta(j),\rho(k-1)) + \Delta(\varepsilon(j),\rho(k+1)) \\
& + \Delta(\rho(j+1),\Gamma(k+1))\},
\end{aligned}$$

as $\Delta(\delta(j),\varepsilon(j)) = 0$. The following three lines are elementary applications of Lemma 5.5.2.

$$\begin{aligned}
\Delta(\delta(j),\rho(k-1)) &\leq \|\delta(j)\|\,\|\rho(k-1)\| \leq c_0\|\delta(j)\|\Delta(\delta(k),\varepsilon(k-2)), \\
\Delta(\varepsilon(j),\rho(k+1)) &\leq c_0\|\varepsilon(j)\|\Delta(\delta(k+2),\varepsilon(k)), \\
\Delta(\rho(j+1),\Gamma(k+1)) &\leq c_0\|\Gamma(k+1)\|\Delta(\delta(j+2),\varepsilon(j)).
\end{aligned}$$

Bringing all these together, we arrive at

$$Q(n+1) \leq Q(n) + \Lambda(n)\big(2c_0 L(n) + c_0^2\Lambda(n) - 1\big). \tag{5.14}$$

The 'good' functional, which permits us to estimate *a priori* the approximate solution if $\mathrm{TV}(\bar{u}) + \|\bar{u} - \hat{u}\|_\infty$ is sufficiently small, is $\mathscr{L} := L + 2c_0 Q$. We recall that, $\hat{u}$ being fixed, ω, ω_1 and ω_2 have been chosen *a priori*, but compact, and c_0 depends only on this construction. Taking (5.11), (5.14) and (5.12) into account we obtain the upper bound

$$\mathscr{L}(n+1) \leq \mathscr{L}(n) + c_0\Lambda(n)\{4c_0\mathscr{L}(n) - 1\}, \tag{5.15}$$

provided that $(u_{n,k})_{k\in n+2\mathbb{Z}}$ has values in ω_2. It is important to note that we do not suppose that the states $u_{n+1,j}$ are in ω_2.

The induction

We now choose a number α which satisfies the following two conditions:

$\alpha < \frac{1}{8c_0 C}$

the ball $B(\hat{u}; (1 + 5C^2/4)\alpha)$ is included in ω_2.

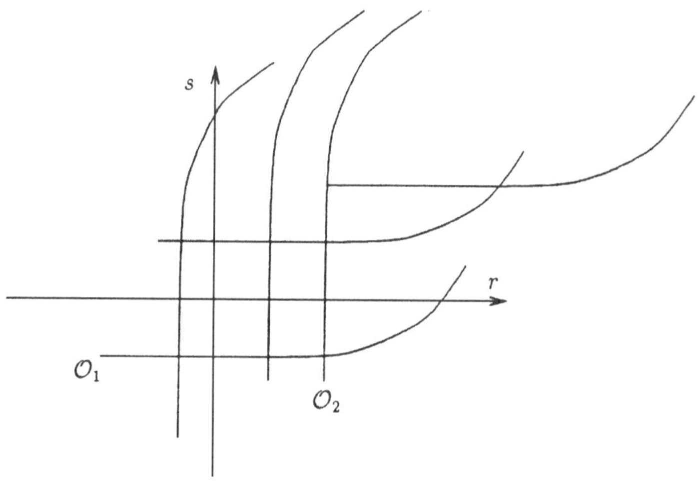

Fig. 5.2: Wave curves of a system 'à la Nishida'.

Finally, we suppose that the initial condition is close to the constant state $\hat{u}$, in the sense that

$$\|\bar{u} - \hat{u}\|_\infty + \mathrm{TV}(\bar{u}) \le \alpha. \tag{5.16}$$

We have $L(0) \le C\alpha$ and hence $\mathscr{L}(0) \le L(0)(1 + 2c_0 L(0)) \le \frac{5}{4}C\mathrm{TV}(\bar{u}) \le \frac{5}{4}C\alpha$ because $Q(n) \le L(n)^2$ since $\Delta(\delta, \varepsilon) \le \|\delta\|\,\|\varepsilon\|$. In particular $\mathscr{L}(0) \le 1/4c_0$.

Let us make the induction hypothesis $\mathscr{L}(n) \le \mathscr{L}(0)$. Then $\mathscr{L}(n) \le 1/4c_0$ and therefore

$$\sum_{k \in n+1+2\mathbb{Z}} \|u_{n,k+1} - u_{n,k-1}\| \le CL(n) \le C\mathscr{L}(n) \le C\mathscr{L}(0) \le \frac{5C^2\alpha}{4}.$$

As $\lim_{k\to-\infty} u_{n,k} = \lim_{x\to-\infty} \bar{u}(x)$ and $\|\bar{u}(-\infty) - \hat{u}\| \le \alpha$, we have $\|u_{n,k} - \hat{u}\| \le (1 + 5C^2/4)\alpha$ and hence $u_{n,k} \in \omega_2$. The approximate solution is thus again defined up to $(n+1)\Delta t$ and the upper bound (5.15) is valid. As $\mathscr{L}(n) \le 1/4c_0$, we have $\mathscr{L}(n+1) \le \mathscr{L}(0)$, which proves the induction.

Under the hypothesis (5.16), we thus have $\mathscr{L}(n) \le \mathscr{L}(0)$ for all $n \ge 0$, from which we derive

$$\mathrm{TV}\big(u_a^h(n\Delta t)\big) \le \frac{5C^2}{4}\mathrm{TV}(\bar{u}),$$
$$\big\|u_a^h(n\Delta t) - \hat{u}\big\|_\infty \le \left(1 + \frac{5C^2}{4}\right)(\|\bar{u} - \hat{u}\|_\infty + \mathrm{TV}(\bar{u})).$$

The analogous estimates for $n\Delta t < t < (n+1)\Delta t$ are elementary and are left to the reader. This completes the proof of Theorem 5.2.2 and hence that of Theorem 5.2.1.

5.6 The example of Nishida

This section is devoted to the study of certain systems for which we can show the stability of Glimm's scheme without supposing that the initial datum is small. Since the stability involves the convergence for almost all $a \in A$, we deduce a global existence theorem for a weak solution for the Cauchy problem without other hypothesis than $\mathrm{TV}(\bar{u}) < +\infty$.

Hypotheses and theorem

Since the initial datum is of arbitrarily large variation, the wave curves must be defined globally in order that we can solve all the Riemann problems.

We only consider 2×2 strictly hyperbolic systems ($n = 2$), with characteristic velocities $u \mapsto \lambda_j(u)$, $j = 1, 2$, $\lambda_1 < \lambda_2$. We introduce the Riemann invariants $u \mapsto r(u)$ and $u \mapsto s(u)$ which are the solutions without critical points of the differential equations

$$\mathrm{d}r\,(\mathrm{d}f - \lambda_1) = 0,$$
$$\mathrm{d}s\,(\mathrm{d}f - \lambda_2) = 0.$$

We suppose that $u \mapsto (r, s)$ is a diffeomorphism of $\mathscr{U}$ on $\mathbb{R}^2$. The class of systems which we consider is such that the wave curves $\mathscr{O}_j(v)$, $j = 1, 2$, are connected, allowing the unique solvability of the Riemann problem and, last but not least, deducing the one from the other by the translations $(r, s) \mapsto (r + r_0, s + s_0)$ and the symmetries $(r, s) \mapsto (s + r_0, r + s_0)$.

To ensure the connectedness of the wave curves, we are led to suppose the characteristic fields are genuinely non-linear or else linearly degenerate (the case of a scalar equation is sufficiently instructive). Each wave curve $\mathscr{O}_j(v)$ contains thus a half straight-line ending in v, parallel to the axis $s = 0$ if $j = 1$, $r = 0$ otherwise. From the uniqueness of the solution of the Riemann problem, we deduce that $\mathscr{O}_1(v)$ is transversal to the straight lines $r =$ const. In fact, being given a and b in $\mathscr{O}_1(v)$ with $r_a = r_b$, these two states would be linked by a 2-rarefaction-wave, for example in the direction $a \mapsto b$ (or by a 2-contact-discontinuity-wave). The Riemann problem between v and b would therefore have two distinct solutions

$$v \overset{\text{1-S}}{\longmapsto} b,$$

and

$$v \overset{1\text{-S}}{\longmapsto} a \overset{2\text{-R}}{\longmapsto} b,$$

which contradicts the uniqueness (here S stands for shock, and R for rarefaction waves).

The curves $\mathcal{O}_1$ are thus parametrised by r and the property of translation shows that there exists a smooth function $f\colon \mathbb{R} \to \mathbb{R}$ with for example $f(\mathbb{R}^-) = \{0\}$ such that $\mathcal{O}_1(v)$ is given by the equation

$$s - s_v = f(r - r_v).$$

By symmetry the 2-wave curves $\mathcal{O}_2(v)$ are given by the equation

$$r - r_v = f(s - s_v).$$

Again, the uniqueness of the solution of the Riemann problem implies $|f(\tau) - f(\sigma)| \neq |\tau - \sigma|$ for $\tau \neq \sigma$. In fact, if $f(\tau) - f(\sigma) = \tau - \sigma$, then the Riemann problem between u_L and u_R ($r_L = s_R = 0$, $s_L = r_R = \tau - f(\tau)$) admits two solutions

$$u_L \overset{1\text{-S}}{\longmapsto} u_m(r = s = \tau) \overset{2\text{-S}}{\longmapsto} u_R,$$
$$u_L \overset{1\text{-S}}{\longmapsto} u^m(r = s = \sigma) \overset{2\text{-S}}{\longmapsto} u_R.$$

Similarly, if $f(\tau) - f(\sigma) = \sigma - \tau$, the Riemann problem between u_L and u_R ($r_L = s_R = 0$, $s_L = r_R = \tau - f(\sigma)$) admits two solutions

$$u_L \overset{1\text{-S}}{\longmapsto} u_m(r = \tau, s = \sigma) \overset{2\text{-S}}{\longmapsto} u_R,$$
$$u_L \overset{1\text{-S}}{\longmapsto} u^m(r = \sigma, s = \tau) \overset{2\text{-S}}{\longmapsto} u_R.$$

Since $f(\mathbb{R}^-) = \{0\}$ we deduce by continuity that $|f(\tau) - f(\sigma)| < |\tau - \sigma|$ for $\tau \neq \sigma$. Putting $w = \frac{1}{2}(r + s)$ and $z = \frac{1}{2}(r - s)$, we can rewrite the equations of 1-wave curves in the form

$$w - w_- = F(z - z_-)$$

and similarly the 2-wave curves as

$$w - w_- = F(z_- - z).$$

We have $F' = (1 + f')/(1 - f')$, with the result that F is strictly increasing. The solution of the Riemann problem leads to eliminating the component w_m of the median state u_m, and to searching for the unique root z_m of the equation

$$F(z - z_R) + F(z - z_L) = w_R - w_L.$$

The last hypothesis is the inequality

$$F(z_1 + z_2) \geq F(z_1) + F(z_2), \qquad \forall z_1, z_2 \geq 0. \tag{5.17}$$

Remarking that $F(z) = z$ for $z \leq 0$, we have also

$$F(z_1 + z_2) \leq F(z_1) + F(z_2), \qquad \forall z_1, z_2 \leq 0. \tag{5.18}$$

We are now able to state the result which Nishida [80] has obtained for the isothermal model of gas dynamics. This example will be the object of §5.6.

Theorem 5.6.1 *For a system of two conservation laws such as are described above, the scheme of Glimm is stable in* BV($\mathbb{R}$) *for every given initial condition* $\bar{u} \in$ BV($\mathbb{R}$)^2. *The families* $(u_a^h)_{h>0}$ *of approximate solutions are relatively compact in* $L^1_{\text{loc}}(\mathbb{R}^+ \times \mathbb{R})$ *and their limiting values are for almost all* $a \in A$ *weak entropy solutions of the Cauchy problem.*

Of course, only the stability has to be proved.

A distance in $\mathcal{U}$

Being given two states $a, b \in \mathcal{U}$, we define a 'distance' $d(a, b) := |z_{\text{m}} - z_a| + |z_{\text{m}} - z_b|$ where $u_{\text{m}} = (w_{\text{m}}, z_{\text{m}})$ is the median state in the Riemann problem between a and b:

$$F(z_{\text{m}} - z_a) + F(z_{\text{m}} - z_b) = w_b - w_a. \tag{5.19}$$

In fact, d is not symmetric. Of the two properties of d of which we shall have need, the first is the triangular inequality.

Lemma 5.6.2 *If* $a_j \in \mathcal{U}$, $j = 1, 2, 3$, *we have*

$$d(a_1, a_3) \leq d(a_1, a_2) + d(a_2, a_3).$$

The proof of this lemma has remained complex for a long time, necessitating the study of fifteen or so cases, rarely presented in an exhaustive way, until done so elegantly by F. Poupaud [84].

Proof Let a_{ij} be the median state between a_i and a_j. Eliminating w from the equations (5.19), we obtain

$$F(z_{13} - z_1) + F(z_{13} - z_3) = F(z_{12} - z_1) + F(z_{12} - z_2) + F(z_{23} - z_2) + F(z_{23} - z_3).$$

We let $x = z_{13} - z_1$, $y = z_{13} - z_3$, $\alpha = z_{12} - z_1$, $\beta = z_{12} - z_2$, $\gamma = z_{23} - z_2$,

$\delta = z_{23} - z_3$. The following equalities arise:

$$x - y = \alpha - \beta + \gamma - \delta, \tag{5.20}$$

$$F(x) + F(y) = F(\alpha) + F(\beta) + F(\gamma) + F(\delta). \tag{5.21}$$

If $xy \leq 0$, then

$$\begin{aligned} d(a_1, a_3) &= |x| + |y| = |x - y| = |\alpha - \beta + \gamma - \delta| \\ &\leq |\alpha| + |\beta| + |\gamma| + |\delta| = d(a_1, a_2) + d(a_2, a_3). \end{aligned}$$

If x and y are positive, we write that $F(\alpha^+) + F(\alpha^-) = F(|\alpha|) + F(0) \geq F(\alpha)$ with $\alpha^+ = \max(\alpha, 0)$, and $\alpha^- = \max(-\alpha, 0)$. We deduce from (5.21) that one of the two following inequalities is true:

$$F(x) \leq F(\alpha^+) + F(\beta^-) + F(\gamma^+) + F(\delta^-)$$

or

$$F(y) \leq F(\alpha^-) + F(\beta^+) + F(\gamma^-) + F(\delta^+).$$

If the first holds (the two cases are similar) then $F(x) \leq F(\alpha^+ + \beta^- + \gamma^+ + \delta^-)$ because of (5.17). Since F is strictly increasing, we have $x \leq \alpha^+ + \beta^- + \gamma^+ + \delta^-$. Finally,

$$\begin{aligned} d(a_1, a_3) &= x + y = 2x - (x - y) \\ &\leq 2(\alpha^+ + \beta^- + \gamma^+ + \delta^-) - \alpha + \beta - \gamma + \delta \\ &= |\alpha| + |\beta| + |\gamma| + |\delta| = d(a_1, a_2) + d(a_2, a_3). \end{aligned}$$

The case $x \leq 0$, $y \leq 0$ is treated in the same way using (5.18).

Simpler is the following case of an equality. ■

Lemma 5.6.3 *If a_2 is an intermediate value in the solution of the Riemann problem between a_1 and a_3, then $d(a_1, a_3) = d(a_1, a_2) + d(a_2, a_3)$.*

In fact, $d(a, b)$ is nothing but the total variation of $x/t \mapsto z$ in the solution of the Riemann problem between a and b and we can express it by Chasles' relation.

Finally, d is equivalent to the usual distance of $\mathcal{U}$.

Lemma 5.6.4 *Let K be a compact set of $\mathbb{R}^2$ and C its image by $(w, z) \mapsto u$. There exists a number $c_K \geq 1$ such that for all $a, b \in C$, we have*

$$c_K^{-1} \|b - a\| \leq d(a, b) \leq c_K \|b - a\|.$$

Proof First of all, C is a compact set and

$$0 < m = \min F'(z) \leq M = \max F'(z) < +\infty$$

for the zs of the form $z_+ - z_-$ with $u_\pm \in C$. In each wave, we thus have

$$N^{-1}|z_+ - z_-| \leq |w_+ - w_-| \leq N|z_+ - z_-|,$$

with $N = \max(M, m^{-1})$. We therefore have

$$d(a, b) \geq \frac{1}{1 + N_1}(|z_{\mathrm{m}} - z_a| + |w_{\mathrm{m}} - w_a| + |z_{\mathrm{m}} - z_b| + |w_{\mathrm{m}} - w_b|)$$

where N_1 is the same number but associated with a compact set C_1 which contains all the median states of the Riemann problems between the points a and b of C. Thus

$$d(a, b) \geq \frac{1}{1 + N_1}(|z_b - z_a| + |w_b - w_a|) \geq \text{const.}\|b - a\|$$

since $(w, z) \mapsto u$ is Lipschitz of constant K in C.

Conversely, the mapping $(r_a, s_a, r_b, s_b) \mapsto (r_{\mathrm{m}}, s_{\mathrm{m}})$ is of class $\mathscr{C}^2$ with $(r_{\mathrm{m}}, s_{\mathrm{m}}) = (r_b, s_a) + \mathrm{O}(\|b - a\|^2)$ when b tends to a, from Lax's theorem. We thus have $z_{\mathrm{m}} = \frac{1}{2}(r_b - s_a) + \mathrm{O}(\|b - a\|^2)$ and

$$d(a, b) = \frac{1}{2}(|r_b - r_a| + |s_b - s_a|) + \mathrm{O}(\|b - a\|^2) \leq \text{const.}\|b - a\|^2$$

on the compact set C. ∎

Stability

We make use of a functional slightly different from that of the general case:

$$M(n) := \sum_{k \in n+1+2\mathbb{Z}} d(u_{n,k-1}, u_{n,k+1}).$$

By the triangle inequality, we have

$$M(n + 1) \leq \sum_{k \in n+2\mathbb{Z}} (d(u_{n+1,k-1}, u_{n,k}) + d(u_{n,k}, u_{n+1,k+1})).$$

Let us reorder the terms of this sum:

$$M(n + 1) \leq \sum_{k \in n+1+2\mathbb{Z}} (d(u_{n,k-1}, u_{n+1,k}) + d(u_{n+1,k}, u_{n,k+1})).$$

But as $u_{n+1,k}$ is an intermediate value in the solution of the Riemann problem between $u_{n,k-1}$ and $u_{n,k+1}$, the right-hand side takes the value $\sum_k d(u_{n,k-1}, u_{n,k+1})$

from Lemma 5.6.3. We therefore have

$$M(n+1) \leq M(n).$$

By induction, while the approximate solution is defined, we have $M(n) \leq M(0)$, that is $\mathrm{TV}(z_a^h(\cdot, n\Delta t + 0)) \leq \mathrm{TV}(\bar{z})$. As $z_{n,-\infty} = \bar{z}(-\infty)$, we deduce

$$|z_{n,k}| \leq |\bar{z}(-\infty)| + \mathrm{TV}(\bar{z}) := \tilde{z}.$$

Let us write $\tilde{c} = \max\{|F'(z)|; |z| \leq \tilde{z}\}$. In each j-wave, we have $\mathrm{TV}(w) \leq \tilde{c}\,\mathrm{TV}(z)$ and hence $\mathrm{TV}(w_a^h(\cdot, n\Delta t + 0)) \leq \tilde{c}\,\mathrm{TV}(\bar{z})$. Thus

$$|w_{n,k}| \leq |\bar{w}(-\infty)| + \tilde{c}\,\mathrm{TV}(\bar{z}),$$

which shows that u_a^h takes its values in a compact set C which does not depend on h or on a or on n or on ρ. We can thus choose ρ *a priori* for the approximate solution to be defined for all time. In addition, the mapping $u \mapsto (w, z)$ being bi-Lipschitz on the compact set in question, we have

$$\begin{aligned}\mathrm{TV}\big(u_a^h(\cdot, n\Delta t + 0)\big) &\leq c_C\big(\mathrm{TV}\big(w_a^h\big) + \mathrm{TV}\big(z_a^h\big)\big)\\ &\leq (1+\tilde{c})c_C\mathrm{TV}(\bar{z}) \leq c\mathrm{TV}(\bar{u}),\end{aligned}$$

which completes the proof of Nishida's theorem.

The isothermal model of gas dynamics

An isothermal gas has for equation of state, with the standard notation, $p(\rho) = c^2\rho$ where c is the constant speed of sound. By a change of scale, we can suppose that $c = 1$. In lagrangian coordinates, the one-dimensional motion is governed by the equations

$$\left.\begin{aligned} v_t &= q_x,\\ q_t + \left(\frac{1}{v}\right)_x &= 0, \end{aligned}\right\} \tag{5.22}$$

v being the specific volume and q the velocity of the gas. The characteristic speeds are $\lambda = \pm v^{-1}$ while the Riemann invariants are

$$r = q + \log v, \qquad s = q - \log v.$$

Here, $w = q$ and $z = \log v$. In a shock the Rankine–Hugoniot condition is written

$$[q] + \sigma[v] = 0, \qquad \left[\frac{1}{v}\right] = \sigma[q]$$

and hence $v_-v_+\sigma^2 = 1$. In a 1-shock-wave, we have $\sigma = -(v_-v_+)^{-\frac{1}{2}}$ and therefore

$$w_+ - w_- = \sqrt{\frac{v_+}{v_-}} - \sqrt{\frac{v_-}{v_+}}$$

with $v_+ < v_-$ as the entropy condition. On the other hand, in a 1-rarefaction-wave, s is constant, that is to say that $w_+ - w_- = \log(v_+/v_-)$ and we have $v_2 > v_1$. Thus the 1-wave curve is given by $w_+ - w_- = F(z_+ - z_-)$ with

$$F(z) = \begin{cases} z, & z \geq 0, \\ 2\sinh\dfrac{z}{2}, & z \leq 0. \end{cases}$$

In a 2-shock-wave, we have $\sigma = (v_-v_+)^{-\frac{1}{2}}$,

$$w_+ - w_- = \sqrt{\frac{v_-}{v_+}} - \sqrt{\frac{v_+}{v_-}}$$

and $v_+ > v_-$ as the entropy condition. Similarly, in a 2-rarefaction-wave, r is constant, that is to say $w_+ - w_- = \log(v_-/v_+)$ and we have $v_+ < v_-$ and thus again $w_+ - w_- = F(z_- - z_+)$. The wave curves have thus the form demanded and we have clearly $F' > 0$. In fact $f' = (F' - 1)/(F' + 1) \in (-1, 1)$. It remains to verify that $F(a+b) \geq F(a) + F(b)$ for $a, b \geq 0$ which is trivial, and $F(a+b) \leq F(a) + F(b)$ for $a, b \leq 0$, which is classic:

$$\begin{aligned} F(a+b) &= 2\left(\sinh\frac{a}{2}\cosh\frac{b}{2} + \sinh\frac{b}{2}\cosh\frac{a}{2}\right) \\ &\leq 2\left(\sinh\frac{a}{2} + \sinh\frac{b}{2}\right) = F(a) + F(b), \end{aligned}$$

where we have used $\sinh \leq 0$, $\cosh \geq 1$.

As an application of Theorem 5.6.1, we therefore have

Theorem 5.6.5 **(Nishida)** *Let* $\bar{v} \in \mathrm{BV}(\mathbb{R})$ *and* $q \in \mathrm{BV}(\mathbb{R})$ *be such that* $\inf_x v(x) > 0$. *Then the Cauchy problem for the system (5.22) possesses a weak entropy solution which satisfies* $v(x,t) \geq v^*$ *where* v^* *is an explicitly calculable constant.*

In fact, $v^* = \exp(-\bar{z})$ with the notation of the preceding section. More precisely,

$$v^* = \bar{v}(-\infty)\exp(-\mathrm{TV}(\log\,\bar{v})).$$

Of course, this value is not, in general, an optimal bound.

5.7 2 × 2 Systems with diminishing total variation

Description

We shall consider strictly hyperbolic systems, whose wave curves have for equations $r = \text{const.}$ (respectively $s = \text{const.}$), r and s being the Riemann invariants. These systems were studied for the first time by B. Temple [103]. This property is satisfied at least by the wave curves which correspond to a linearly degenerate field. As the appropriate Riemann invariant is likewise constant across a rarefaction wave, we see that the hypothesis concerns only shock waves.

The solution of the Riemann problem is particularly simple in this case. Let us consider a characteristic quadrilateral $K \subset \mathcal{U}$ defined by

$$K = \{u \in \mathcal{U} : r_- \leq r(u) \leq r_+, s_- \leq s(u) \leq s_+\}$$

and which is complete, that is to say that

$$K \to [r_-, r_+] \times [s_-, s_+], \qquad u \mapsto (r(u), s(u)),$$

is a diffeomorphism. Then for u_{L} and u_{R} in K, the Riemann problem has a unique solution with values in K, for which the median state u_{m} is determined by $r_{\mathrm{m}} = r_{\mathrm{R}}$, $s_{\mathrm{m}} = s_{\mathrm{L}}$.

In particular, the remark of Hoff (see §5.2) shows that if the Cauchy datum $\bar{u}$ has values in K, there exists a value ρ of the ratio $\rho = \Delta t / \Delta x$ such that u_a^h is defined for all time and has values in K.

Stability

To study the stability of Glimm's scheme, we consider the functionals

$$V_1(t) = \mathrm{TV}\big(r \circ u_a^h(t); \mathbb{R}\big),$$
$$V_2(t) = \mathrm{TV}\big(s \circ u_a^h(t); \mathbb{R}\big).$$

If $n\Delta t < s, t < (n+1)\Delta t$, $u_a^h(s)$ and $u_a^h(t)$ differ only by a diffeomorphism of $\mathbb{R}$,

$$u_a^h(s, x) = u_a^h(t, \psi_{s,t}(x))$$

with the result that $V_j(s) = V_j(t)$. In addition, sampling is an operation which diminishes the total variation:

$$V_j(n\Delta t) \leq V_j(n\Delta t - 0).$$

After these two remarks which do not make use of the particular structure of the system we calculate $V_j(n\Delta t + 0)$. For $j = 1$, this is the sum of the variations of

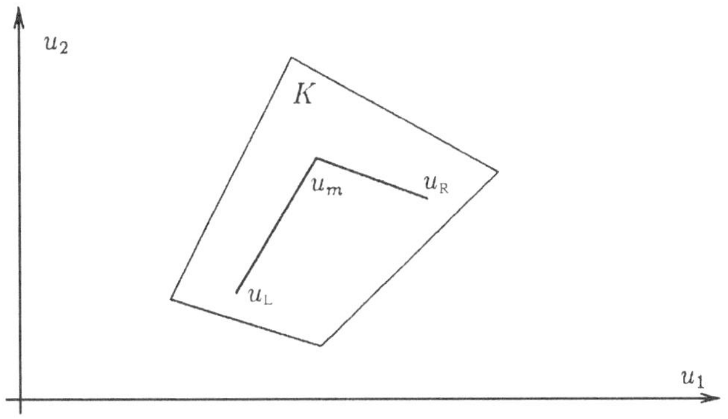

Fig. 5.3: The Riemann problem for a system 'of B. Temple'.

$r \circ u_a^h$ in the Riemann problems solved at the instant $n\Delta t$. As r is constant in the 2-waves, $V_1(n\Delta t + 0)$ is the sum of the variations of r in the 1-waves. Finally r is monotonic in the 1-waves (this is true for very general systems), for example in supposing that a 1-wave is either a shock wave, or a rarefaction wave, or a contact discontinuity, as r is monotonic in a rarefaction, Finally, the variation of r in a Riemann problem is $|r_R - r_L|$ and we deduce that $V_1(n\Delta t + 0) = V_1(n\Delta t)$ and similarly for V_2. Thus $t \mapsto V_j(t)$ is decreasing:

$$V_j(t) \leq V_j(0) \in \{\mathrm{TV}(r \circ \bar{u}), \mathrm{TV}(s \circ \bar{u})\}.$$

Finally, C being a Lipschitz constant on K of the diffeomorphism $u \mapsto (r, s)$ and its inverse, we have the estimate

$$\begin{aligned} \mathrm{TV}\big(u_a^h(t)\big) &\leq C\mathrm{TV}\big((r, s) \circ u_a^h(t)\big) \\ &\leq C\mathrm{TV}((r, s) \circ \bar{u}) \leq C^2\mathrm{TV}(\bar{u}), \end{aligned}$$

which shows the stability of Glimm's scheme in $\mathrm{BV}(\mathbb{R})$. We have seen that this entails the convergence. Let us state the result.

Theorem 5.7.1 (Leveque and Temple, Serre) *We suppose that the integral curves of the eigenvector fields of* $\mathrm{d}f$ *are the wave curves of the system (5.1). Let* K *be a complete characteristic quadrilateral in* $\mathcal{U}$.

For all $\bar{u} \in \mathrm{BV}(\mathbb{R})^2$ *with values in* K*, the Cauchy problem has a weak entropy solution with values in* K *and which satisfies*

$$\mathrm{TV}(r \circ u(t)) \leq \mathrm{TV}(r \circ \bar{u}),$$
$$\mathrm{TV}(s \circ u(t)) \leq \mathrm{TV}(s \circ \bar{u}).$$

Example 5.7.2 The following system has been considered by numerous authors, for example [52]:

$$u_t + (\varphi(u)u)_x = 0, \tag{5.23}$$

where $u = (u_1, u_2)$ and $\varphi \in \mathscr{C}^2(\mathscr{U} = \mathbb{R}^+ \times \mathbb{R}; \mathbb{R})$. We write $r = \|u\|$ (the Riemann invariants will be denoted differently) and we suppose that $\varphi_r > 0$ (but $\varphi_r < 0$ could also arise) with $r\partial_r := u_1\partial_1 + u_2\partial_2$. Finally, we suppose that $\lim_{r\to+\infty} \varphi(u) = +\infty$. Thus, (θ, φ): $\mathscr{U} \to [-\frac{1}{2}\pi, \frac{1}{2}\pi] \times (\varphi(0), +\infty)$ realises a diffeomorphism.

The system (5.23) is strictly hyperbolic, with characteristic speeds $\lambda_1 = \varphi$, $\lambda_2 = \varphi + r\varphi_r$. The first characteristic field is linearly degenerate while the second is genuinely non-linear when $(r\varphi)_{rr} \neq 0$.

The Riemann invariants of this system are the angle $\theta = \arctan(u_2/u_1)$ and the function φ. We have seen that only the shock waves have to be considered. These are relative to the second characteristic field and satisfy the Rankine–Hugoniot condition

$$[(\varphi(u) - \sigma)u] = 0,$$

from which we deduce either $u_{\mathrm{L}} \wedge u_{\mathrm{R}} = 0$, or $\varphi_{\mathrm{L}} = \varphi_{\mathrm{R}} = \sigma$. But this latter case is that of a contact discontinuity. The shocks therefore satisfy $u_{\mathrm{L}} \wedge u_{\mathrm{R}} = 0$, that is to say $\theta_{\mathrm{L}} = \theta_{\mathrm{R}}$. The system thus satisfies the hypotheses of the preceding section. If the given initial condition is of bounded variation and with values in $\mathscr{U}^*$,

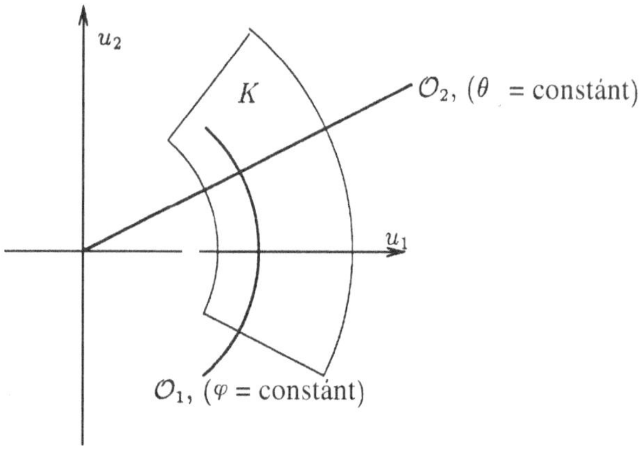

Fig. 5.4: Wave curves and invariant domain for the system (5.23).

if besides $\inf_x \bar{r}(x) > 0$, there exists a complete characteristic quadrilateral K defined by

$$|\theta| \leq \frac{1}{2}\pi, \qquad \inf_x \varphi(u) \leq \varphi \leq \sup_x \varphi(u),$$

which contains the values of $\bar{u}$ (cf. Fig. 5.4). The compact set K is invariant for the Riemann problem and thus for Glimm's scheme. By Theorem 5.7.1, this one converges. The Cauchy problem therefore has a weak entropy solution with values in K for almost all $(t, x) \in \mathbb{R}^+ \times \mathbb{R}$.

In $\inf_x r(x) = 0$, the situation is more delicate as the hypothesis does not ensure that $\bar{\theta}$ is of bounded variation, $u \mapsto \theta$ not being Lipschitz. On the other hand, as $(\theta, \varphi) \mapsto u$ is Lipschitz, it is sufficient to consider a given initial function $\bar{u}$ for which $\bar{\theta}$ and $\bar{\varphi}$ are of bounded variation to obtain the convergence of Glimm's scheme and the existence of a solution of the Cauchy problem.

5.8 Technical lemmas

Proof of Lemma 5.5.2

We begin by stating an intermediary result, the inspiration for which comes from a course at Rennes given by G. Métivier.

Lemma 5.8.1 *Let I be a part of $\{1, \ldots, m\} \times \{1, \ldots, n\}$ and $f \in \mathscr{C}_b^2([0, +\infty)^{m+n})$. If f is identically zero when $\Pi_I(x, y) := \sum_{(i,j)\in I} x_i y_j$ is identically zero, then we have the inequality*

$$|f(x; y)| \leq C_{m,n} \Pi_I(x, y) \sup_{a,b,i,j} \left| \frac{\partial^2 f}{\partial x_i \partial y_j}(a; b) \right|, \qquad \forall x, y > 0.$$

The constant $C_{m,n}$ depends only on m and n.

Proof By linearity, we can suppose that

$$\sup_{a,b,i,j} \left| \frac{\partial^2 f}{\partial x_i \partial y_j}(a; b) \right| \leq 1.$$

We proceed by induction on (m, n). If $m = 0$ or $n = 0$, I is empty, hence $f \equiv 0$, and the result is obvious. Similarly if $m, n \geq 1$ when I is empty. We suppose therefore that $m \geq 1$, $n \geq 1$, that I is not empty and that the lemma is true for the pairs $(m - 1, n)$ and $(m, n - 1)$. We can suppose that $(1, 1) \in I$.

We apply the lemma (as an induction hypothesis) to the functions

$$(x', y) \overset{f_0}{\mapsto} f(0, x'; y),$$
$$(x, y') \overset{f_1}{\mapsto} f(x; 0; y'),$$
$$(\mathbb{R}^+)^{m+n-1} \to \mathbb{R}.$$

For example, f_0 is identically zero on the zeros of Π_J where $J = I \cap (\{2, \ldots, m\} \times \{1, \ldots, n\})$. We thus have

$$|f_0| \le C_{m-1,n} \Pi_J \le C_{m-1,n} \Pi_I.$$

Now let us fix a point (x', y') of $[0, +\infty)^{m+n-2}$ and let us define

$$g(x_1, y_1) := f(x_1, x'; y_1; y') - f(0, x'; y_1, y') - f(x_1, x'; 0, y') + f(0, x'; 0, y').$$

We have $g(0, y) = g(x, 0) = 0$ and so

$$g(x_1, y_1) = \int_0^{x_1} \int_0^{y_1} \frac{\partial^2 g}{\partial a \partial b}(a, b) \, \mathrm{d}a \, \mathrm{d}b,$$

from where we derive $|g(x_1, y_1)| \le x_1 y_1$. Finally,

$$\begin{aligned} |f(x, y)| &\le x_1 y_1 + (C_{m-1,n} + C_{m-1,n-1} + C_{m,n-1}) \Pi_I(x, y) \\ &\le C_{m,n} \Pi_I(x, y). \end{aligned}$$

The constant $C_{m,n}$ is calculated by the recurrence relation

$$C_{m,n} : = 1 + C_{m-1,n} + C_{m-1,n-1} + C_{m,n-1}.$$

■

We now apply Lemma 5.8.1 to the function $(\delta, \varepsilon) \mapsto f(\delta, \varepsilon) := \gamma(\delta, \varepsilon, u_\mathrm{m}) - \delta - \varepsilon$ considering separately the 2^{2n} sectors of the form $J_1 \times \cdots \times J_{2n}$ with $J_j = \mathbb{R}^\pm$. Clearly, f must be defined over the whole of $\mathbb{R}^{2n}$ by truncation and extension by zero, or else quite simply remark that Lemma 5.8.1 remains true when we replace $[0, \infty]$ by $[0, A]$ where $A > 0$ is an arbitrary finite number. Noting that $\Delta(\delta, \varepsilon, u_\mathrm{m})$ is of the form $\Pi_I(\delta, \varepsilon)$ in each of the sectors, the inequality $|\gamma(\delta, \varepsilon, u_\mathrm{m}) - \delta - \varepsilon| \le c_0 \Delta(\delta, \varepsilon, u_\mathrm{m})$ is thus established, this with the constant c_0 depending on the base point u_m. But from Lemma 5.8.1 it depends only on the upper bound, on a neighbourhood of $(0, 0)$ of the second derivatives of $\gamma(\cdot, \cdot, u_\mathrm{m})$. This bound is itself bounded above by a constant when u_m remains in a neighbourhood of $\hat{u}$ and this completes the proof of Lemma 5.5.2. ■

Proof of Theorem 5.1.3

Let us fix a compact set $K = [0, T] \times [-L, L]$ of $\mathbb{R}_t^+ \times \mathbb{R}_x$. The sequence $(a_m(t))_{m \in \mathbb{N}}$ of the restrictions of $v_m(t)$ to $(-L, L)$ is relatively compact in $L^1(-L, L)$ for all

$t \in [0, T]$ by Helly's theorem. Let Q be a dense subset of $[0, T]$ (for example the rational numbers). We can extract, making use of the diagonal procedure, a sequence $(a_{m(k)}(t))_{k\in\mathbb{N}}$ such that $m(k) \to +\infty$ and $(a_{m(k)}(t))_{k\in\mathbb{N}}$ converges in $L^1(-L, L)$ for all t belonging to Q. We denote this limit by $a(t)$.

Passing to the limit in the inequality

$$\int_{-L}^{L} |a_{m(k)}(t, x) - a_{m(k)}(s, x)| \,\mathrm{d}x \leq \varepsilon_{m(k)} + M|t - s|, \quad \forall t, s \in Q,$$

it becomes

$$\int_{-L}^{L} |a(t, x) - a(s, x)| \,\mathrm{d}x \leq M|t - s|, \quad \forall t, s \in Q.$$

This shows that a is the restriction to Q of a Lipschitz function defined on $[0, T]$ with values in $L^1(-L, L)$. We again denote this function by $a(t)$, which is clearly unique.

Being given $\varepsilon > 0$ and $t \in [0, T]$, there exists $s \in Q$ such that $2M|t - s| < \varepsilon$. Then there exists $l \in \mathbb{N}$ such that $\varepsilon_{m(k)} < \varepsilon$ and $\int_{-L}^{L} |a_{m(k)}(s) - a(s)| \,\mathrm{d}x < \varepsilon$ for all $k > l$. For these indices we therefore have

$$\begin{aligned}
\int_{-L}^{L} |a_{m(k)}(t) - a(t)| \,\mathrm{d}x &\leq \int_{-L}^{L} |a_{m(k)}(t) - u_{m(k)}(s)| \,\mathrm{d}x \\
&\quad + \int_{-L}^{L} |a_{m(k)}(s) - a(s)| \,\mathrm{d}x + \int_{-L}^{L} |a(t) - a(s)| \,\mathrm{d}x \\
&\leq 2M|t - s| + \varepsilon_{m(k)} + \int_{-L}^{L} |a_{m(k)}(s) - a(s)| \,\mathrm{d}x \\
&\leq 3\,\varepsilon,
\end{aligned}$$

that is to say that

$$\lim_{k\to+\infty} a_{m(k)}(t) = a(t), \qquad \forall t \in [0, T],$$

for the norm of $L^1((-L, L))$.

Now let us define $H_k(t) = \int_{-L}^{L} |a_{m(k)}(t) - a(t)| \,\mathrm{d}x$. We have

$$H_k(t) \leq 2ML + \sup_{t\in[0,T]} \int_{-L}^{L} |a| \,\mathrm{d}x \leq 4ML, \qquad \forall t \in [0, T].$$

We have seen also that $(H_k(t))_{k\in\mathbb{N}}$ tends to zero for all $t \in [0, T]$. The theorem of dominated convergence thus ensures that

$$\lim_{k\to+\infty} \int_0^T H_k(t) \,\mathrm{d}t = 0,$$

that is to say that $(a_{m(k)}(t))_{k\in\mathbb{N}}$ tends to a in $L^1(K)$. The sequence $(a_m)_{m\in\mathbb{N}}$ is thus relatively compact in $L^1(K)$.

Finally, as $\mathbb{R}^+ \times \mathbb{R}$ is the denumerable union of such blocks, the diagonal procedure allows us to find a sequence again denoted by $(m(k))_{k\in\mathbb{N}}$ such that $(a_{m(k)})_{k\in\mathbb{N}}$ converges in $L^1(\omega)$ for every bounded open set ω of $\mathbb{R}^+ \times \mathbb{R}$.

Remark The proof of Theorem 5.1.3 is simpler when $\varepsilon_m = 0$, for all m. This is then a consequence of the theorems of Helly and of Ascoli and Arzelà.

5.9 Supplementary remarks

Other numerical schemes

Several other numerical schemes are very close in conception to that of Glimm, especially that of Lax and Friedrichs. In the latter, only the sampling disappears, to be replaced by an averaging:

$$u_{n,k} := \frac{1}{2h}\int_{(k-1)h}^{(k+1)h} u_a^h(n\Delta t - 0, y)\,dy.$$

The scheme of Godunov differs from that of Lax and Friedrichs only by the position of the meshes. In place of being in alternate rows, they are aligned according to the rectangular network $\mathbb{Z} \times 2\mathbb{Z}$. These three schemes are particular cases of those defined and studied by H. Gilquin [29], where the definition of u_n depends on a probability measure $d\nu_n$:

$$u_{n,k} := \int_{-1}^{1} u_a^h(n\Delta t - 0, (k+s)\Delta x) d\nu_n(s).$$

For all these schemes, which are *monotonic* (that is to say preserve the order) in the scalar case, convergence takes place provided that the sequence $(d\nu_n)_{n\in\mathbb{N}}$ is equi-distributed in $(-1, 1)$ and that the approximation is stable in BV (exercise). Unfortunately, we do not in general know how to prove this stability, except for the scalar conservation laws and their natural generalisations, the Temple systems (see Chapter 13). In the case of certain systems of two conservation laws, called 2×2 systems, the method of compensated compactness has enabled us to obtain theorems of convergence to weak solutions which we do not know to be of bounded variation [10, 14].

The rich case

The systems of conservation laws called *rich* (or semi-hamiltonian according to the terminology of Tsarev [106, 107]) will be studied in greater detail in Chapter 12. The system (5.1) is called rich if it is strictly hyperbolic and if there exists a complete system of Riemann invariants, that is to say a system of curvilinear coordinates $w_1(u), \ldots, w_n(u)$ satisfying

$$\mathrm{d}w_j(u)(\mathrm{d}f(u) - \lambda_j(u)) \equiv 0, \quad j = 1, \ldots, n.$$

The 2×2 systems are rich as long as they are strictly hyperbolic. In addition, we still suppose that each characteristic field is genuinely non-linear or else linearly degenerate in order to be able to use Lax's theorem.

Lemma 5.9.1 *If the system (5.1) is rich in ω, then there exists a constant c_0 such that for all u_L, u_R, and u_m in ω_1, we have*

$$\|\gamma - \varepsilon - \delta\| \le c_0(\|\varepsilon\| + \|\delta\|)\Delta(\varepsilon, \delta),$$

with the notation $u_m = \varphi(\varepsilon; u_L)$ and $u_R = \varphi(\delta; u_m) = \varphi(\gamma; u_L)$.

In making use of this estimate, Glimm [32] improved the stability result in weakening the condition of smallness on the given Cauchy condition: for every number $V_0 > 0$, there exists a number α such that if $\mathrm{TV}(\bar{u}) < V_0$ and $\|\bar{u} - \hat{u}\| \le \alpha$, then Glimm's scheme is stable in $\mathrm{BV}(\mathbb{R})$ (hence the Cauchy problem admits a weak entropy solution).

Proof We use the expansion (4.15) for three Riemann problems.
We obtain

$$\begin{aligned}
u_m - u_L &= \sum_j \varepsilon_j r_j^L + \frac{1}{2}\sum_j \varepsilon_j^2 (\mathrm{d}r_j \cdot r_j)_L + \sum_{j<k} \varepsilon_j \varepsilon_k (\mathrm{d}r_k \cdot r_j)_L + O(\|\varepsilon\|^3),\\
u_R - u_m &= \sum_j \delta_j r_j^m + \frac{1}{2}\sum_j \delta_j^2 (\mathrm{d}r_j \cdot r_j)_m + \sum_{j<k} \delta_j \delta_k (\mathrm{d}r_k \cdot r_j)_m + O(\|\delta\|^3)\\
&= \sum_j \delta_j r_j^L + \frac{1}{2}\sum_j \delta_j^2 (\mathrm{d}r_j \cdot r_j)_L + \sum_{j<k} \delta_j \delta_k (\mathrm{d}r_k \cdot r_j)_L\\
&\quad + \sum_{j,k} \delta_k \varepsilon_j (\mathrm{d}r_k \cdot r_j)_L + O(\|\delta\|^3 + \|\varepsilon\|^3),\\
u_R - u_L &= \sum_j \gamma_j r_j^L + \frac{1}{2}\sum_j \gamma_j^2 (\mathrm{d}r_j \cdot r_j)_L + \sum_{j<k} \gamma_j \gamma_k (\mathrm{d}r_k \cdot r_j)_L + O(\|\gamma\|^3).
\end{aligned}$$

We know that $\gamma = \varepsilon + \delta + O(\|\varepsilon\|^2 + \|\delta\|^2)$. Combining the above equalities and making use of the dual basis $(l_p(u_L))_{1 \le p \le n}$ to that of the eigenvectors r_j, we have therefore

$$\gamma_p - \varepsilon_p - \delta_p = l_p \cdot \sum_{j<k} \delta_k \varepsilon_j (\mathrm{d}r_k \cdot r_j - \mathrm{d}r_j \cdot r_k)_L + O(\|\varepsilon\|^3 + \|\delta\|^3).$$

But the existence of the pth Riemann invariant w_p is equivalent to the geometric condition of Frobenius:

$$l_p \cdot (\mathrm{d}r_k \cdot r_j - \mathrm{d}r_j \cdot r_k) = 0, \qquad \forall 1 \le j, k \le n.$$

Thus $\gamma_p - \varepsilon_p - \delta_p = \mathrm{O}(\|\varepsilon\|^3 + \|\delta\|^3)$. Finally, the fact that the right-hand side is zero when $\Delta(\varepsilon, \delta) = 0$ leads by an analogous calculation to that in the general case to the upper bound

$$|\gamma_p - \varepsilon_p - \delta_p| \le \text{const.}(\|\varepsilon\| + \|\delta\|)\Delta(\varepsilon, \delta).$$

■

'Continuous' Glimm functional

M. Schatzman [89, 90] has adapted the calculation of Glimm for an exact solution of a system of conservation laws, when it is of class $\mathscr{C}^1$ outside a denumerable family of shocks which statisfy Lax's condition. Without going into details, let us say that in the absence of shocks, the linear term is replaced by

$$V(u(t)) = \sum_i \int_{\mathbb{R}} |l_i(u) \cdot u_x| \mathrm{d}x,$$

the l_i being the linear eigenforms of $\mathrm{d}f$. The quadratic form becomes then

$$Q(u(t)) = \iint_{x<y} D(Z(x), Z(y))\, \mathrm{d}x\mathrm{d}y, \qquad Z := (l_1(u) \cdot u_x, \ldots, l_n(u) \cdot u_x).$$

In the general case, these functionals contain supplementary terms to take into account shock waves and can even be defined for a general spatial curve σ; we write then $V(u; \sigma)$ and $Q(u; \sigma)$. For a given initial condition of small total variation, there exists a number $K > 0$ such that $\sigma \mapsto (V + KQ)(u; \sigma)$ is decreasing, which furnishes an *a priori* estimate of $u(t)$ in $\mathrm{BV}(\mathbb{R})$, uniform in time.

5.10 Exercises

5.1 (1) Let θ be a real number and $a = n\theta - E(n\theta)$ the fractional part of $n\theta$. Show that the sequence $(a_n)_{n\in\mathbb{N}}$ is equi-distributed in $(0, 1)$ if and only if θ is irrational. When $\theta \in \mathbb{R}\backslash\mathbb{Q}$ we shall apply the criterion (5.7) to functions judiciously chosen, then we shall proceed by a density argument.

(2) Find a real number $\theta > 1$ for which the sequence $a_n = \theta^n - E(\theta^n)$ is not equi-distributed in $(0, 1)$.

(3) For $n \in \mathbb{N}$, let $a_n = 10^{-k}2^n$, where $k = E(\log_{10} 2^n)$. It is thus an element of the interval $(\frac{1}{10}, 1)$. Show that $(a_n)_{n\in\mathbb{N}}$ is not equi-distributed in $(\frac{1}{10}, 1)$.

5.2 (1) Show that if an entropy E (not necessarily convex) of flux F satisfies $E_t + F_x \leq 0$, for all simple admissible waves, then the weak solutions obtained by Glimm's scheme also satisfy this inequality.

(2) For the system (5.23), show that these solutions satisfy

$$(rg(\theta))_t + (r\varphi(u)g(\theta))_x = 0$$

for all $g \in \mathscr{C}[-\frac{1}{2}\pi, \frac{1}{2}\pi]$. Verify that these equations contain that of the system (5.23). Finally show the inequality

$$r_t + (r\varphi(u))_x \leq 0.$$

5.3 The dynamics of an isothermal gas in one dimension and in eulerian variables and governed by the system

$$\left.\begin{aligned} \rho_t + (\rho u)_x &= 0, \\ (\rho u)_t + (\rho u^2 + c^2\rho)_x &= 0, \end{aligned}\right\} \qquad (5.24)$$

where $\mathscr{U} = \mathbb{R}^{+^*} \times \mathbb{R}$ and $c > 0$ is the (constant) speed of sound. Show that this system belongs to the class described in §5.6. The Cauchy problem thus has a weak solution.

5.4 We consider the so-called 'Leroux' system

$$\left.\begin{aligned} \rho_t + (\rho u)_x &= 0, \\ u_t + (u^2 + \rho)_x &= 0, \end{aligned}\right\}$$

with $\mathscr{U} = \mathbb{R}^+ \times \mathbb{R}$.

(1) Show that it is strictly hyperbolic in $\mathscr{U}^*$. Show that the Riemann invariants are the slopes of the straight lines which pass through (ρ, u) and which are tangents to the parabola Γ of equation $u^2 + 4\rho = 0$.

(2) Show that these straight lines are the wave curves of the system and that $\mathscr{U}$ is an invariant domain for the Riemann problem.

(3) Deduce that if the given initial condition $(\bar{\rho}, \bar{u})$ is of bounded variation and if $\inf_x(\bar{\rho} + \bar{u}) > 0$, then the Cauchy problem has a weak solution.

(4) Let Δ be one of the tangents of Γ, with equation $\alpha\rho + \beta u = \gamma$. Show that the weak solution which we have constructed satisfies

$$((\alpha\rho + \beta u - \gamma)^+)_t + \left(\left(u + \frac{\beta}{\alpha}\right)(\alpha\rho + \beta u - \gamma)^+\right)_x \leq 0.$$

5.5 We consider a strictly hyperbolic 2×2 system of 'Temple' type. Let $K \subset \mathcal{U}$ be a complete characteristic quadrilateral. We denote the Riemann invariants by r and s.

(1) Show that if K is 'small enough', then

$$\sup_{u \in K} \lambda_1(u) < \inf_{u \in K} \lambda_2(u).$$

(2) We suppose from now on that

$$\sup_{u \in K} \lambda_1(u) := c_1 < c_2 := \inf_{u \in K} \lambda_2(u).$$

We consider a given initial datum $\bar{u}$, such that

$$\bar{u}(\bar{x}) = \begin{cases} u_{\mathrm{L}}, & \forall x \leq \alpha, \\ u_{\mathrm{R}}, & \forall x \geq \beta. \end{cases}$$

Show that $s(u_{n,j}) = s(u_{\mathrm{L}})$ for all $j \leq J_n$, where

$$J_n = 2E\left(\frac{\alpha}{2h}\right) - n + 2\,\mathrm{card}\{m\colon 1 \leq m \leq n, a_{\mathrm{m}} < \rho c_2\}.$$

(3) Deduce that the weak solutions of the Cauchy problem, obtained by Glimm's scheme, satisfy

$$(x < \alpha + c_2 t) \Rightarrow (s(u(x,t)) = s(u_{\mathrm{L}}))$$

and similarly

$$(x > \beta + c_1 t) \Rightarrow (r(u(x,t)) = r(u_{\mathrm{R}})).$$

(4) Show that there exists a time T of *uncoupling*, beyond which the evolution proceeds according to independent scalar conservation laws. More precisely, there exists a point X such that we have three zones for $t > T$. A zone is defined by $c_1(t - T) < x - X < c_2(t - T)$ where $u(t,x)$, constant, is determined by $r(u) = r(u_{\mathrm{R}})$ and $s(u) = s(u_{\mathrm{L}})$. A zone is defined by $x < X + c_1(t - T)$ where $u(t,x) = \varphi_1(\varepsilon(t,x); u_{\mathrm{L}})$ and ε is a solution of a scalar conservation law. Finally a zone is defined by $x > X + c_2(t - T)$ where $u(t,x) = \varphi_2(\delta(t,x); u_{\mathrm{R}})$ and δ is the solution of another scalar conservation law.

5.6 We consider a strictly hyperbolic 2×2 system whose characteristic fields are genuinely non-linear. To fix our ideas, the speeds, expressed as functions of the Riemann invariants, satisfy

$$\frac{\partial \lambda_1}{\partial r} > 0, \quad \frac{\partial \lambda_2}{\partial s} > 0.$$

We suppose that the Riemann problem has a unique solution. We are given a bounded initial condition $\bar{u}$ such that $r \circ u$ and $s \circ u$ are increasing and have values in a characteristic quadrilateral $K \subset \mathscr{U}$.

(1) Show that $\bar{u} \in \mathrm{BV}(\mathbb{R})^2$.

(2) Let u_{L} and u_{R} be the data of a Riemann problem. Show that if $r(u_{\mathrm{L}}) \leq r(u_{\mathrm{R}})$ and $s(u_{\mathrm{L}}) \leq s(u_{\mathrm{R}})$ the waves of the Riemann problem are rarefaction waves. We might begin by studying the case of equality.

(3) Show by induction on n that the sequences $(r(u_{n,j}))_{j \in n+2\mathbb{Z}}$ and $(s(u_{n,j}))_{j \in n+2\mathbb{Z}}$ are increasing and that the Riemann problems which are solved by putting into effect Glimm's scheme only make use of rarefaction waves.

(4) Deduce that u_a^h is defined for all time $t \geq 0$ and that

$$\mathrm{TV}\big(r \circ u_a^h(t)\big) = \mathrm{TV}(r \circ \bar{u}),$$
$$\mathrm{TV}\big(s \circ u_a^h(t)\big) = \mathrm{TV}(s \circ \bar{u}).$$

Conclude that the Cauchy problem possesses a weak entropy solution on $\mathbb{R}^+ \times \mathbb{R}$.

6

Second order perturbations

We are interested in this chapter in perturbations of hyperbolic systems of conservation laws

$$\partial_t u + \sum_{\alpha=1}^{d} \partial_\alpha f^\alpha(u) = 0 \tag{6.1}$$

by diffusion terms of second order. To adhere closely to physical examples, we restrict ourselves to perturbations of conservative form:

$$\partial_t u + \sum_{\alpha=1}^{d} \partial_\alpha f^\alpha(u) = \varepsilon \sum_{\alpha,\beta=1}^{d} \partial_\alpha(B^{\alpha\beta}(u)\partial_\beta u). \tag{6.2}$$

In these models, $B^{\alpha\beta}(u)$, for $1 \leq \alpha, \beta \leq d$, denotes matrices whose coefficients are smooth functions defined on the state space $\mathcal{U}$. The coefficient $\varepsilon > 0$ is ultimately to tend to zero. It may come from a change of scale $(x, t) \mapsto (\eta x, \eta t)$ in which the process of diffusion seems secondary with respect to the transport phenomena. For example, for a gas, we generally admit that the thermal conduction and, especially, the viscosity have rather weak effects.

For a given initial state $u_0(x)$, the solution of the Cauchy problem for (6.2), when it is well-posed, will be denoted by $u^\varepsilon(t, x)$. The natural question is to know if u^ε converges, in a sense to be made precise, when ε tends to zero, to an entropy solution u of the Cauchy problem for (6.1). We are also interested in the behaviour of u^ε in the neighbourhood of a shock wave of u. Let us say immediately that the convergence is proved only in very few cases.

To be complete, we should hope equally to know that u^ε converges when the given initial state u_0^ε also depends on ε and converges to a u_0. This will lead us to consider particular solutions in the form of a progressive wave $U((\nu \cdot x - ct)/\varepsilon)$. Their analysis, relatively simpler than that of the Cauchy problem, allows another approach to the stability of shocks.

6.1 Dissipation by viscosity

Not every perturbation of the form (6.2) leads to a well-posed Cauchy problem. A natural requirement is that it is linearly well-posed in the neighbourhood of a constant state. By the change of variables $(x,t) \mapsto (\varepsilon x, \varepsilon t)$, we are led to the case $\varepsilon = 1$. The linearised system

$$\partial_t v + \sum_{\alpha=1}^{d} A^\alpha(\bar{u})\partial_\alpha v = \sum_{\alpha,\beta=1}^{d} B^{\alpha\beta}(\bar{u})\partial_\alpha\partial_\beta v \tag{6.3}$$

possesses particular solutions of the form $v(x,t) = \exp(\omega t + i\xi \cdot x)V$, $V \in \mathbb{R}^n$, if and only if V is an eigenvector of the matrix

$$M(\xi) := \sum_{\alpha,\beta=1}^{d} \xi_\alpha \xi_\beta B^{\alpha\beta}(\bar{u}) + i\sum_{\alpha=1}^{d} \xi_\alpha A^\alpha(\bar{u})$$

associated with the eigenvalue $-\omega$. The *dispersion relation*, which links ξ and ω, is thus

$$\det(\omega I_n + M(\xi)) = 0. \tag{6.4}$$

A necessary condition for the Cauchy problem to be well-posed for (6.3) is that the real part of the solutions ω of the equation (6.4) retains an upper bound when ξ ranges over $\mathbb{R}^d$. In particular, making $\|\xi\|$ tend to $+\infty$, we see that the eigenvalues of the matrices $B(\bar{u};\xi) := \sum_{\alpha,\beta} \xi_\alpha\xi_\beta B(\bar{u})$ must all have their real parts non-negative, for $\xi \in S^{d-1}$. Quite evidently, none of these conditions is sufficient. Not only are they not sufficient for the Cauchy problem for (6.3) to be well-posed for each $\varepsilon > 0$ (after all, we have not excluded that the matrices $B^{\alpha\beta}$ are singular, even zero), but they ensure still less the convergence of u^ε when ε tends to zero.

Non-dissipative case

To understand why the convergence of u^ε demands stronger hypotheses than the existence, let us look at the case of a physical system, where (6.1) is compatible with a strongly convex entropy E ($\mathrm{D}_u^2 E > 0$), of flux F. Let us suppose that the tensor $B(u)$ satisfies the following condition:

$$\sum_{\alpha,\beta,i,j,k} \frac{\partial^2 E}{\partial u_i \partial u_k} B^{\alpha\beta}_{kj}(u) m_{\alpha j} m_{\beta i} = 0, \quad \forall m \in M_{d\times n}(\mathbb{R}). \tag{6.5}$$

This condition excludes neither the Cauchy problem being well-posed nor that it produces a smoothing effect. For example, the linear system

$$\partial_t \begin{pmatrix} v \\ w \end{pmatrix} = \varepsilon \begin{pmatrix} 0 & 1 \\ -1 & 0 \end{pmatrix} \partial_x^2 \begin{pmatrix} v \\ w \end{pmatrix}$$

(take $E = v^2 + w^2$ to satisfy (6.5)) leads to two uncoupled Schrödinger equations *via* the change of unknowns $(v, w) \mapsto (v + iw, v - iw)$, an equation whose semi-group is smoothing for initial data with rapid decay.

For such systems, we have the identity

$$\partial_t E(u^\varepsilon) + \operatorname{div} F(u^\varepsilon) = \varepsilon \sum_{\alpha,\beta} \partial_\alpha((d_u E \cdot B^{\alpha\beta})(u^\varepsilon)\partial_\beta u^\varepsilon), \tag{6.6}$$

valid at least for smooth solutions, let us say of class $\mathscr{C}^2$. For these, when they decay quickly enough at infinity, we deduce from (6.6) the conservation of energy (or of entropy, according to the context)

$$\int_{\mathbb{R}^d} E(u^\varepsilon(t, x))\,\mathrm{d}x = \int_{\mathbb{R}^d} E(u_0(x))\,\mathrm{d}x, \quad \forall t > 0.$$

This shows that the sequence $(u^\varepsilon)_{\varepsilon>0}$ is bounded in a certain Lebesgue–Orlicz space associated with E. If in addition u^ε and $f(u^\varepsilon)$ converge simultaneously to u and $f(u)$ in the sense of distributions,[1] then this convergence will occur in general for the strong topology of a Lebesgue space because of the non-linearity of f (see Exercise 6.1). Free to extract a sub-sequence, pointwise convergence will occur almost everywhere. Finally, we can think that $(u^\varepsilon)_\varepsilon$ remains localized (the speed of propagation is finite when ε is zero), at least enough to be able to apply the theorem of dominated convergence. We then obtain

$$\int_{\mathbb{R}^d} E(u(t, x))\,\mathrm{d}x = \int_{\mathbb{R}^d} E(u_0(x))\,\mathrm{d}x, \quad \forall t > 0.$$

The conservation of energy when $\varepsilon = 0$ contradicts the development of shock waves for which we hope that the non-positive measure $\partial_t E(u) + \operatorname{div} F(u)$ is not identically zero.

Dissipation or production of entropy

Another way to see that the condition (6.5) is incompatible with the convergence of u^ε to an entropy solution of (6.1) when that contains a shock wave is an asymptotic analysis in a neighbourhood of a point of discontinuity (t_0, x_0) of this wave. If the

[1] with the result that u is a weak solution of (6.1).

wave front is tangent to the hyperplane with equation $y := \nu \cdot (x - x_0) - c(t - t_0) = 0$, a change of scale suggests that u^ε has an asymptotic expansion of the form

$$u^\varepsilon(t, x) = U\left(\frac{y}{\varepsilon}, t, x\right) + O(\varepsilon).$$

The function $z \mapsto U(z, t, x)$ is called a *profile* of the shock wave. With values in $\mathscr{U}$, it is smooth and tends when $z \to \pm\infty$ to the states $u_\pm(t, x)$ situated on one side or other of the shock. We shall see in Chapter 7 that U is the heteroclinic orbit of a very simple vector field. On the right-hand side of (6.6), in the form $\partial_\alpha e^\alpha$, the terms e^α obey the following asymptotic expansion:

$$e^\alpha = \sum_\beta (d_u E \cdot B^{\alpha\beta})(U)U' + O(\varepsilon),$$

where the first term tends to zero in L^p_{loc} for all finite p since it is bounded and localized in a zone of size ε round about the shock wave. Therefore $e^\alpha \rightharpoonup 0$ in $\mathscr{D}'(\mathbb{R}^{d+1})$ and a passage to the limit in (6.6) must give the entropy equality $\partial_t E(u) + \operatorname{div} F(u) = 0$ instead of the entropy inequality expected.

We remark that the situation remains the same when we replace the condition (6.5) by the weaker hypothesis

$$\sum_{\alpha,\beta,i,j,k} \xi_\alpha \xi_\beta \frac{\partial^2 E}{\partial u_i \partial u_k} B^{\alpha\beta}_{kj}(u)\eta_i \eta_j = 0, \quad \forall \xi \in \mathbb{R}^d, \ \forall \eta \in \mathbb{R}^n.$$

The calculation is the same but the right-hand side of the balance of the entropy is now written

$$\sum_\alpha \partial_\alpha e^\alpha + \varepsilon \sum_{\alpha,\beta,i,j} Q^{\alpha\beta}_{ij}(u^\varepsilon)\big(\partial_\alpha u^\varepsilon_i \partial_\beta u^\varepsilon_j - \partial_\alpha u^\varepsilon_j \partial_\beta u^\varepsilon_i\big).$$

We must verify that the second sum tends to zero in the sense of distributions by making use of the asymptotic expansion of u^ε. Now the coefficient of ε^{-2} in $\partial_\alpha u^\varepsilon_i \partial_\beta u^\varepsilon_j - \partial_\alpha u^\varepsilon_j \partial_\beta u^\varepsilon_i$ is identically zero (this is due to the structure of codimension 1 of the shock waves), with the result that the terms of this sum are of the form $L(U)U' + O(\varepsilon)$ where $L(U)$ is bounded. This term thus tends to zero in $\mathscr{D}'(\mathbb{R}^{d+1})$.

This discussion shows that it is essential that the perturbation is strictly dissipative for the entropy of the system. However, if we demand that it satisfies the Legendre–Hadamard condition

$$\sum_{\alpha,\beta,i,j,k} \xi_\alpha \xi_\beta \frac{\partial^2 E}{\partial u_i \partial u_k} B^{\alpha\beta}_{kj}(\bar{u})\eta_i \eta_j \geq c(\bar{u})\|\xi\|^2\|\eta\|^2, \quad \forall \xi \in \mathbb{R}^d, \forall \eta \in \mathbb{R}^n, \qquad (6.7)$$

we shall miss most of the perturbations of physical origin. In fact, each time that the system (6.1) contains a conservation law such as that of mass, $\rho_t + \operatorname{div}(\rho v) = 0$, this will not be perturbed, that is to say that the matrices $B(\bar{u};\xi) := \sum_{\alpha,\beta} \xi_\alpha \xi_\beta \times B^{\alpha\beta}(u)$ will be singular, having a line of zeros. The left-hand side of (6.7) will thus be zero for $\eta \in \ker B(\bar{u};\xi)$ and also for $D_u^2 E \cdot \eta \in \ker B(\bar{u};\xi)^{\mathrm{T}}$. Since the quadratic form $\eta \mapsto (B(u;\xi)\eta \mid D_u^2 E \cdot \eta)$ must be positive semi-definite, a natural hypothesis is that there exists a continuous function $c(u) > 0$ such that

$$\big(B(\bar{u};\xi)\eta \mid D_u^2 E\eta\big) \geq c(\bar{u})\|B(\bar{u};\xi)\eta\|^2, \quad \forall \xi \in S^{d-1}, \ \forall \eta \in \mathbb{R}^n. \tag{6.8}$$

We say then that the tensor B is *dissipative* with respect to the entropy E.

Now let us see a formal consequence of (6.8). The entropy balance for the perturbed problem is now

$$\partial_t E(u^\varepsilon) + \operatorname{div} F(u^\varepsilon) + \varepsilon \sum_{\alpha,\beta,i,j,k} \frac{\partial^2 E}{\partial u_i \partial u_k}(u^\varepsilon) B_{kj}^{\alpha\beta}(u^\varepsilon) \partial_\alpha u_i^\varepsilon \partial_\beta u_j^\varepsilon = \sum_\alpha \partial_\alpha e^\alpha. \tag{6.9}$$

If $d = 1$ (the multi-dimensional case is not as clear but the reader will treat it without difficulty where $E = \|u\|^2$ when B is constant), we deduce

$$\partial_t E(u^\varepsilon) + \partial_x F(u^\varepsilon) + \varepsilon c(u^\varepsilon)\|B(u^\varepsilon)\partial_x u^\varepsilon\|^2 \leq \partial_x e^\varepsilon.$$

Supposing that the solution is sufficiently smooth (so that the above inequality is correct) and that it decays rapidly at infinity, the integration with respect to x yields

$$\frac{\mathrm{d}}{\mathrm{d}t} \int_{\mathbb{R}} E(u^\varepsilon)\,\mathrm{d}x + \varepsilon \int_{\mathbb{R}} c(u^\varepsilon)\|B(u^\varepsilon)\partial_x u^\varepsilon\|^2\,\mathrm{d}x \leq 0.$$

Finally

$$\int_{\mathbb{R}} E(u^\varepsilon(T,x))\,\mathrm{d}x + \varepsilon \int_0^T \mathrm{d}t \int_{\mathbb{R}} c(u^\varepsilon)\|B(u^\varepsilon)\partial_x u^\varepsilon\|^2\,\mathrm{d}x \leq \int_{\mathbb{R}} E(u_0(x))\,\mathrm{d}x,$$

which furnishes *a priori* estimates. If now the sequence $(u^\varepsilon)_\varepsilon$ is bounded in $L^\infty(\omega)$ and converges almost everywhere to $u(t,x)$ in an open set ω of $\mathbb{R}^{d+1}$, then u^ε and $f(u^\varepsilon)$ converge to u and $f(u)$ in $L^1_{\mathrm{loc}}(\omega)$, hence in $\mathscr{D}'(\omega)$, and $u_t^\varepsilon + f(u^\varepsilon)_x$ tends to $u_t + f(u)_x$ in the sense of distributions. In addition, the hypothesis above shows that $\varepsilon^{1/2}\partial_x(\varepsilon^{1/2}B(u^\varepsilon)\partial_x u^\varepsilon)$ tends to zero in the sense of distributions. Passing to the limit in (6.2), we obtain $u_t + f(u)_x = 0$. Passing also to the limit in

$$E(u^\varepsilon)_t + F(u^\varepsilon)_x \leq \varepsilon\big(\mathrm{d}E(u^\varepsilon) \cdot B(u^\varepsilon)u_x^\varepsilon\big)_x,$$

we obtain this time the desired entropy inequality $E(u)_t + F(u)_x \leq 0$.

Let us note again that the asymptotic analysis of a shock wave no longer contradicts the production of entropy. In fact, the dominant term of

$$\sum_{\alpha} \partial_\alpha e^\alpha - \varepsilon \sum_{\alpha,\beta,i,j,k} \frac{\partial^2 E}{\partial u_i \partial u_k}(u^\varepsilon) B_{kj}^{\alpha\beta}(u^\varepsilon)\partial_\alpha u_i^\varepsilon \partial_\beta u_j^\varepsilon$$

is $-\varepsilon^{-1}(B(U;\nu)U' \mid \mathrm{D}_u^2 E(U)U')$ which is of the form $-\varepsilon^{-1}V(\varepsilon^{-1}y)$ where V is positive, zero at infinity. This term has a mass independent of ε, strictly positive. The set of the terms considered therefore tends to a negative singular measure carried by the front of the shock wave, which is nothing but the measure of entropy dissipation (the opposite of the measure of entropy production).

Example 6.1.1 Let us illustrate the criterion (6.8) by the Navier–Stokes equations of the dynamics of a compressible, viscous, heat-conducting fluid. The case $d = 1$ is as usual the easiest to treat since we can use lagrangian coordinates. Denoting by τ, v, $p(\tau, e)$, $T(\tau, e)$, e the specific volume, the velocity, the pressure, the temperature, the specific internal energy, we express the relative viscosity and the thermal conduction as functions of τ and of e. The Navier–Stokes equations are written as

$$\begin{aligned} \tau_t - v_x &= 0, \\ v_t + p_x &= \varepsilon(b v_x)_x, \\ \left(e + \frac{1}{2}v^2\right)_t + (pv)_x &= \varepsilon(b v v_x + k T_x)_x. \end{aligned}$$

The coefficients b and k are positive. There are two points of view, according as we neglect or not the effects due to the viscosity compared with those due to the thermal transfers (which is realistic in the case of a gas). In one case, we shall have $b = 0$ and $k > 0$, in the other $b, k > 0$. The diffusion tensor takes the value

$$B = \begin{pmatrix} 0 \\ b\,\mathrm{d}v \\ bv\,\mathrm{d}v + k\,\mathrm{d}T \end{pmatrix}.$$

Its kernel is the plane $\mathrm{d}T = 0$ in the case without viscosity, the straight line $\mathrm{d}v = \mathrm{d}T = 0$ in the viscous case.

The mathematical entropy is the opposite $E = -S(\tau, e)$ of the physical entropy. This satisfies the relation $T\,\mathrm{d}S = \mathrm{d}e + p\,\mathrm{d}\tau$. We have $T > 0$. For the smooth solutions of the Navier–Stokes equations, we find

$$S_t = \varepsilon\left(\frac{kT_x}{T}\right)_x + \varepsilon\frac{b}{T}v_x^2 + \varepsilon\frac{kT_x^2}{T^2},$$

that is to say that

$$(B\eta \mid \mathrm{D}^2 E\eta) = \frac{b}{T}(\mathrm{d}v \cdot \eta)^2 + \frac{k}{T^2}(\mathrm{d}T \cdot \eta)^2.$$

In both cases, the condition (6.8) is satisfied, with $c = 1/(kT^2)$ in the inviscid case and a more complicated expression in the viscous case.

Partially hyperbolic systems

The Navier–Stokes equations give the occasion to remark that the diffusion effect may or may not have a smoothing effect.

In the inviscid case, with only thermal conduction, Ling Hsiao and Dafermos [16] have shown that shock waves can develop from very smooth initial data. For such solutions, τ, v and T_x are discontinuous while T is continuous (even if its initial value T_0 is not). The discontinuities satisfy the Rankine–Hugoniot relations:

$$[v] + s[\tau] = 0, \quad [p] = s[v],$$
$$[T] = 0, \quad [pv] = s\left[e + \frac{1}{2}v^2\right] + \varepsilon[kT_x].$$

The shock waves are therefore similar to those of an isothermal gas, which explains the importance given in the literature to this model.

We note however that to a sufficiently small and smooth initial condition there can correspond a global smooth solution as has been shown by Slemrod [95]. On the other hand, in the viscous case, the diffusion is powerful enough for the solution of the Cauchy problem to be smooth for all time if the given initial condition is (see for example [51]). In fact a discontinuity should have to satisfy the Rankine–Hugoniot conditions

$$[v] + s[\tau] = 0, \qquad [v] = 0, \qquad [T] = 0.$$

We should have $s = 0$. We show easily (see [43]) that $\mathrm{d}[\tau]/\mathrm{d}t = \mathrm{O}([\tau])$ with the result that no discontinuity can appear if it did not exist before the initial instant (and similarly no discontinuity can disappear). In fact, as the tensor B is not invertible, the Navier–Stokes system is not parabolic and its semi-group is not smoothing. For a measurable bounded given initial condition, the velocity v and the temperature T are a little smoothed in the sense that $v_x, T_x \in L^2_{\mathrm{loc}}(\mathbb{R}^+ \times \mathbb{R})$, but the smoothness of the specific volume τ is simply propagated. For example, if $0 \leq s < 1$ and $1 \leq p < \infty$,

$$\tau(\,\cdot\,, t_1) \in W^{s,p}_{\mathrm{loc}} \iff \tau(\,\cdot\,, t_2) \in W^{s,p}_{\mathrm{loc}}, \quad \forall t_1, t_2 \geq 0.$$

The viscous isentropic case gives way to an analogous phenomenon. The Navier–Stokes equations are reduced to the conservation of mass and to Newton's law:

$$\tau_t - v_x = 0,$$
$$v_t + p(\tau)_x = \varepsilon(b(\tau)v_x)_x.$$

A discontinuity satisfies the conditions $[v] = 0$ and $[v] + s[\tau] = 0$, hence the conditions $s = 0$. Again, there is a discontinuity in τ at (t_0, x_0) if $\tau(0, \cdot)$ is discontinuous at x_0. The smoothness of τ is propagated without it improving.

We see the fact that the diffusion tensors are not invertible allow that discontinuous solutions exist for the perturbed problems. However, the diffusion is often sufficient for the discontinuities not to appear spontaneously. When they do so even then, we must see there a genuinely non-linear hyperbolic behaviour and anticipate that the equation (6.9) will not be satisfied. The solution of the Cauchy problem will not be unique. We must then select the physically admissible solution by imposing the entropy inequality

$$\partial_t E(u) + \operatorname{div} F(u) + \varepsilon \sum_{\alpha,\beta,i,j,k} B_{kj}^{\alpha\beta} \frac{\partial^2 E}{\partial u_i \partial u_k} \partial_\alpha u_i^\varepsilon \partial_\beta u_j^\varepsilon \leq \sum_\alpha \partial_\alpha e^\alpha \tag{6.10}$$

while giving a convenient sense to the third and fourth terms.

6.2 Global existence in the strictly dissipative case

The parabolic case is that where the Legendre–Hadamard condition (6.7) is satisfied. Following Hoff and Smoller [43, 44], we restrict ourselves to the case of a 'diagonal' perturbation and with constant coefficients

$$B_{ij}^{\alpha\beta} = \delta_i^j b_i^{\alpha\beta}, \quad \forall u \in \mathcal{U}.$$

The hypothesis (6.7) amounts to saying that the differential operators $Q_i := -\sum_{\alpha,\beta} b_i^{\alpha\beta} \partial_\alpha \partial_\beta$ are elliptic. The equation of the heat type $v_t + Q_i v = 0$ possesses a fundamental solution K_i of the form

$$K_i(t, x) = t^{-d/2} k_i\left(\frac{x}{\sqrt{t}}\right)$$

where k_i belongs to $S(\mathbb{R}^d)$ (the Schwarz class), $k_i > 0$ and $\int k_i(y)\,dy = 1$. The solution of the Cauchy problem for $v_t + Q_i v = g$ is given by the Duhamel formula

$$v(t, \cdot) = K_i(t) * v_0 + \int_0^t K_i(t-s) * g(s)\,ds$$

where $*$ denotes the convolution product in $\mathbb{R}^d$ and v_0 the initial condition.

Local existence in L^∞

Let us return to the non-linear system that we have to solve. In view of later applications, we also take into account external forces $g(x,t) = (g_1, \dots, g_n)$ which are *a priori* given:

$$\partial_t u_i + \operatorname{div} f^i(u) + Q_i u_i = g_i, \quad 1 \le i \le n. \tag{6.11}$$

Being given an initial condition $u_0 := \mathbb{R}^d \mapsto \mathcal{U}$, a solution of the Cauchy problem is a solution of the non-linear integral equation

$$u_i(t) = K_i(t) * u_{0i} - \sum_{\alpha=1}^{d} \int_0^t \partial_\alpha K_i(t-s) * f_\alpha^i(u(s))\,\mathrm{d}s \\ + \int_0^t K_i(t-s) * g_i(s)\,\mathrm{d}s, \quad 1 \le i \le n, \tag{6.12}$$

when the terms have a meaning. We shall suppose that there exists a point $\bar{u}$ such that $u_0 - \bar{u}$ is bounded and square-integrable.

The existence and uniqueness (local in time) of a solution of (6.12) are obtained by writing it as a fixed point of the mapping

$$u \mapsto Lu,$$
$$(Lu)_i(t) := K_i(t) * u_{0i} - \sum_{\alpha=1}^{d} \int_0^t \partial_\alpha K_i(t-s) * f_\alpha^i(u(s))\,\mathrm{d}s \\ + \int_0^t K_i(t-s) * g_i(s)\,\mathrm{d}s,$$

and by showing that this is a contraction in a suitable complete metric space.

Norms

For $v \in \mathbb{R}^n$ we write $\|v\| = \max_{1\le i\le n} |v_i|$. We have chosen this norm because it defines invariant balls $B(a;s)$ for the semi-group associated with the system of non-coupled linear equations $\partial_t v_i + Q_i v_i = 0$, $1 \le i \le n$ (when $Q_1 = \dots = Q_n$, then any norm of $\mathbb{R}^n$ can be used), thanks to the maximum principle. For $m \in \mathbb{N}^*$ and $1 \le p \le \infty$ we write

$$\|v\|_{\infty p} = \operatorname*{ess\,sup}_{0\le t\le T} \|v(t)\|_p$$

the usual norm of $L^\infty(0,T;(L^p(\mathbb{R}^d))^m)$ and

$$\|v\|_{1p} = \int_0^T \|v(t)\|_p\,\mathrm{d}t$$

that of $L^1(0, T; (L^p(\mathbb{R}^d))^m)$. We shorten $\|\cdot\|_{\infty\infty}$ to $\|\cdot\|_\infty$ (it is the norm of $(L^\infty((0, T)\times\mathbb{R}^d))^m$). We remark that these norms depend on the choice of $T > 0$ if v is defined in a time interval containing $(0, T)$. For example, $\lim_{T\to 0+} \|v\|_{1p} = 0$ when v belongs to $L^1(0, T_1; (L^p(\mathbb{R}^d))^m)$. Finally we shall define, similarly, the norms in $L^q(0, T; X)$ where X is a Banach space, for example a Sobolev space.

Hypotheses

We are given two numbers r_0 and r such that $0 < r_0 < r$ and a point $\bar{u}$ such that $\overline{B(\bar{u}; r)} \subset \mathcal{U}$. We suppose that the initial datum has values in $B(\bar{u}; r_0)$ (which is restrictive only if $\mathcal{U} \neq \mathbb{R}^n$) and that $u_0 - \bar{u}$ is square-integrable. As far as the forces g_i are concerned their smoothness will be made precise in each statement. We shall denote by G_T the ball of $(L^\infty(0, T) \times \mathbb{R}^d)^n$ defined by $\|u - \bar{u}\|_\infty \leq r$. It is a complete metric space. We are going to consider L as a mapping defined on G_T. The essential result concerning local existence in time is the following.

Lemma 6.2.1 *Let $g \in L^1(0, S; (L^\infty(\mathbb{R}^d))^n)$ with $S > 0$. There exists a time $T > 0$ such that L is a contracting mapping of G_T into itself for the norm $\|\cdot\|_\infty$. If in addition $g \in L^1(0, T; (L^2(\mathbb{R}^d))^n)$, then L is equally contracting for the norm $\|\cdot\|_{\infty 2}$.*

We deduce then from Picard's theorem the statement of existence (in which we do not suppose that f is the flux of a hyperbolic system).

Corollary 6.2.2 *Let g and T be as above. The mapping L possesses a unique fixed point $u \in G_T$. In addition $u \in L^\infty(0, T; (L^2(\mathbb{R}^d))^n)$. In fact, there exists a number $C_1 = C_1(r_0, r, T)$ such that we have*

$$\|u - \bar{u}\|_\infty \leq C_1(\|u_0 - \bar{u}\|_\infty + \|g\|_{1\infty}), \tag{6.13}$$

$$\|u - \bar{u}\|_{\infty 2} \leq C_1(\|u_0 - \bar{u}\|_{\infty 2} + \|g\|_{12}). \tag{6.14}$$

Proof Let us first show that we can choose T with the result that $L(G_T) \subset G_T$. Using the fact that $\int_{\mathbb{R}^d} K_i(t, x)\,\mathrm{d}x \equiv 1$, we have

$$Lu(t) - \bar{u} = K(t) * (u_0 - \bar{u}) - \int_0^t \nabla_x K(t - s) * (f(u(s)) - f(\bar{u}))\,\mathrm{d}s$$
$$+ \int_0^t K(t - s) * g(s)\,\mathrm{d}s.$$

We have used a vector notation to simplify the equations. Let us denote by $M(r)$ a Lipschitz constant for the function f in $B(\bar{u}; r)$. As $K_i \geq 0$, we have $\|K_i(t)\|_1 = 1$, therefore (and it is there that we use the fact that $B(\bar{u}; r_0)$ has its edges parallel to the axes)

$$\|Lu(t) - \bar{u}\|_\infty \le \|u_0 - \bar{u}\|_\infty + M(r)\|u - \bar{u}\|_\infty \int_0^t \|\nabla_x K(t-s)\|_1 \, \mathrm{d}s + \int_0^t \|g(s)\|_\infty \, \mathrm{d}s.$$

However, since $\nabla_x K_i(t) = t^{-\frac{1}{2}(d+1)} l_i(t^{-\frac{1}{2}}x)$ where $l_i \in \mathscr{S}(\mathbb{R}^d)$, we have $\|\nabla_x K(t)\| = Ct^{-\frac{1}{2}}$. Thus, for $u \in G_T$.

$$\|Lu(t) - \bar{u}\|_\infty \le r_0 + CrM(r) \int_0^t \frac{\mathrm{d}s}{\sqrt{t-s}} + \|g\|_{1\infty},$$

hence

$$\|Lu(t) - \bar{u}\|_\infty \le r_0 + 2CrM(r)\sqrt{T} + \|g\|_{1\infty}.$$

Choosing T small enough that

$$r_0 + 2CrM(r)\sqrt{T} + \|g\|_{1\infty} \le r, \tag{6.15}$$

we have that $Lu \in G_T$. For $u, v \in G_T$, $q = 2$ and $q = \infty$, we then have

$$\|Lv(t) - Lu(t)\|_q \le M(r) \int_0^t \|\nabla_x K(t-s)\|_1 \|v(s) - u(s)\|_q \, \mathrm{d}s$$

and therefore

$$\|Lv - Lu\|_{\infty q} \le 2CM(r)\sqrt{T}\|v - u\|_{\infty q}.$$

From (6.15) the ratio $k = 2CM(r)T^{1/2}$ is strictly less than 1. The mapping L is thus contracting in G_T for the norm $\|\cdot\|_{\infty 2}$. ∎

To show that the fixed point u of L is in $L^\infty(0, T; (L^2(\mathbb{R}^d))^n)$, we note that the sequence of iterates (we take $u^{m+1} = Lu^m$ and $u^0 \equiv \bar{u}$) is Cauchy in this space, hence converges for the norm $\|\cdot\|_{\infty 2}$ to the limit $\bar{u}$. But there exists a sub-sequence which converges almost everywhere and hence $\bar{u} = u$ is simultaneously in $L^\infty(0, T; (L^2(\mathbb{R}^d))^n)$ and in $(L^\infty((0, T) \times \mathbb{R}^d)^n)$. Finally, the constant C_1 has the value $1/(1-k)$.

Estimate of the derivatives

To show that $u(t)$ has partial derivatives of order p which are in $L^2 \cap L^\infty$, we show that each iterate u^m has and that the corresponding norms remain bounded when m tends to infinity. For that, we shall suppose that in the first instance $u_0 - \bar{u} \in$

$H^{p-1} \cap W^{p-1,\infty}$. Using the smoothing property of the semi-group we shall remove this hypothesis later. Let us begin with the case[2] $p = 1$.

Lemma 6.2.3 *Let* $g \in L^1(0, T; (H^1 \cap W^{1,\infty}(\mathbb{R}^d))^n)$ *and* T *be as above. There exist* $T_0 \in (0, T]$ *and* $C_2 > 1$ *such that the solution of (6.12) satisfies*

$$u - \bar{u} \in L^\infty(0, T_0; L^2),$$
$$t^{1/2} \nabla_x u \in L^\infty(0, T_0; L^2 \cap L^\infty)$$

with the upper bounds

$$\|u - \bar{u}\|_{\infty 2} \le C_2(\|u_0 - \bar{u}\|_2 + \|g\|_{12}),$$
$$\|t^{1/2} \nabla_x u\|_{\infty 2} \le C_2(\|u_0 - \bar{u}\|_2 + \|\nabla_x g\|_{12}),$$
$$\|t^{1/2} \nabla_x u\|_\infty \le C_2(\|u_0 - \bar{u}\|_\infty + \|\nabla_x g\|_{1\infty}).$$

Remark In this statement as in those that follow, the time of existence T_0 depends only on r_0, r and the norm of g in $L^1(0, S; X)$ where X is an appropriate Banach space (here, $X = H^1 \cap W^{1,\infty}$). The constants C_1, C_2 depend only on r, while T_0 is bounded below by a number $(2CM(r))^{-2}$ which depends only on r.

Proof of lemma

The first inequality has already been proved. It is sufficient then to show that L preserves the above inequalities, that is to say that if v (given in G_T) satisfies them, then Lv satisfies them. Let $v \in G_{T_0} \cap L^\infty(0, T; L^2)$ where T_0 has still to be made precise. We have

$$\nabla_x Lv(t) = \nabla_x K(t) * (u_0 - \bar{u}) - \int_0^t \nabla_x K(t-s) * \nabla_x f(v(s))\, ds$$
$$+ \int_0^t K(t-s) * \nabla_x g(s)\, ds.$$

Hence

$$\|\nabla_x Lv(t)\|_2 \le Ct^{-1/2}\|u_0 - \bar{u}\|_2 + C\int_0^t \|\nabla_x f(v(s))\|_2 \frac{ds}{\sqrt{t-s}} + \|\nabla_x g\|_{12}.$$

But

$$\|\nabla_x f(v(s))\|_q \le M(r)\|\nabla_x v(s)\|_q$$

[2] From here on, we differ from the analysis of Hoff and Smoller, who do not estimate the L^∞-norms of the derivatives. It does not seem possible to perform an induction argument using only the L^2-norms and the Lemma 2.1 of [44] appears to be false.

for all $1 \le q \le \infty$. The above equation therefore becomes

$$\|\nabla_x Lv(t)\|_2 \le Ct^{-1/2}\|u_0 - \bar{u}\|_2 + CM(r)\int_0^t \|\nabla_x v(s)\|_2 \frac{\mathrm{d}s}{\sqrt{t-s}} + \|\nabla_x g\|_{12}.$$

Let us choose $T_0 \in (0, T]$ with the result that

$$l := CM(r)\sqrt{T_0}\int_0^1 \frac{\mathrm{d}\sigma}{\sqrt{\sigma(1-\sigma)}} < 1. \tag{6.16}$$

Then

$$\|\nabla_x Lv(t)\|_2 \le Ct^{-1/2}\|u_0 - \bar{u}\|_2 + lT_0^{-1/2}\|t^{1/2}\nabla_x v\|_{\infty 2} + \|\nabla_x g\|_{12},$$

from which

$$\|t^{1/2}\nabla_x Lv\|_{\infty 2} \le C\|u_0 - \bar{u}\|_2 + l\|t^{1/2}\nabla_x v\|_{\infty 2} + T_0^{1/2}\|\nabla_x g\|_{12}.$$

As $u^0 \equiv \bar{u}$, the repeated use of the preceding argument implies

$$\|t^{1/2}\nabla_x u^m\|_{\infty 2} \le \frac{1}{1-l}\big(C\|u_0 - \bar{u}\|_2 + T_0^{1/2}\|\nabla_x g\|_{12}\big).$$

Finally

$$\|\nabla_x Lv(t)\|_\infty \le Ct^{-1/2}\|u_0 - \bar{u}\|_\infty + C\int_0^t \|\nabla_x f(v(s))\|_\infty \frac{\mathrm{d}s}{\sqrt{t-s}} + \|\nabla_x g\|_{1\infty},$$

which leads in a similar way to

$$\|t^{1/2}\nabla_x Lv\|_\infty \le C\|u_0 - \bar{u}\|_\infty + l\|t^{1/2}\nabla_x v\|_\infty + T_0^{1/2}\|\nabla_x g\|_{1\infty}$$

and to

$$\|t^{1/2}\nabla_x u^m\|_\infty \le \frac{1}{1-l}\big(C\|u_0 - \bar{u}\|_\infty + T_0^{1/2}\|\nabla_x g\|_{1\infty}\big).$$

■

We continue with the derivatives of higher order.

Lemma 6.2.4 *Let $p \ge 2$, $g \in L^1(0, S; (H^p \cap W^{p,\infty}(\mathbb{R}^d))^n)$ and T_0 be as above. There exists a polynomial P depending on p, with positive coefficients depending on r, s, and on the norm of $\mathrm{D}^2 f$ in $\mathscr{C}^{p-2}(B(\bar{u}; r))$, such that, if also $u_0 - \bar{u} \in (H^{p-1} \cap W^{p-1,\infty}(\mathbb{R}^d))^n$, then the solution u of (6.12) satisfies*

$$u - \bar{u} \in L^\infty(0, T_0; (H^{p-1} \cap W^{p-1,\infty}(\mathbb{R}^d))^n),$$
$$t^{1/2}(u - \bar{u}) \in L^\infty(0, T_0; (H^p \cap W^{p,\infty}(\mathbb{R}^d))^n),$$

with

$$\left\|\nabla_x^{p-1}u\right\|_{\infty 2}+\left\|\nabla_x^{p-1}u\right\|_{\infty}\leq P(\|u_0-\bar{u}\|+G_{p-1}),$$

$$\left\|t^{1/2}\nabla_x^{p}u\right\|_{\infty 2}+\left\|t^{1/2}\nabla_x^{p}u\right\|_{\infty}\leq P(\|u_0-\bar{u}\|+G_{p}),$$

where $\|u_0-\bar{u}\|$ *is the norm in* $H^{p-1}\cap W^{p-1,\infty}$ *and* G_p *that of* g *in* $L^1(0,T;(H^p\cap W^{p,\infty}(\mathbb{R}^d))^n)$.

Proof We proceed by induction on p. We show that L preserves the inequalities that $u^0\equiv\bar{u}$ satisfies trivially. We deduce that u^m, then u also, satisfies them, which is the substance of the lemma. In the first place we have

$$\nabla_x^{p-1}Lv(t)=K(t)*\nabla_x^{p-1}u_0-\int_0^t\nabla_x K(t-s)*\nabla_x^{p-1}f(v(s))\,\mathrm{d}s$$
$$+\int_0^t K(t-s)*\nabla_x^{p-1}g(s)\,\mathrm{d}s.$$

Hence, if $q=2$ or $q=\infty$,

$$\left\|\nabla_x^{p-1}Lv(t)\right\|_q\leq\left\|\nabla_x^{p-1}u_0\right\|_q+C\int_0^t\left\|\nabla_x^{p-1}f(v(s))\right\|_q\frac{\mathrm{d}s}{\sqrt{t-s}}$$
$$+\int_0^t\left\|\nabla_x^{p-1}g(s)\right\|_q\mathrm{d}s.$$

However, $\nabla_x^{p-1}f(v)-\mathrm{d}f(v)\nabla_x^{p-1}v$ is a polynomial without constant term in the variables $\nabla_x v,\ldots,\nabla_x^{p-2}v$, whose coefficients are the derivatives of f of orders between 2 and $p-1$, and hence are bounded on $B(\bar{u};r)$. From the induction hypothesis, the iterates u^m therefore satisfy

$$\left\|\nabla_x^{p-1}f(v)-\mathrm{d}f(v)\nabla_x^{p-1}v\right\|_q\leq Q(\|u_0-\bar{u}\|+G_{p-2})$$

where in fact Q depends only on the norm of $u_0-\bar{u}$ in $H^{p-2}\cap W^{p-2,\infty}$, on G_{p-2} and on the norm of D^2f in $C^{p-3}(B(\bar{u};r))$. Thus

$$\left\|\nabla_x^{p-1}Lu^m(t)\right\|_q\leq\left\|\nabla_x^{p-1}u_0\right\|_q+G_{p-1}+CM(r)\int_0^t\left\|\nabla_x^{p-1}u^m(s)\right\|_q\frac{\mathrm{d}s}{\sqrt{t-s}}$$
$$+C\int_0^t Q(\|u_0-\bar{u}\|+G_{p-2})\frac{\mathrm{d}s}{\sqrt{t-s}}$$

and hence

$$\left\|\nabla_x^{p-1}Lu^m(t)\right\|_{\infty q}\leq Q_1(\|u_0-\bar{u}\|+G_{p-1})+k\left\|\nabla_x^{p-1}u^m\right\|_{\infty q},$$

from which

$$\left\|\nabla_x^{p-1}u^m(t)\right\|_{\infty q}\leq\frac{1}{1-k}Q_1(\|u_0-\bar{u}\|+G_{p-1}).\tag{6.17}$$

Similarly, applying the induction hypothesis and using (6.17) to bound the term $\|\nabla_x^p f(u^m) - \mathrm{d}f(u^m)\nabla_x^p u^m\|_q$, we obtain

$$\begin{aligned}\left\|\nabla_x^p Lu^m(t)\right\|_q &\le Ct^{-1/2}\left\|\nabla_x^{p-1}u_0\right\|_q + C\int_0^t \left\|\nabla_x^p f(u^m(s))\right\|_q \frac{\mathrm{d}s}{\sqrt{t-s}} \\ &\quad + \int_0^t \left\|\nabla_x^p g(s)\right\|_q \mathrm{d}s \\ &\le Ct^{-1/2}\left\|\nabla_x^{p-1}u_0\right\|_q + G_p \\ &\quad + C\int_0^t \left(M(r)\left\|\nabla_x^p u^m(s)\right\|_q + Q_2(\|u_0-\bar u\| + G_{p-1})\right)\frac{\mathrm{d}s}{\sqrt{t-s}}.\end{aligned}$$

Thus

$$\left\|t^{1/2}\nabla_x^p Lu^m\right\|_{\infty q} \le Q_3(\|u_0-\bar u\| + G_p) + l\|t^{1/2}\nabla_x^p u^m\|_{\infty q},$$

from which it follows that

$$\left\|t^{1/2}\nabla_x^p u^m\right\|_{\infty q} \le \frac{1}{1-l}Q_3(\|u_0-\bar u\| + G_p).$$

■

Now let us show that the solution is smooth when $t > 0$, this being true even for non-smooth data. If $u_0 - \bar u$ belongs only to $L^2(\mathbb{R}^d)$ (with still $\|u_0-\bar u\|_\infty \le r_0$), then the iteration converges to the unique solution u in $L^\infty(0, T; (L^2 \cap L^\infty(\mathbb{R}^d))^n)$. In addition $t^{1/2}u^m \in L^\infty(0, T_0; H^1 \cap W^{1,\infty})$. Let $t_0 > 0$. Making use of the semi-group property of K, we have

$$Lv(t) = K(t-t_0)*(Lv)(t_0) - \int_{t_0}^t \nabla_x K(t-s)*f(v(s))\,\mathrm{d}s + \int_{t_0}^t K(t-s)*g(s)\,\mathrm{d}s.$$

Applying this to $v = u^{m-1}$ and using the fact that $Lv(t_0) \in H^1 \cap W^{1,\infty}$, we deduce that if also $g \in L^1(0, S; (H^2 \cap W^{2,\infty}(\mathbb{R}^d))^n)$, we have for all $t_0 < t < T_0$ and $q = 2$ or ∞,

$$\left\|(t-t_0)^{1/2}\nabla_x^2 u^m\right\|_{\infty q} \le P\left(\|u_0-\bar u\|_2, \|u_0-\bar u\|_\infty, G_1, G_2, \frac{1}{\sqrt{t_0}}\right)$$

for a suitable polynomial P. In proceeding by induction on the order of the derivatives, we state that if g is still more smooth, there exists a polynomial P_p of $p+3$ variables such that if $t^* \in (0, T_0]$, then

$$\left\|(t-t^*)^{1/2}\nabla_x^p u^m\right\|_{\infty q} \le P_p\left(\|u_0-\bar u\|_2, \|u_0-\bar u\|_\infty, G_1, \dots, G_p, \frac{1}{\sqrt{t^*}}\right) \qquad (6.18)$$

for $q = 2$, $q = \infty$, $p \ge 2$ and for all m. The convergence towards the solution of (6.12) thus confirms that $u \in L^\infty_{\mathrm{loc}}(0, T_0; (H^p \cap W^{p,\infty})^n)$ when $g \in L^1(0, S; (H^p \cap W^{p,\infty}(\mathbb{R}^d))^n)$. Let us sum up this in the following theorem.

Theorem 6.2.5 *Let $u_0 \in (L^2 \cap L^\infty(\mathbb{R}^d))^n$ be such that u_0 takes its values in a block $\prod_{i=1}^n [a_i, b_i]$ strictly included in $\mathscr{U}$. Let $g \in L^1(0, S; (H^p \cap W^{p,\infty}(\mathbb{R}^d))^n)$. Then there exists $T_0 \in (0, S]$ such that the system (6.11) possesses a unique solution in $\mathscr{C}([0, T_0]; (L^2(\mathbb{R}^d))^n) \cap L^\infty((0, \infty) \times \mathbb{R}^d)$ with $u(0, \cdot) = u_0$. In addition there exists $C > 1$ such that this solution satisfies*

$$u \in L^\infty_{\mathrm{loc}}(0, T_0; (H^p \cap W^{p,\infty})^n),$$

$$\|u - \bar{u}\|_{\infty 2} + \|t^{1/2}\nabla_x u\|_{\infty 2} \leq C(\|u_0 - \bar{u}\|_2 + \|g\|_{12}),$$

$$\|u - \bar{u}\|_\infty + \|t^{1/2}\nabla_x u\|_\infty \leq C(\|u_0 - \bar{u}\|_\infty + \|g\|_{1\infty}).$$

Finally, for all $t^* \in (0, T_0]$, there exists a polynomial P_p such that for all $t \in (t^*, T_0]$, we have

$$\left\|(t - t^*)^{1/2}\nabla_x^p u\right\|_{\infty 2} \leq P_p(\|u_0 - \bar{u}\|_2, \|u_0 - \bar{u}\|_\infty, G_1, \ldots, G_p),$$

$$\left\|(t - t^*)^{1/2}\nabla_x^p u\right\|_\infty \leq P_p(\|u_0 - \bar{u}\|_2, \|u_0 - \bar{u}\|_\infty, G_1, \ldots, G_p).$$

Proof If $u \in \mathscr{C}([0, T]; (L^2(\mathbb{R}^d))^n) \cap (L^\infty((0, T) \times \mathbb{R}^d))^n$ is the solution of (6.2) and satisfies $u(0, \cdot) = u_0$, then this is a solution of (6.12) and we have seen that this exists (continuity with values in L^2 comes from the fact that the u^m have this property since $K(t) * u_0$ has it) and is unique. We have already shown all the other stated properties. ■

Remark This theorem is not optimal. We can for example weaken the hypotheses concerning g. If, in addition, g is somewhat smooth with respect to the time (for example $\partial_t g \in (L^2 \cap L^\infty)^n$), the solution itself is also somewhat smooth when $t > 0$. That is shown as previously, by differentiating the integral equation as often as is necessary with respect to the time. As an example, we can state

Theorem 6.2.6 *Let $g \in (\mathscr{D}(\mathbb{R}^{d+1}))^n$ and $u_0 \in (L^2 \cap L^\infty(\mathbb{R}^d))^n$. Then there exists $T > 0$ such that the (unique) solution of the Cauchy problem (6.11) is of class $\mathscr{C}^\infty$ on $(0, T] \times \mathbb{R}^d$. If moreover u_0 is of class $\mathscr{C}^\infty$, then the solution is of class $\mathscr{C}^\infty$ on $[0, T] \times \mathbb{R}^d$.*

Extension of the solution (case $g \equiv 0$)

In what has gone before we have not had the use of an entropy. In fact, the analysis has not used the hyperbolicity of the inviscid system (when we suppress the viscosity). However, the entropy plays an essential rôle in the extension of the solution to $\mathbb{R}^+ \times \mathbb{R}^d$. We suppose that the inviscid system has a strongly convex entropy, denoted by E, of flux F (hence this system is symmetrisable hyperbolic by

Theorem 3.4.2). Free to replace the entropy E by $u \mapsto E(u)-E(\bar{u})-\mathrm{d}E(\bar{u})\cdot(u-\bar{u})$, we can suppose that E is positive, and zero at $\bar{u}$. Hence, there exists a number $\delta > 0$ such that, for all $a \in B(\bar{u};r)$, we have

$$\delta\|a-\bar{u}\|^2 \le E(a) \le \delta^{-1}\|a-\bar{u}\|^2.$$

The solution of the Cauchy problem, since it is smooth, satisfies

$$E(u)_t + \operatorname{div} F(u) + c(u)\|B\nabla_x u\|^2 \le \sum_\alpha \partial_\alpha e^\alpha.$$

Similarly, on $B(\bar{u};r)$, $c(u)$ satisfies $c(u)\|Bm\|^2 \ge \gamma\|m\|^2$, $\forall m \in M_{d\times n}$ where γ is a positive constant. The expressions e^α, of the form $\sum_\beta(\mathrm{d}E\cdot B^{\alpha\beta})(u)\partial_\beta u$, tend to zero at infinity since $u \in L^\infty \cap H^p$ for p large enough (let us say $p > \frac{1}{2}d+1$). Similarly for $F(u)$ which satisfies $F(\bar{u})=0$ (we are allowed this choice) and $\mathrm{d}F(\bar{u}) = \mathrm{d}E(\bar{u})\,\mathrm{d}f(\bar{u}) = 0$. The integration over $\mathbb{R}^d$ thus gives

$$\frac{\mathrm{d}}{\mathrm{d}t}\int_{\mathbb{R}^d} E(u)\,\mathrm{d}x + \gamma\int_{\mathbb{R}^d}\|\nabla_x u\|^2\,\mathrm{d}x \le 0.$$

In particular, $t \mapsto \int_{\mathbb{R}^d} E(u)\,\mathrm{d}x$ decreases on $(0,T_0]$ and hence also on $[0,T_0]$ since u is continuous in $L^2(\mathbb{R}^d)$. Thus

$$\int_{\mathbb{R}^d} E(u(t,x))\,\mathrm{d}x \le \int_{\mathbb{R}^d} E(u_0(x))\,\mathrm{d}x,$$

from which it follows that

$$\|u(t)-\bar{u}\|_2 \le \frac{1}{\delta}\|u_0-\bar{u}\|_2.$$

Also, if $0 < t^* < t_1 \le T_0$, the upper bound (6.18) and Sobolev's inequality which corresponds to the injection $H^m \subset \mathscr{C}^0$ for $m > \frac{1}{2}n$ give

$$\begin{aligned}\|u(t_1)-\bar{u}\|_\infty &\le \|u(t_1)-\bar{u}\|_2^{1-\theta}\left\|\nabla_x^m u(t_1)\right\|_2^\theta\\ &\le \delta^{\theta-1}\|u_0-\bar{u}\|_2^{1-\theta}P_m\left(\|u_0-\bar{u}\|_2, \frac{1}{\sqrt{t^*}}\right)^\theta (t_1-t^*)^{-\theta/2}\end{aligned}$$

with $\theta = \theta(m,n) \in (0,1]$. The numbers r_0, r, t^* and t_1 being fixed, there exists a number $r_1 > 0$ such that if $\|u_0-\bar{u}\| \le r_1$, then the right-hand side is less than r_0. But then the Cauchy problem, made up of the system (6.2) and the initial condition $u(t_1)$ at $t = t_1$, possesses a smooth solution in the interval (t_1, t_1+T_0) which has all the properties stated in Theorem 6.2.5. The solution sought is therefore defined at least on the interval $(0, t_1+T_0)$. We can take $t^* = \frac{1}{4}T_0$ and $t_1 = \frac{1}{2}T_0$. The number r_1 depends only on the choice of r_0 and of r, with the result that we can extend the solution of the Cauchy problem to all the intervals of the form $(0, \frac{1}{4}qT_0)$ by repeating the same argument. The final result is therefore the following.

Theorem 6.2.7 *($g \equiv 0$) Let $\bar{u} \in \mathcal{U}$ and the numbers $r > r_0 > 0$ be such that the block $B(\bar{u}; r)$ is contained in $\mathcal{U}$. There exists $r_1 > 0$ such that if $u_0 \in (L^2 \cap L^\infty(\mathbb{R}^d))^n$ and if $\|u_0 - \bar{u}\|_\infty \le r_0$, $\|u_0 - \bar{u}\|_2 \le r_1$, then the Cauchy problem for (6.11) (where $g \equiv 0$) possesses a global solution $u \in \mathcal{C}_b(\mathbb{R}^+; (L^2(\mathbb{R}^d))^n) \cap L^\infty(\mathbb{R}^+ \cap \mathbb{R}^d)^n$. This solution is unique in this class and satisfies the estimates of Theorem 6.2.5 with in addition*

$$\|u - \bar{u}\|_{\infty 2} \le \delta(r)\|u_0 - \bar{u}\|_2.$$

Existence with a small diffusion

The above theorem provides global existence for small data and Theorem 6.2.5 ensures the local existence for all smooth data when $\varepsilon > 0$ is fixed. In practice, as we consider the sequence $(u^\varepsilon)_{\varepsilon>0}$ when ε tends to zero, we wish to know that these solutions are defined in a common strip $(0, S) \times \mathbb{R}^d$. We ought therefore to show that the time of existence T_ε of u^ε does not tend to zero.

In fact, neither of the two above theorems leads to this conclusion. To apply them, we consider a fixed value of ε, let us say $\varepsilon = 1$, by the change of variables $(t, x) \mapsto (\varepsilon t, \varepsilon x)$. Hence, we apply them to the given initial condition

$$\tilde{u}_{0\varepsilon}(x) := u_0(\varepsilon x).$$

This always satisfies $\|\tilde{u}_{0\varepsilon} - \bar{u}\| \le r_0$ and the local theorem yields a solution $\tilde{u}^\varepsilon$ in the strip $(0, T_0(r)) \times \mathbb{R}^d$. But the solution $u^\varepsilon(t, x) = \tilde{u}(t/\varepsilon, x/\varepsilon)$ is only defined for $0 < t \le T_\varepsilon = \varepsilon T_0$. Besides, the global theorem 6.2.7 does not apply for small ε as $\|\tilde{u}_{0\varepsilon} - \bar{u}\|_2 = \varepsilon^{-d/2}\|u_0 - \bar{u}\|_2 > r_1$.

We shall see in the following section a sharper estimate which makes use of the regular solution of the hyperbolic problem (6.1) (which is clearly hyperbolic since it has a convex entropy) and which permits us to prove the existence of u^ε and the convergence of the sequence $(u^\varepsilon)_{\varepsilon>0}$ to that value in a strip $(0, S) \times \mathbb{R}^d$ for an $S > 0$. However, we shall restrict ourselves to the case of a single space dimension.

6.3 Smooth convergence as $\varepsilon \to 0^+$

In this section, we shall prove a convergence result when $d = 1$, if the diffusion has constant coefficients.

We suppose still that the system (6.1) possesses a strongly convex entropy E of flux F. Hence, it is symmetrisable hyperbolic. Let $0 < r_0 < r$, $\bar{u}$ and $u_0 \in L^\infty(\mathbb{R}; B(\bar{u}; r_0))$ as in the preceding section. There exists a unique local smooth solution of the Cauchy problem for (6.2). We denote this solution by u^ε and its time of existence by $T_\varepsilon > 0$. We have seen that $T_\varepsilon \geq \varepsilon T_0(r)$. We define

$$S_\varepsilon = \inf\{\tau \leq T_\varepsilon; \|u^\varepsilon(t) - \bar{u}\|_\infty > r_0, \ \forall t \in [\tau, \tau + \varepsilon T_0] \cap [\tau, T_\varepsilon)\}.$$

There exist times arbitrarily close to S_ε for which $\|u^\varepsilon(t) - \bar{u}\|_\infty \leq r_0$. From the local existence theorem, we thus have that $T_\varepsilon \geq S_\varepsilon + \varepsilon T_0$ and all the estimates that we have stated are valid for u^ε when $t \in (0, \varepsilon T_0 + S_\varepsilon)$ as $u(t)$ can be considered as the solution of the Cauchy problem after a time less than εT_0 and for a given initial condition with values in $B(\bar{u}; r_0)$. In fact, we shall use only the estimate $\|u^\varepsilon(t) - \bar{u}\|_\infty \leq r$.

Finally, suppose that $u_0 \in H^2(\mathbb{R})^n$. Then Theorem 3.6.1 assures us that the Cauchy problem for the system (6.1) possesses a unique regular solution u in a strip $(0, T) \times \mathbb{R}$ which satisfies $u \in \mathscr{C}([0, T); H^2(\mathbb{R})) \cap \mathscr{C}^1([0, T) \times \mathbb{R})$. The aim of this section is to prove the convergence of u^ε to u when ε tends to zero.

For that, we begin by establishing and energy estimate.

The energy estimate

This estimate having an interest in its own right, we present it for very general diffusion tensors.

Theorem 6.3.1 *Let $v \mapsto B(v)$ be a diffusion tensor satisfying the inequality*

$$(\mathrm{D}^2 E(v)\xi \mid B(v)\xi) \geq c(v)\|B(v)\xi\|^2, \quad \forall \xi \in \mathbb{R}^n,$$

where $v \mapsto c(v) > 0$ is a continuous function.

We suppose that the Cauchy problem for the system (6.2) has a smooth solution u^ε with values in $B(\bar{u}; r)$ for all $(t, x) \in [0, t_\varepsilon] \times \mathbb{R}$. We suppose finally that u, the solution of the Cauchy problem for (6.1), has values in a compact set in $\mathscr{U}$ (this is true, even if it entails reducing the value of T).

Then there exists a constant $C > 0$ such that the upper bounds below are valid for all $t \in [0, \min(T, t_\varepsilon)]$:

$$\|u^\varepsilon(t) - u(t)\|_2 \leq C\sqrt{\varepsilon t}, \tag{6.19}$$

$$\int_0^t \int_{\mathbb{R}} \left\| B(u^\varepsilon)u_x^\varepsilon \right\|^2 \mathrm{d}x \, \mathrm{d}s \leq Ct. \tag{6.20}$$

It is clear that these estimates cannot remain valid if u is a weak solution with discontinuities. For example (6.20), when B is invertible, implies, that u_x^ε remains in a bounded set of $(L^2((0, t) \times \mathbb{R}))^n$. By equation (6.2), we deduce that u_t^ε remains in a compact set of $L^2(0, t; (H_{\mathrm{loc}}^{-1}(\mathbb{R}))^n)$. By a classical compactness lemma, it follows that $(u^\varepsilon(t))_\varepsilon$ is a relatively compact sequence for the topology of uniform convergence, for almost all t, and this prevents the convergence (even in $L^\infty(0, t; L^2(\mathbb{R}))$) to a discontinuous function, in contradiction to the first upper bound (6.19).

We shall note also that this theorem does not require that B is invertible. It constitutes a *uniform* estimate with respect to the diffusion.

Proof For every function g defined on $\mathscr{U}$, we write $g = g(u)$ and $g^\varepsilon = g(u^\varepsilon)$. We introduce the expressions

$$\begin{aligned}\Delta(t,x) &:= E^\varepsilon - E - \mathrm{d}E\cdot(u^\varepsilon - u),\\ \delta(t,x) &:= F^\varepsilon - F - \mathrm{d}E\cdot(f^\varepsilon - f).\end{aligned}$$

We have

$$\begin{aligned}\Delta_t + \delta_x &= \varepsilon \mathrm{d}E^\varepsilon \cdot \left(B^\varepsilon u_x^\varepsilon\right)_x - \mathrm{d}E\cdot((u^\varepsilon - u)_t + (f^\varepsilon - f)_x)\\ &\quad -(\mathrm{d}E)_t\cdot(u^\varepsilon - u) - (\mathrm{d}E)_x\cdot(f^\varepsilon - f)\\ &= \varepsilon(\mathrm{d}E^\varepsilon - \mathrm{d}E)\cdot\left(B^\varepsilon u_x^\varepsilon\right)_x - \mathrm{D}^2E(u_t, u^\varepsilon - u) - \mathrm{D}^2E(u_x, f^\varepsilon - f)\\ &= \varepsilon\left((\mathrm{d}E^\varepsilon - \mathrm{d}E)\cdot B^\varepsilon u_x^\varepsilon\right)_x - \varepsilon\left(\mathrm{D}^2E^\varepsilon u_x^\varepsilon - \mathrm{D}^2E u_x \mid B^\varepsilon u_x^\varepsilon\right)\\ &\quad + \mathrm{D}^2E(\mathrm{d}f\cdot u_x, u^\varepsilon - u) - \mathrm{D}^2E(u_x, f^\varepsilon - f).\end{aligned}$$

Let $\gamma > 0$ be a lower bound of $c(v)$ on $B(\bar{u}; r)$, then

$$\begin{aligned}\Delta_t + \delta_x + \gamma\varepsilon\left\|B^\varepsilon u_x^\varepsilon\right\|^2 \le{}& \varepsilon\left((\mathrm{d}E^\varepsilon - \mathrm{d}E)\cdot B^\varepsilon u_x^\varepsilon\right)_x + \varepsilon\|\mathrm{D}^2E\, u_x\|\,\left\|B^\varepsilon u_x^\varepsilon\right\|\\ &- \mathrm{D}^2E(u_x, f^\varepsilon - f - \mathrm{d}f\cdot(u^\varepsilon - u)),\end{aligned}$$

where we have also used the symmetry of $\mathrm{d}f$ relatively to the quadratic form D^2E (Theorem 3.4.2).

On the compact set $B(\bar{u}; r)$, the expression Δ is greater than $c_1(r)\|u^\varepsilon - u\|^2$ where $c_1(r) > 0$. As $f^\varepsilon - f - \mathrm{d}f\cdot(u^\varepsilon - u) = \mathrm{O}(\|u^\varepsilon - u\|^2)$, we therefore have

$$\begin{aligned}\Delta_t + \delta_x + \gamma\varepsilon\left\|B^\varepsilon u_x^\varepsilon\right\|^2 \le{}& \varepsilon\left((\mathrm{d}E^\varepsilon - \mathrm{d}E)\cdot B^\varepsilon u_x^\varepsilon\right)_x + c_2(r)\|u_x\|_\infty\Delta\\ &+ \varepsilon\|\mathrm{D}^2E\, u_x\|\,\left\|B^\varepsilon u_x^\varepsilon\right\|.\end{aligned}$$

Using the Cauchy–Schwarz inequality, we find that this gives

$$\Delta_t + \delta_x + \frac{\gamma\varepsilon}{2}\left\|B^\varepsilon u_x^\varepsilon\right\|^2 \le \varepsilon\left((\mathrm{d}E^\varepsilon - \mathrm{d}E)\cdot B^\varepsilon u_x^\varepsilon\right)_x + c_2(r)\|u_x\|_\infty\Delta + \frac{\varepsilon}{2\gamma}\|\mathrm{D}^2E\, u_x\|^2.$$

Let us integrate this inequality over $\mathbb{R}$. As $u^\varepsilon - u \in L^2(\mathbb{R})$ and $u_x \in L^2(\mathbb{R})$, the integrals of δ_x and of $\varepsilon((\mathrm{d}E^\varepsilon - \mathrm{d}E)\cdot B^\varepsilon u_x^\varepsilon)_x$ are zero. There thus remains

$$\frac{\mathrm{d}}{\mathrm{d}t}\int_{\mathbb{R}}\Delta\,\mathrm{d}x + \frac{\gamma\varepsilon}{2}\int_{\mathbb{R}}\left\|B^\varepsilon u_x^\varepsilon\right\|^2\mathrm{d}x \le c_3\left(\varepsilon + \int_{\mathbb{R}}\Delta\,\mathrm{d}x\right). \tag{6.21}$$

Since $\Delta(0,\cdot) \equiv 0$, the Gronwall inequality yields

$$\int_{\mathbb{R}} \Delta \, \mathrm{d}x \le e^{c_3 t} c_3 \varepsilon t \le c_4 \varepsilon t,$$

from which $\|u^\varepsilon(t) - u(t)\|_2^2 \le c_4 \varepsilon t / c_1$. Finally, integrating (6.21) from 0 to t, we obtain

$$\frac{\gamma\varepsilon}{2}\int_0^t \int_{\mathbb{R}} \|B^\varepsilon u_x^\varepsilon\|^2 \, \mathrm{d}x \, \mathrm{d}s \le c_3 \varepsilon \left(t + \frac{1}{2} c_4 t^2\right) \le c_5 \varepsilon t,$$

which completes the proof of the theorem. ■

The essential point of this theorem is that the constant C depends only on r and on the norms of u in $H^1(\mathbb{R})^n$ and in $W^{1,\infty}(\mathbb{R})^n$. But it does not depend on $\varepsilon > 0$ or on the time of existence $S_\varepsilon + \varepsilon T_0$. In particular, we deduce immediately that u^ε converges to u in $L^\infty(0, S; L^2(\mathbb{R})^n)$ where $S := \min(T, \liminf_{\varepsilon \to 0} S_\varepsilon)$. For this result to have a significance, it must be shown that $S > 0$. That will be shown later on. But first we examine two particular cases.

Two most favourable cases

We return to the estimates of the preceding section to prove the following equality:

$$\Delta_t + \delta_x = \varepsilon(\mathrm{d}E^\varepsilon - \mathrm{d}E)\cdot\left(B^\varepsilon u_x^\varepsilon\right)_x + \mathrm{D}^2E(u_x, f^\varepsilon - f - \mathrm{d}f\cdot(u^\varepsilon - u)). \quad (6.22)$$

Now we use differently the dissipation:

$$\begin{aligned} &\Delta_t + \delta_x + \varepsilon(\mathrm{D}^2E^\varepsilon(u^\varepsilon - u)_x | B^\varepsilon(u^\varepsilon - u)_x) \\ &\quad = \varepsilon\left((\mathrm{d}E^\varepsilon - \mathrm{d}E)\cdot B^\varepsilon u_x^\varepsilon\right)_x + \varepsilon((\mathrm{D}^2E - \mathrm{D}^2E^\varepsilon)u_x | B^\varepsilon(u^\varepsilon - u)_x) \\ &\qquad + \varepsilon(\mathrm{d}E - \mathrm{d}E^\varepsilon)_x \cdot B^\varepsilon u_x + \mathrm{D}^2E(u_x, f^\varepsilon - f - \mathrm{d}f\cdot(u^\varepsilon - u)). \end{aligned}$$

Thus

$$\begin{aligned} &\Delta_t + \delta_x + \gamma\varepsilon\|B^\varepsilon(u^\varepsilon - u)_x\|^2 \le \varepsilon\left((\mathrm{d}E^\varepsilon - \mathrm{d}E)B^\varepsilon u_x^\varepsilon\right)_x + \varepsilon(\mathrm{d}E - \mathrm{d}E^\varepsilon)_x \cdot B^\varepsilon u_x \\ &\quad + \varepsilon((\mathrm{D}^2E - \mathrm{D}^2E^\varepsilon)u_x | B^\varepsilon(u^\varepsilon - u)_x) + c_2(r)\|u_x\|_\infty \Delta. \end{aligned}$$

Using Young's inequality, we have

$$\begin{aligned} &\Delta_t + \delta_x + \frac{\gamma\varepsilon}{2}\|B^\varepsilon(u^\varepsilon - u)_x\|^2 \le \varepsilon[(\mathrm{d}E^\varepsilon - \mathrm{d}E)\cdot B^\varepsilon(u^\varepsilon - u)_x]_x \\ &\quad + c_2(r)\|u_x\|_\infty\Delta + c_3\|u_x\|_\infty^2 \varepsilon\Delta + \varepsilon(\mathrm{d}E - \mathrm{d}E^\varepsilon)\cdot\left(\mathrm{d}B^\varepsilon \cdot u_x^\varepsilon\right)u_x. \end{aligned} \quad (6.23)$$

Let us decompose the last term into two parts. The first,

$$\varepsilon(\mathrm{d}E - \mathrm{d}E^\varepsilon)\cdot(\mathrm{d}B^\varepsilon u_x)u_{x'}$$

is integrable since $u \in L^\infty(0, T; H^2(\mathbb{R}))$ and its integral is bounded above by

$\int_{\mathbb{R}} \Delta\,\mathrm{d}x + c_4(\|u\|_{H^2(\mathbb{R})})\varepsilon^2$. The second is

$$\varepsilon(\mathrm{d}E - \mathrm{d}E^\varepsilon)\cdot(\mathrm{d}B^\varepsilon\cdot(u^\varepsilon - u_x))u_{x'}$$

which is a bad one, in general. There are, however, two favourable cases. On the one hand, when the tensor B has *constant coefficients*, since this term is then zero. On the other hand the *parabolic* case (B is thus invertible), since then

$$\begin{aligned}\varepsilon(\mathrm{d}E - \mathrm{d}E^\varepsilon)\cdot(\mathrm{d}B^\varepsilon\cdot(u^\varepsilon - u)_x)u_x &\leq \varepsilon c\|u_x\|\,\|u^\varepsilon - u\|\,\|B^\varepsilon(u^\varepsilon - u)_x\| \\ &\leq \frac{\gamma\varepsilon}{4}\|B^\varepsilon(u^\varepsilon - u)_x\|^2 + \varepsilon c_5(\|u_x\|_\infty)\Delta.\end{aligned}$$

Finally, the inequality (6.23), integrated with respect to x, leads to

$$\frac{\mathrm{d}}{\mathrm{d}t}\int_{\mathbb{R}}\Delta\,\mathrm{d}x + \frac{\gamma\varepsilon}{4}\int_{\mathbb{R}}\|B^\varepsilon(u^\varepsilon - u)_x\|^2 \leq c_6\int_{\mathbb{R}}\Delta\,\mathrm{d}x + c_4\varepsilon^2.$$

The estimate of Theorem 6.3.1 is therefore improved in

Theorem 6.3.2 *We suppose that*

(1) *either the diffusion tensor has constant coefficients,*
(2) *or the perturbation is parabolic (B is invertible and $\mathrm{D}^2E\cdot B$ is positive definite).*

We suppose that the Cauchy problem for the system (6.2) has a smooth solution u^ε with values in $B(\bar{u}, r)$ for all $(t, x) \in [0, t_\varepsilon] \times \mathbb{R}$. We suppose, finally, that the solution of the Cauchy problem for (6.1) has values in a compact set of $\mathcal{U}$ (which is true even if it means a reduction in the value of T).

Then, there exists a constant $C > 0$ such that the upper bounds below are valid for all $t \in [0, \min(T, t_\varepsilon)]$:

$$\|u^\varepsilon(t) - u(t)\|_2 \leq C\varepsilon\sqrt{t}, \tag{6.24}$$

$$\int_0^t\int_{\mathbb{R}}\|B(u^\varepsilon)(u^\varepsilon - u)_x\|^2\,\mathrm{d}x\,\mathrm{d}s \leq C\varepsilon t. \tag{6.25}$$

Uniformity of the existence times

We return to the case where the diffusion B is invertible and with constant coefficients. The inequality (6.20) is therefore an estimate or $u^\varepsilon_x - u_x$ in $L^2((0, t) \times \mathbb{R})$.

Even if we have to replace r_0 by a number $r_1 \in (r_0, r)$, we can suppose that $\|u_0 - \bar{u}\|_\infty < r_0$. Then, we denote by $T_1 > 0$ the time during which the solution u of the hyperbolic problem remains with its values in $B(\bar{u}; \frac{1}{2}(r_0 + \|u_0 - \bar{u}\|_\infty))$. If $\liminf_{\varepsilon\to 0} S_\varepsilon < T_1$, then $S_\varepsilon + \varepsilon T_0 < T_1$ for arbitrarily small values of ε. On the interval $[S_\varepsilon, S_\varepsilon + \varepsilon T_0]$, we have $\|u^\varepsilon(t) - \bar{u}\|_\infty \geq r_0$ and therefore $\|u^\varepsilon(t) - u(t)\|_\infty \geq r_0 - \|u(t) - \bar{u}\|_\infty \geq \frac{1}{2}(r_0 - \|u_0 - \bar{u}\|_\infty)$ which is a strictly positive constant. In addition, the classical inequality $\|v\|_\infty^2 \le 2\|v\|_2\|v_x\|_2$, valid for all v in $H^1(\mathbb{R})$, gives $\|u^\varepsilon(t) - u(t)\|_\infty^4 \le 4c(r)\varepsilon t\|(u^\varepsilon - u)_x\|_2^2$. Finally,

$$\begin{aligned}\varepsilon T_0\left(\frac{r_0 - \|u_0 - \bar{u}\|_\infty}{2}\right)^4 &\le \int_{S_\varepsilon}^{S_\varepsilon+\varepsilon T_0} \|u^\varepsilon - u\|_\infty^4 \,\mathrm{d}t \\ &\le c_1(r)\varepsilon(S_\varepsilon + \varepsilon T_0)\int_{S_\varepsilon}^{S_\varepsilon+\varepsilon T_0} \|(u^\varepsilon - u)_x\|_2^2 \,\mathrm{d}t \\ &\le 2c_1(r)\varepsilon(S_\varepsilon + \varepsilon T_0)\int_{S_\varepsilon}^{S_\varepsilon+\varepsilon T_0} \left(\|u_x^\varepsilon\|_2^2 + \|u_x\|_2^2\right) \mathrm{d}t \\ &\le c_2(r)\varepsilon(S_\varepsilon + \varepsilon T_0)^2,\end{aligned}$$

where we have used Theorem 6.3.1 and the $\mathscr{C}(0, T; H^1)$ smoothness of u. From this inequality, we deduce an explicit lower bound of the time during which u^ε exists and $\|u^\varepsilon - \bar{u}\|_\infty$ stays less than r:

$$S_\varepsilon + \varepsilon T_0 \ge \min\left(\left(\frac{T_0}{c_2(r)}\right)^{1/2}\left(\frac{r_0 - \|u_0 - \bar{u}\|_\infty}{2}\right)^2, T_1\right).$$

The final result is therefore the following.

Theorem 6.3.3 *Let $v_t + f(v)_x = 0$ be a system of conservation laws equipped with a strongly convex entropy E (thus, it is symmetrisable hyperbolic). Let $B \in M_n(\mathbb{R})$ be a matrix with constant coefficients satisfying $(\mathrm{D}^2E(u)\eta \mid B\eta) \ge c(u)\|\eta\|^2$, $\eta \in \mathbb{R}^n$, with $u \mapsto c(u) > 0$ continuous. Finally, let $u_0 \in H^2(\mathbb{R})^n$, with values in a compact set K of $\mathscr{U}$, this compact set being invariant for the equation $v_t = Bv_{xx}$.*

We denote by u the local smooth solution of the Cauchy problem

$$\begin{cases} u_t + f(u)_x = 0, \\ u(0, \cdot) \quad = u_0. \end{cases}$$

For $\varepsilon > 0$, we denote by u^ε the local smooth solution of the Cauchy problem

$$\begin{cases} u_t^\varepsilon + f(u^\varepsilon)_x = \varepsilon B u_{xx}^\varepsilon, \\ u^\varepsilon(0, \cdot) \quad = u_0. \end{cases}$$

Then there exist a time $T > 0$ and a constant $c(K) > 0$ such that u and u^ε (for $0 < \varepsilon < 1$) are defined on $[0, T) \times \mathbb{R}$ and satisfy

$$\|u^\varepsilon(t) - u(t)\|_2 \le C(K)\sqrt{\varepsilon t}, \quad \forall t \in [0, T),$$

$$\int_0^t \|u_x^\varepsilon\|_2^2 \,\mathrm{d}s \le C(K)t, \quad \forall t \in [0, T).$$

In particular, $u = \lim_{\varepsilon\to 0^+} u^\varepsilon$ for the norm of $L^\infty(0, T; (L^2(\mathbb{R}))^n)$.

Comments It is not, in general, clear that T_c, the time during which we have the convergence of u^ε to u, is equal to the time of the existence T_e of u. But as the sole obstacle to the energy estimates is the growth of u^ε in $L^\infty(\mathbb{R})^n$, we clearly have $T_c = T_e$ once a maximum principle yields a set K of $\mathcal{U}$ in which u^ε remains indefinitely. The most obvious case is that of a scalar equation. Let us take also as an example the Keyfitz and Kranzer system

$$u_t + (\varphi(\|u\|)u)_x = 0,$$

which we perturb in a diagonal manner:

$$u_t^\varepsilon + (\varphi(\|u^\varepsilon\|)u^\varepsilon)_x = \varepsilon u_{xx}^\varepsilon.$$

The expression $\rho^\varepsilon := \frac{1}{2}\|u^\varepsilon\|^2$ satisfies the inequatility

$$\rho_t^\varepsilon + (A(\rho^\varepsilon))_x \leq \varepsilon \rho_{xx}^\varepsilon$$

for a suitable function $A\colon \mathbb{R}^+ \mapsto \mathbb{R}$. Thus $\rho^\varepsilon(t, x) \leq \frac{1}{2}\|u_0\|_\infty^2$ and we deduce that u^ε tends to u in $L^\infty([0, T_e); (L^2(\mathbb{R}))^n)$.

We are restricted to invertible diffusion tensors with constant coefficients because this allows the use of Duhamel's formula and the kernel of a parabolic linear equation, with all the explicit estimates which result. It is, however, plausible from Theorem 6.3.1 that an existence result, uniform with respect to $\varepsilon \in (0, 1]$, must arise for the diffusions $v \mapsto B(v)$ satisfying the inequality

$$(\mathrm{D}^2 E(v)\eta \mid B(v)\eta) \geq c(v)\| B(v)\eta\|^2, \quad \forall \eta \in \mathbb{R}^n,$$

where $c(v) > 0$. The difficulty in the general case is that we must proceed with the estimates of derivatives of order α for $0 \leq |\alpha| \leq m$ with $m > 1 + \frac{1}{2}d$, exactly as in the proof of Theorem 3.6.1, treating in addition the diffusion term.

Even in the case of an invertible diffusion tensor with constant coefficients, we are restricted to the one-dimensional case because of the inequality $\|v\|_\infty^2 \leq 2\|v\|_2\|v_x\|_2$ which is precisely what we need to obtain a lower bound for S_ε. Here also, we should need the estimates of higher order derivatives when $d \geq 2$.

The convergence of u^ε to a weak entropy solution of the system (6.1) is a much more delicate question. On the one hand, the estimates, if they exist, must be valid for sufficiently weak norms, for example L^p norms. On the other hand, we do not have a theorem giving *a priori* the existence of entropy solutions (that of Glimm, restricted to small data, is not satisfactory). It is just this convergence which has been used to construct such solutions. The main method used to establish this procedure has been that of compensated compactness (see the fundamental articles by Tartar [101] and DiPerna [17, 18]). This method is restricted to 2×2 systems (more generally to the systems called *rich*) and gives no information concerning the smoothness of the entropy solution.

The arguments of this section have been used in a more complex body of problems, that of a problem with boundary conditions of Dirichlet type, in [30], [31]. See Chapter 15.

6.4 Scalar case. Accuracy of approximation

In the scalar case, we are given the very strong properties such as the maximum principle, the uniqueness of the Cauchy problem and the contraction property in $L^1(\mathbb{R}^d)$ (see Theorem 2.3.5): for two solutions u and v of the same equation $u_t + \operatorname{div} f(u) = 0$, we have

$$\|u(t) - v(t)\|_1 \leq \|u(0) - v(0)\|_1.$$

These properties and their generalisations to solutions of the perturbed equation

$$u_t^\varepsilon + \operatorname{div} f(u^\varepsilon) = \varepsilon \Delta u^\varepsilon \tag{6.26}$$

allow us (cf. [58]) to bound the error $\|u^\varepsilon(t) - u(t)\|_1$ due to the approximation.

Theorem 6.4.1 (Kuznetsov) *There exists a constant $C > 0$ such that, if $u_0 \in \mathrm{BV}(\mathbb{R}^d)$ and if $u(0) = u^\varepsilon(0) = u_0$, then*

$$\|u^\varepsilon(t) - u(t)\|_1 \leq C\sqrt{\varepsilon t}\,\mathrm{TV}(u_0).$$

This statement can be improved in the genuinely non-linear case in.

Theorem 6.4.2 *We suppose that $d = 1$ and that $\inf_{\mathbb{R}} f'' > 0$. Then, for $u_0 \in \mathrm{BV}(\mathbb{R})$ and with compact support, we have*

$$\|u(t) - u^\varepsilon(t)\|_1 \leq C(u_0)\varepsilon^{1/2}t^{1/4}.$$

Remark (1) Of course, this last estimate is only better than that of Kuznetsov for $t \gg 1$. In addition, if $u_0 \in L^1(\mathbb{R}^d)$, the two estimates are only useful when they are better than the trivial bound

$$\|u(t) - u^\varepsilon(t)\|_1 \leq \|u(t)\|_1 + \|u^\varepsilon(t)\|_1 \leq 2\|u_0\|.$$

The interesting times are therefore

(a) $t \ll \varepsilon^{-1}$, in the general case,
(b) $1 \ll t \ll \varepsilon^{-2}$, in the one-dimensional genuinely non-linear case.

(2) Neither of these two results is uniform with respect to the time and indeed they could not be. In the linear case, with $f \equiv 0$, we have $u(t, x) = u_0(x)$ while

$\|u^\varepsilon(t)\|_\infty$ tends to zero when t tends to infinity. We thus have that

$$\liminf_{t\to+\infty} \|u(t) - u^\varepsilon(t)\|_1 \geq \|u_0\|_1,$$

which is independent of ε.

The genuinely non-linear case (in which $d = 1$ and $u_0 \in L^1$) is subtler and is supported by the asymptotic description of u and of u^ε with $\varepsilon > 0$ and fixed. We can suppose that $f'(0) = 0$ and $f''(0) = 1$. First of all, $u(t)$ is asymptotic in L^1 to an N-wave (see [19], Theorem 9.1):

$$N(x,t) = \begin{cases} x/t, & -\sqrt{(2pt)} \leq x \leq \sqrt{(2qt)}, \\ 0, & \text{otherwise}, \end{cases}$$

where

$$p := -\inf_{x\in\mathbb{R}} \int_{-\infty}^{x} u_0(\xi)\,d\xi, \qquad q := \sup_{x\in\mathbb{R}} \int_{x}^{+\infty} u_0(\xi)\,d\xi.$$

In addition, $u^\varepsilon(t)$ is asymptotic in L^1 to a non-linear diffusion wave of the form

$$v(t,x) = \frac{1}{\sqrt{t}} V\left(\frac{x}{\sqrt{t}}\right),$$

where V is *positive* (if $\int u_0\,dx > 0$, negative otherwise). Thus

$$\liminf_{t\to+\infty} \|u(t) - u^\varepsilon(t)\|_1 \geq \lim_{t\to+\infty} \int_{-\sqrt{2pt}}^{0} \frac{|x|}{t}\,dx = p,$$

which is independent of ε.

(3) As $\|u^\varepsilon(t) - u(t)\|_\infty \leq 2\|u_0\|_\infty \leq 2\,\mathrm{TV}(u_0)$ if $\inf u_0 \leq 0 \leq \sup u_0$, we can deduce from Kuznetsov's theorem and the Hölder inequality the following estimate:

$$\|u^\varepsilon(t) - u(t)\|_p \leq C(\varepsilon t)^{1/2p}\,\mathrm{TV}(u_0)$$

for $p \geq 1$.

(4) If we measure the error in a norm other than $L^1(\mathbb{R})$, we can obtain a power of ε different from $\frac{1}{2}$; the above remark is an illustration of this. But the exponent can approach the optimal value 1 in the favorable cases. Tadmor [100] has shown that if $\inf f'' > 0$ and if the initial condition has a Lipschitz increasing part, that is if

$$\exists M;\quad x < y \Longrightarrow \frac{u_0(y) - u_0(x)}{y - x} \leq M,$$

then

$$\|(u^\varepsilon(t) - u(t)) * \varphi\|_\infty \leq K(t, u_0)\varepsilon\|\varphi_x\|_\infty$$

for every test function $\varphi \in \mathscr{D}(\mathbb{R})$.

Before proving these theorems, we are going to state some properties of the parabolic equation (6.26). First of all, the Cauchy problem has a unique locally smooth solution. This satisfies the maximum principle since the equation can also be written as a transport–diffusion equation $v_t + f'(v) \cdot \nabla v = \varepsilon \Delta v$. Thus, u^ε remains with values in the interval $I = [\inf_{x \in \mathbb{R}} u_0(x), \sup_{x \in \mathbb{R}} u_0(x)]$ which entails that the smooth solution is defined for all time $t \geq 0$. If v is another solution of the same equation (6.26), we have

$$(u^\varepsilon - v)_t + \operatorname{div}(f(u^\varepsilon) - f(v)) = \varepsilon \Delta(u^\varepsilon - v). \tag{6.27}$$

Now, we use the following formulae. If $\varphi \colon \mathbb{R}^2 \to \mathbb{R}$ is a Lipschitz function and if a, b are two functions belonging to $W^{1,1}_{\text{loc}}(\mathbb{R}^{d+1})$ (in particular if a and b are smooth) then $\varphi(a, b) \in W^{1,1}_{\text{loc}}(\mathbb{R}^{d+1})$ and we have

$$\frac{\partial}{\partial x_j}\varphi(a, b) = \frac{\partial \varphi}{\partial a}\frac{\partial a}{\partial x_j} + \frac{\partial \varphi}{\partial b}\frac{\partial b}{\partial x_j}, \quad 0 \leq j \leq d. \tag{6.28}$$

In addition if $\psi \colon \mathbb{R} \to \mathbb{R}$ is convex and Lipschitz then we have

$$\Delta(\psi \circ a) \geq \psi'(a)\Delta a. \tag{6.29}$$

Multiplying equation (6.27) by $\operatorname{sgn}(u^\varepsilon - v)$ and applying the preceding formulae with $\varphi(a, b) := |a - b|$, $\varphi(a, b) := \operatorname{sgn}(a - b)(f^\alpha(a) - f^\alpha(b))$ and $\psi(a) := |a|$, we deduce

$$|u^\varepsilon - v|_t + \operatorname{div}\{\operatorname{sgn}(u^\varepsilon - v)(f(u^\varepsilon) - f(v))\} \leq \varepsilon \Delta |u^\varepsilon - v|.$$

Integrating over $[0, T] \times \mathbb{R}^d$, we obtain

$$\|u^\varepsilon(t) - v(t)\|_1 \leq \|u^\varepsilon(0) - v(0)\|_1, \quad \forall t \geq 0.$$

In particular, if $u^\varepsilon(0) = u_0$ and $v(0, x) = u_0(x - h)$, then v is nothing but a translation of u^ε: $v(t, x) = u^\varepsilon(t, x - h)$, with the result that

$$\|u^\varepsilon(t) - u^\varepsilon(t, \cdot - h)\|_1 \leq \|u_0 - u_0(\cdot - h)\|_1.$$

Dividing by $h > 0$ and letting h tend to zero, we arrive at the decay of the total variation of u^ε,

$$\operatorname{TV}(u^\varepsilon(t)) \leq \operatorname{TV}(u_0). \tag{6.30}$$

We can now proceed with the proof of Theorem 6.4.1.

Proof From the entropy inequality for u we have for all $(s, y) \in \mathbb{R}^+ \times \mathbb{R}^d$

$$|u - u^\varepsilon(s, y)|_t + \operatorname{div}_x\{\operatorname{sgn}(u - u^\varepsilon(s, y))(f(u) - f(u^\varepsilon(s, y)))\} \leq 0. \quad (6.31)$$

In fact, it is necessary to see this inequality in its integral form, including the initial condition which uses test functions. Similarly, again making use of the formulae (6.28) and (6.29), we have for all $(s, y) \in \mathbb{R}^+ \times \mathbb{R}^d$

$$\begin{aligned}&|u^\varepsilon - u(s, y)|_t + \operatorname{div}_x(\operatorname{sgn}(u^\varepsilon - u(s, y))(f(u^\varepsilon) - f(u(s, y))))\\&\quad \leq \varepsilon\Delta|u^\varepsilon - u(s, y)|.\end{aligned} \quad (6.32)$$

Let $\alpha > 0$ and $\beta > 0$ be two parameters which we shall adjust in a moment. We use a smoothing kernel $\omega_\alpha(z) = \alpha^{-1}\omega(z/\alpha)$ where $\omega \in \mathscr{D}^+(\mathbb{R})$ is even and satisfies $\int_{\mathbb{R}} \omega \,\mathrm{d}z = 1$. We also use the smoothing kernel $\Omega_\beta(x) := \omega_\beta(x_1)\ldots\omega_\beta(x_d)$ on R^d. We put

$$g_{\alpha\beta}(s, \tau, x, y) = \omega_\alpha(s - \tau)\Omega_\beta(x - y).$$

Let us define for every smooth function $h \in \mathscr{D}(\mathbb{R}^{d+1})$ and every $a \in \mathbb{R}$

$$\begin{aligned}\theta^t(h, u, a) := &\iint_{(0,t)\times\mathbb{R}^d} \left\{\frac{\partial h}{\partial s}(s, x)|u(s, x) - a|\right.\\&\qquad\left. + \nabla_x h \cdot \operatorname{sgn}(u - a)(f(u) - f(a))\right\} \mathrm{d}x\,\mathrm{d}s\\&+ \int_{\mathbb{R}^d} h(0, x)|u_0(x) - a|\,\mathrm{d}x - \int_{\mathbb{R}^d} h(t, x)|u(t, x) - a|\,\mathrm{d}x.\end{aligned}$$

Since u is an entropy solution of the Cauchy problem of the unperturbed equation, we have $\theta^t(h, u, a) \geq 0$, provided that $h \geq 0$. Let us substitute in this inequality $g(\tau, y, \cdot, \cdot)$ for h and $u^\varepsilon(\tau, y)$ for a. Then let us integrate the resulting expression with respect to τ and y in the strip $(0, t) \times \mathbb{R}^d$. Using the formulae $\partial g/\partial\tau = -\partial g/\partial s$ and $\nabla_y g = -\nabla_x g$ we find that $\delta^t(u, u^\varepsilon) \leq 0$ where

$$\begin{aligned}\delta^t(u, u^\varepsilon) := &\iiiint_{0<s,\tau<t} \left\{\frac{\partial g}{\partial\tau}|u(s, x) - u^\varepsilon(\tau, y)| + \operatorname{sgn}(u(s, x)\right.\\&\qquad\left. - u^\varepsilon(\tau, y))\nabla_y g \cdot (f(u(s, x)) - f(u^\varepsilon(\tau, y)))\right\} \mathrm{d}x\,\mathrm{d}y\,\mathrm{d}s\,\mathrm{d}\tau\\&- \iiint_{0<\tau<t} g(\tau, 0, x, y)|u_0(x) - u^\varepsilon(\tau, y)|\,\mathrm{d}x\,\mathrm{d}y\,\mathrm{d}\tau\\&+ \iiint_{0<\tau<t} g(\tau, t, x, y)|u(t, x) - u^\varepsilon(\tau, y)|\,\mathrm{d}x\,\mathrm{d}y\,\mathrm{d}\tau.\end{aligned}$$

Now, making use of the inequality (6.32) we obtain

$$
\begin{aligned}
0 \geq \delta^t(u, v) \geq -\varepsilon & \iiiint_{0<s,\tau<t} \nabla_y g \cdot \nabla_y |u(s, x) - u^\varepsilon(\tau, y)| \,\mathrm{d}x\,\mathrm{d}y\,\mathrm{d}s\,\mathrm{d}\tau \\
& - \iiint_{0<\tau<t} g(\tau, 0, x, y)|u_0(x) - u^\varepsilon(\tau, y)| \,\mathrm{d}x\,\mathrm{d}y\,\mathrm{d}\tau \\
& + \iiint_{0<\tau<t} g(\tau, t, x, y)|u(t, x) - u^\varepsilon(\tau, y)| \,\mathrm{d}x\,\mathrm{d}y\,\mathrm{d}\tau \\
& - \iiint_{0<s<t} g(0, s, x, y)|u(s, x) - u_0(y)| \,\mathrm{d}x\,\mathrm{d}y\,\mathrm{d}s \\
& + \iiint_{0<s<t} g(t, s, x, y)|u(s, x) - u^\varepsilon(t, y)| \,\mathrm{d}x\,\mathrm{d}y\,\mathrm{d}s.
\end{aligned}
$$

Now, making α tend to zero, we derive

$$
\begin{aligned}
\varepsilon & \iiint_{0<s<t} \nabla_y \Omega_\beta(x-y) \cdot \nabla_y |u(s, x) - u^\varepsilon(s, y)| \,\mathrm{d}x\,\mathrm{d}y\,\mathrm{d}s \\
& \geq - \iint \Omega_\beta(x-y)|u_0(x) - u_0(y)| \,\mathrm{d}x\,\mathrm{d}y \\
& \quad + \iint \Omega_\beta(x-y)|u(t, x) - u^\varepsilon(t, y)| \,\mathrm{d}x\,\mathrm{d}y. \qquad (6.33)
\end{aligned}
$$

We find an upper bound for the initial integral as follows:

$$
\begin{aligned}
\iint \Omega_\beta(x-y)|u_0(x) - u_0(y)| \,\mathrm{d}x\,\mathrm{d}y &= \iint \Omega_\beta(z)|u_0(x+z) - u_0(x)| \,\mathrm{d}x\,\mathrm{d}z \\
&\leq \int \Omega_\beta(z)\|z\| \,\mathrm{TV}(u_0)\,\mathrm{d}z = C\beta\,\mathrm{TV}(u_0).
\end{aligned}
$$

Similarly, we have

$$
\begin{aligned}
& \iint \Omega_\beta(x-y)|u(t, x) - u^\varepsilon(t, y)| \,\mathrm{d}x\,\mathrm{d}y \\
& \geq \iint \Omega_\beta(x-y)(|u(t, y) - u^\varepsilon(t, y)| - |u(t, x) - u(t, y)|) \,\mathrm{d}x\,\mathrm{d}y \\
& = \|u(t) - u^\varepsilon(t)\|_1 - \int \Omega_\beta(x-y)|u(t, x) - u(t, y)| \,\mathrm{d}x\,\mathrm{d}y \\
& \geq \|u(t) - u^\varepsilon(t)\|_1 - C\beta\,\mathrm{TV}(u(t)) \\
& \geq \|u(t) - u^\varepsilon(t)\|_1 - C\beta\,\mathrm{TV}(u_0),
\end{aligned}
$$

since $t \mapsto \mathrm{TV}(u(t))$ is decreasing. Thus

$$\|u(t) - u^\varepsilon(t)\|_1 \le 2C\beta\,\mathrm{TV}(u_0) + \varepsilon \iiint_{0<s<t} \nabla_y \Omega_\beta(x-y) \cdot \nabla_y |u^\varepsilon(s,y) - u(s,x)|\,\mathrm{d}x\,\mathrm{d}y\,\mathrm{d}s.$$

Let us accept temporarily the following result.

Lemma 6.4.3 *Let $p\colon \mathbb{R}^{2d} \to \mathbb{R}$ be a measurable function and μ a positive measure of finite mass on $\mathbb{R}^d$. We suppose on the one hand that $y \mapsto p(x,y)$ is bounded and smooth, and on the other hand that $x \mapsto p(x,y)$ is of bounded variation and that $|\nabla_x p(x,y)| \le d\mu(x)$ in the sense of the measures.*

Then

$$\left| \iint_{\mathbb{R}^{2d}} \nabla_y \Omega_\beta(x-y) \cdot \nabla_y p(x,y)\,\mathrm{d}x\,\mathrm{d}y \right| \le \frac{C}{\beta} \int_{\mathbb{R}^d} \mathrm{d}\mu(x).$$

Let us apply the lemma with $p(x,y) := |u(s,x) - u^\varepsilon(s,y)|$ and $\mathrm{d}\mu_x := |\nabla_x u(x,s)|$ to obtain

$$\begin{aligned} \|u(t) - u^\varepsilon(t)\|_1 &\le 2C\beta\,\mathrm{TV}(u_0) + \frac{C\varepsilon}{\beta} \int_0^t \mathrm{TV}(u(s))\,\mathrm{d}s \\ &\le 2C\beta\,\mathrm{TV}(u_0) + \frac{C\varepsilon t}{\beta} \mathrm{TV}(u_0), \end{aligned} \tag{6.34}$$

since $\mathrm{TV}(u(s)) \le \mathrm{TV}(u_0)$. Choosing $\beta = \surd(\varepsilon t)$, we obtain the result sought:

$$\|u^\varepsilon(t) - u(t)\|_1 \le C\sqrt{\varepsilon t}\,\mathrm{TV}(u_0).$$

■

Proof of the lemma

If p is smooth with respect to both of its variables, we can carry out two integrations by parts where we have made use of $\nabla_x^2 \Omega_\beta = \nabla_y^2 \Omega_\beta$ to arrive at

$$\iint_{\mathbb{R}^{2d}} \nabla_y \Omega_\beta(x-y) \cdot \nabla_y p(x,y)\,\mathrm{d}x\,\mathrm{d}y = \iint_{\mathbb{R}^{2d}} \nabla_x \Omega_\beta(x-y) \cdot \nabla_x p(x,y)\,\mathrm{d}x\,\mathrm{d}y.$$

We have an upper bound on the absolute value of this integral by

$$\iint_{\mathbb{R}^{2d}} |\nabla_x \Omega_\beta(x-y)|\,\mathrm{d}y\,\mathrm{d}\mu(x) = \iint_{\mathbb{R}^{2d}} |\nabla_z \Omega_\beta(z)|\,\mathrm{d}z\,\mathrm{d}\mu(x) = \frac{C\varepsilon}{\beta} \int_{\mathbb{R}^d} \mathrm{d}\mu(x).$$

In the general case, we approximate p by a sequence $(p_n)_{n\in\mathbb{N}}$ of smooth functions which satisfy the hypotheses uniformly with respect to n, then we pass to the limit in the above inequality when n tends to infinity. It remains to give the

Proof of Theorem 6.4.2

We recall the inequality (6.34) and we make use of the asymptotic estimate of Dafermos [15] (see Exercise 2.19):

$$\mathrm{TV}(u(s)) \leq C(u_0)s^{-1/2},$$

where C depends on $\inf_{\mathbb{R}} f''$. Then

$$\begin{aligned}\|u(t)-u^\varepsilon(t)\|_1 &\leq C\beta\,\mathrm{TV}(u_0)+\frac{C\varepsilon}{\beta}\int_0^t \frac{\mathrm{d}s}{\sqrt{s}}\\ &\leq C\beta\,\mathrm{TV}(u_0)+\frac{C\varepsilon}{\beta}\sqrt{t}.\end{aligned}$$

Then we choose $\beta^2=\varepsilon\sqrt{t}$.

6.5 Exercises

6.1 For the following systems, show that if the sequences $(u^\varepsilon)_{\varepsilon>0}$ and $(f(u^\varepsilon))_{\varepsilon>0}$ are bounded in $L^1(\omega)$ (where ω is a bounded open set in $\mathbb{R}^+\times\mathbb{R}^d$) and converge in the sense of distributions to u and $f(u)$ respectively, then

$$\lim_{\varepsilon\to 0}\int_\omega \|u^\varepsilon-u\|\,\mathrm{d}x\,\mathrm{d}t=0.$$

The Burgers equation.

Isentropic gas dynamics. Here, we suppose that in addition the sequences $(\rho^\varepsilon)_{\varepsilon>0}$ and $((\rho^\varepsilon)^{-1})_{\varepsilon>0}$ are bounded in $L^\infty(\omega)$ and also that the pressure is of the form $p(\rho)=\rho^\gamma$ with $\gamma>1$.

Show that this property is false for the system $u_t+(\varphi(\|u\|)u)_x=0$ even when $\varphi'\neq 0$ and $\varphi''\neq 0$, as well as for the system of isothermal gas dynamics ($p(\rho)=c\rho$, where c is a constant).

6.2 We consider the scalar equation

$$u_t=0,$$

with its perturbation

$$u_t^\varepsilon=\varepsilon u_{xx}^\varepsilon.$$

We choose the initial condition

$$a(x)=-\operatorname{sgn} x=\begin{cases}1, & x<0,\\ -1, & x>0.\end{cases}$$

Show that the viscous solution is of the form $u^\varepsilon(x,t) = v(x/\sqrt{\varepsilon t})$ where we shall make v explicit. Deduce that $\int_{\mathbb{R}} |u^\varepsilon(t) - u(t)|\,\mathrm{d}x = C\sqrt{\varepsilon t}$ where C is a constant. Thus, from the qualitative point of view, Kuznetsov's theorem is optimal.

6.3 Some readers might find the preceding exercise unconvincing since the initial condition is not itself integrable. We recall therefore the equation $u_t = 0$ and its perturbation $u_t = \varepsilon u_{xx}$. The initial condition is now

$$a(x) = K^s(x) = \frac{1}{\sqrt{s}} K\left(\frac{x}{\sqrt{s}}\right),$$

where K is the kernel of the heat equation:

$$\frac{\mathrm{d}K^s}{\mathrm{d}s} = \frac{\mathrm{d}^2 K^s}{\mathrm{d}x^2}.$$

Show that $u^\varepsilon(t) = K^{\varepsilon t+s}$. We choose $s = \varepsilon t$; show that $\|u(t) - u^\varepsilon(t)\|_1$ depends neither on ε nor on t, but that $\mathrm{TV}(u_0) = c(\varepsilon t)^{-1/2}$. Deduce that the best constant C_V in the upper bound $\|u(t) - u^\varepsilon(t)\|_1 \le C_V(\varepsilon, t)\,\mathrm{TV}(u_0)$ is at least equal to $C\sqrt{\varepsilon t}$. Kuznetsov's theorem is therefore optimal.

Similarly, calculate $\|a\|_1$ and conclude that the best constant $C_1(\varepsilon, t)$ in the inequality $\|u(t) - u^\varepsilon(t)\|_1 \le C_1(\varepsilon, t)\|u_0\|_1$ satisfies $C_- \le C_1(\varepsilon, t) \le 2$ where $C_- > 0$ is independent of ε and of t.

6.4 We consider the Burgers equation,

$$u_t + \left(\frac{1}{2}u^2\right)_x = 0,$$

with its perturbation

$$u_t^\varepsilon + u^\varepsilon u_x^\varepsilon = \varepsilon u_{xx}^\varepsilon.$$

We choose the same initial conditions as in Exercise 2.

(1) Calculate the entropy solution u.
(2) Show that the viscous solution is of the form $u^\varepsilon(t, x) = u^1(x/\varepsilon, t/\varepsilon)$. We write $z = u^1$.
(3) Show that the restriction of z to the quarter-plane $(\mathbb{R}^+)^2$ is the solution of the parabolic problem with Dirichlet condition,

$$\left.\begin{aligned} z_t + z z_x &= z_{xx}, \\ z(0, x) &= -1, \\ z(t, 0) &= 0. \end{aligned}\right\}$$

(4) let $x \mapsto Y$ be the solution of the differential equation $Y' = \frac{1}{2}(Y^2 - 1)$ satisfying $Y(0) = 0$. Also, let $b(t) \in [0, 1)$ be the solution of the equation $\log|1 - b| + b = -t$. Show that Y and b are defined on $\mathbb{R}^+$, that $Y < 0$, $Y' < 0$, $b' > 0$, $b(0) = 0$, $b(+\infty) = 1$, $Y(+\infty) = -1$.

(5) We define $y(t, x) := Y(x/b(t))$. Show that $y_t + yy_x \geq y_{xx}$, $y(t, 0) = 0$ and $y(0, x) = -1$. Deduce from the maximum principle that

$$-1 \leq z(t, x) \leq y(t, x).$$

(6) Show that

$$\int_{\mathbb{R}} |u^\varepsilon(t) - u(t)| \, \mathrm{d}x \leq 2 \int_{\mathbb{R}} \left| 1 + Y\left(\frac{\sigma}{b(\frac{t}{\varepsilon})}\right) \right| \mathrm{d}\sigma,$$

then that

$$\int_{\mathbb{R}} |u^\varepsilon(t) - u(t)| \, \mathrm{d}x \leq c\varepsilon b\left(\frac{t}{\varepsilon}\right)$$

where C is a constant.

(7) Compare this result with Kuznetsov's theorem, respectively for $t = \mathrm{o}(\varepsilon)$ and for $t^{-1} = \mathrm{O}(\varepsilon^{-1})$ (it seems natural that the uniform (in time) estimate $\int_{\mathbb{R}} |u^\varepsilon(t) - u(t)| \, \mathrm{d}x = \mathrm{O}(\varepsilon)$ is true for a genuinely non-linear conservation law (that is with $f'' > 0$) when $\lim_{x \to +\infty} u_0(x) < \lim_{x \to -\infty} u_0(x)$).

6.5 We consider the smooth solutions (u, v) of the isentropic Navier–Stokes equation in which the kinematic viscosity $\varepsilon > 0$ is chosen to be constant. The spatial dimension is $d = 1$ and the equations are written in lagrangian coordinates:

$$\left.\begin{aligned} v_t &= u_x, \\ u_t + p(v)_x &= \varepsilon(u_x/v)_x, \end{aligned}\right\} t > 0, \, x \in (0, M).$$

The total mass M is finite. For simplicity, we assume that the fluid spreads out freely in $\mathbb{R}$. The boundary conditions are thus

$$\varepsilon \frac{u_x}{v} = p(v), \quad x = 0, M.$$

The pressure is a given smooth function which satisfies $p(0) = +\infty$, $p(+\infty) = 0$, $p > 0$. We write $e(v) := -\int_1^v p(w) \, \mathrm{d}w$ and we suppose that $e(0) = +\infty$, $e(+\infty) > -\infty$.

(1) Establish the energy estimate

$$\int_0^M \left(\frac{1}{2}u^2 + e(v)\right) \mathrm{d}x + \varepsilon \int_0^t \int_0^M \frac{u_x^2}{v} \, \mathrm{d}x \, \mathrm{d}s = \int_0^M \left(\frac{1}{2}u_0^2 + e(v_0)\right) \mathrm{d}x.$$

(2) Let $h(t, x) := \int_0^x u(t, \xi)\, d\xi - \varepsilon \log v$. Show that $h_t = -p(v)$.
(3) Deduce that $h(t, x) \leq h(0, x)$, then calculate an explicit lower bound K of $v(t, x)/v_0(x)$.
(4) Deduce also that $h(t, x) \geq h(0, x) - tq(K v_0(x))$ where q is a suitable chosen function. Then calculate an explicit upper bound of $v(x, t)$.
(5) Are the above estimates uniform with respect to ε? Can we deduce a bound for the sequence $(u^\varepsilon, v^\varepsilon)_{\varepsilon>0}$?

6.6 We consider again the smooth solutions of the Navier–Stokes equations, but for an isothermal fluid: $p(v) = v^{-1}$. We write $\tau = \varepsilon u - x$. Verify that τ satisfies a diffusion equation $\tau_t = (\alpha \tau_x)_x$ where α will be identified. Deduce from the maximum principle an explicit bound of u. Is this bound uniform with respect to ε?

7
Viscosity profiles for shock waves

This chapter treats of an admissibility condition for discontinuous solutions of a given hyperbolic system. This approach is restricted to a single space dimension ($d = 1$).

We regard the hyperbolic system

$$u_t + f(u)_x = 0 \tag{7.1}$$

as 'the limit' of the perturbed system

$$u_t + f(u)_x = \varepsilon(B(u)u_x)_x \tag{7.2}$$

when ε tends to zero. By 'the limit', we understand that weak solutions of (7.1) are admissible if and only if they are pointwise limits almost everywhere of sequences of solutions $(u^\varepsilon)_{\varepsilon>0}$ of (7.2) which are locally uniformly bounded. The aim of this chapter is the study of the progressive waves for a system of the form (7.2) from the point of view of existence and asymptotic stability. These waves are used as criteria of admissibility for shock waves of the system (7.1).

7.1 Typical example of a limit of viscosity solutions

A simple case of a family of solutions $(u^\varepsilon)_{\varepsilon>0}$ of (7.2) is that of a progressive wave of speed s independent of ε whose profile is the same for each value of ε, within a change of scale:

$$u^\varepsilon(t, x) = U\left(\frac{x - st}{\varepsilon}\right). \tag{7.3}$$

The condition that (7.3) defines a solution of (7.2) is that $U: \mathbb{R} \to \mathscr{U} \subset \mathbb{R}^n$ is a solution of the differential-algebraic system (differential system if $B(U)$ is invertible) $(B(U)U')' = (f(U))' - sU'$, or equivalently

$$B(U)U' = f(U) - sU + \text{const.} \tag{7.4}$$

In another way, to say that $u^\varepsilon(x,t)$ converges almost everywhere and is locally bounded when ε tends to zero means that U is bounded and has limits, which we denote by u_L and u_R, at $\pm\infty$. The limit of u^ε is then a step function with two values:

$$u(t,x) = \begin{Bmatrix} u_\mathrm{L}, & x < st, \\ u_\mathrm{R}, & x > st, \end{Bmatrix} \tag{7.5}$$

where u_L and u_R satisfy the Rankine–Hugoniot condition

$$[f(u)] = s[u]. \tag{7.6}$$

Finally the equation satisfied by U is

$$B(U)U' = f(U) - f(u_\mathrm{L}) - s(U - u_\mathrm{L}) \, (= f(U) - f(u_\mathrm{R}) - s(U - u_\mathrm{R})). \tag{7.7}$$

Definition 7.1.1 We say that a discontinuous solution of (7.1) of the form (7.5) admits a *viscosity profile* U if U is a bounded solution of the system (7.7) (called the *profile equation*) which tends to u_L at $-\infty$ and to u_R at $+\infty$.

We note that this definition depends *a priori* on the viscosity tensor that we have adopted.

Profiles vs. Lax's entropy condition

As we have said above, a discontinuous solution, represented by $(u_\mathrm{L}, u_\mathrm{R}, s)$ and which admits a viscosity profile, will often be considered as admissible (in fact we shall shortly impose a supplementary condition of stability which is frequently satisfied). The existence of a viscosity profile is thus seen here as a *sufficient* condition of admissibility. On the other hand, if E is a strongly convex entropy (of flux F), for which the diffusion B is dissipative, Lax's entropy condition $[F(u)] \leq s[E(u)]$ is a *necessary* condition since it expresses the inequality $E(u)_t + F(u)_x \leq 0$, itself the consequence of the inequality

$$E(u^\varepsilon)_t + F(u^\varepsilon)_x \leq \varepsilon\big(\mathrm{d}E(u^\varepsilon) \cdot B(u^\varepsilon)u_x^\varepsilon\big)_x.$$

Moreover we verify directly that the existence of a profile implies the entropy condition since if $u_\mathrm{R} \neq u_\mathrm{L}$ and if $(D_u^2 E\eta \mid B(u)\eta) \geq c(u)\|B(u)\eta\|^2$, then

$$(F(U) - sE(U))' = \mathrm{d}E(U) \cdot (f(U) - sU)' = \mathrm{d}E(U) \cdot (B(U)U')'$$

and therefore

$$(F(U) - sE(U) - \mathrm{d}E(U) \cdot B(U)U')' \\ = -(D^2E\, U' \mid BU') \leq -c(U)\|BU'\|^2. \tag{7.8}$$

In practice, the last term cannot be identically zero without U being stationary. Thus $\xi \mapsto F(U) - sE(U) - \mathrm{d}E(U) \cdot B(U)U'$ is strictly decreasing and its evaluation at $\pm\infty$ gives $[F(U)] < s[E(U)]$. The existence of a viscosity profile is thus a sufficient condition of admissibility which is not necessary since a contact discontinuity cannot have such a profile (the contact discontinuities satisfy $[F(U)] = s[E(U)]$ for every entropy). We shall see in §7.2 a subtler version of this remark, but right now we can conclude that the truth concerning the admissibility of discontinuous solutions of (7.1) lies somewhere between the existence of a viscosity profile and Lax's entropy condition.

Profile vs. Lax's shock condition

Pushing ahead of the analysis we can ask if a discontinuous wave $(u_{\mathrm{L}}, u_{\mathrm{R}}, s)$ admitting a profile for a viscosity tensor, but whose profile does not persist under a small perturbation of the data (for example $B \mapsto \tilde{B}$, $f \mapsto \tilde{f}$, or $(u_{\mathrm{L}}, u_{\mathrm{R}}; s) \mapsto (\tilde{u}_{\mathrm{L}}, \tilde{u}_{\mathrm{R}}; \tilde{s})$ with the constraint $[\tilde{f}(\tilde{u})] = \tilde{s}[\tilde{u}]$), is admissible. Actually, the system itself, or again the states and the speed of a wave, are never perfectly known and the imprecision of these data contributes to the instability of the profile, and always hinders observing this in practice. In addition, we therefore shall impose on the viscosity profiles the condition that they persist when the profile equation is perturbed. This is the *structural stability* of the profile.

In general, the system (7.7) is a differential-algebraic system since $B(U)$ is not necessarily invertible. It is correct in general to assume that p, the rank of $B(u)$, is constant. Then, we isolate $n - p$ algebraic equations, applying a projection parallel to the image of B. In general, these define a sub-manifold $V(u_{\mathrm{L}}; s)$ in U of dimension p, on which the profile equation consists of the dynamical system

$$v' = g(u_{\mathrm{L}}, s; v) \tag{7.9}$$

provided that $\ker B(u)$ is supplementary to both $\mathrm{Im}\, B(u)$ and the tangent space at u to $V(u_{\mathrm{L}}; s)$. The fact that $\mathbb{R}^n = \ker B(u) \oplus \mathrm{Im}\, B(u)$ follows for example from the existence of strongly convex entropy E for which B is dissipative, in the sense in which there exists a number $c(u) > 0$ such that $(D_u^2 E\xi, B(u)\xi) \geq c(u)\|B(u)\xi\|^2$. Indeed, if $z \in \ker B(u)^2$ and $y = Bz$, then for all $\alpha \in \mathbb{R}$, we have $(D_u^2 E(y + \alpha z), \alpha y) \geq \alpha^2 \|y\|^2$, which, in the limit $\alpha \to 0$, gives $(D_u^2 Ey, y) = 0$, hence $y = 0$: $z \in \ker B(u)$. Finally, $\ker B(u)^2 = \ker B(u)$, which is the expected property. Of course, u_L and u_R are in $V(u_L; s)$ and are stationary points of the reduced system (7.9). Hence, the profile U is a trajectory of this system, which links the critical points u_L and u_R. This is contained in the stable manifold of u_R, denoted by $W^s(u_R)$, and in $W^i(u_L)$, the unstable manifold of u_L. It is structurally stable if and only if $W^s(u_R)$ and $W^i(u_L)$ are transverse to each other, that is to say when at a point $U(\xi)$ of the trajectory their tangent spaces satisfy (in an obvious notation)

$$T_R^s(U(\xi)) + T_L^i(U(\xi)) = T_{U(\xi)}V(u_L; s). \tag{7.10}$$

This condition does not depend on the choice of the point $U(\xi)$ on the trajectory, but only on the trajectory itself. Since these two tangent spaces contain in common the tangent to the trajectory, the condition of tranversality implies in particular an inequality between dimensions,

$$\dim W^s(u_R) + \dim W^i(u_L) \geq p + 1. \tag{7.11}$$

Let us look, in more detail, at the case of an invertible diffusion ($p = n$). We have $V(u_L; s) = \mathcal{U}$. Let us suppose that, in addition, the system (7.1) possesses a strongly convex entropy E for which B is dissipative. This system is therefore hyperbolic and we denote by $\lambda_1(u), \ldots, \lambda_n(u)$ the eigenvalues of $\mathrm{d}f(u)$, arranged in increasing order. Furthermore, we suppose them to be of constant multiplicities.

Let us re-write the profile equation in the form

$$u' = g(U) =: B(U)^{-1}(f(U) - f(u_R) - s(U - u_R)).$$

First of all, we prove a lemma.

Lemma 7.1.2 *We suppose that the speed s satisfies $\lambda_k(u_R) < s < \lambda_{k+1}(u_R)$ for a certain index $1 \leq k \leq n$. Then the endomorphism $\mathrm{d}g(u_R)$ does not have an eigenvalue with real part zero. The sum of the multiplicities of eigenvalues with negative real part is k.*

Proof Let us write $C = \mathrm{d}f(u_R) - sI_n$ which is invertible, and $S = D^2E(u_R)$ which is symmetric and positive definite. Since B is dissipative, we have

$$(S\eta \mid B(u_R)\eta) > 0, \qquad \forall \eta \in \mathbb{R}^n, \ \eta \neq 0.$$

In particular,

$$\Re(S\bar{\eta} \mid B(u_R)\eta) > 0, \qquad \forall \eta \in \mathbb{C}^n, \ \eta \neq 0.$$

Let η be an eigenvector of $\mathrm{d}g(u_R) = B^{-1}C$, associated with an eigenvalue λ. This is not null and we have

$$\lambda(S\bar{\eta} \mid B\eta) = (S\bar{\eta} \mid C\eta).$$

The right-hand side of this formula is real because, E being an entropy of the system (7.1), the matrix SC is symmetric. If $\Re\lambda = 0$ we therefore have $0 = \Re(S\bar{\eta} \mid B\eta)$ which is false. Therefore $\Re\lambda \neq 0$.

This conclusion remains true when we replace B by $I_n + \alpha B$ with $\alpha \in \mathbb{R}^+$ since this is again a dissipative tensor for E. By continuity, the number of eigenvalues $n(\alpha)$ of $(I_n + \alpha B)^{-1}C$ in the half-plane $\Re z < 0$, counted with their orders of multiplicity, does not depend on α. Hence, it is equal to $n(0) = k$. Dividing by α and letting α tend to infinity we obtain the stated result. ■

When s is an eigenvalue of $df(u_R)$, we obtain a similar result by applying the lemma with $s_1 \neq s$ and letting s_1 tend to s: if $\lambda_j(u_R) < s = \lambda_{j+1}(u_R) = \cdots = \lambda_k(u_R) < \lambda_{k+1}(u_R)$ the number of eigenvalues of $dg(u_R)$ in the half-plane $\Re(z) < 0$ (respectively in the half-plane $\Re(z) > 0$) is equal to j (respectively $n - k$). The dimension of $W^s(u_R)$ is therefore between j and k. Similarly, that of $W^i(u_L)$ is bounded above by $n - m$ where $\lambda_m(u_L) < s \leq \lambda_{m+1}(u_L)$. Then, the inequality (7.11) shows that $k - m \geq 1$. We deduce Lax's shock condition.

Theorem 7.1.3 *We consider a hyperbolic system whose characteristic speeds are of constant multiplicities, provided with a strongly convex entropy E. We consider a diffusion tensor B, strictly dissipative for E. Let $(u_L, u_R; s)$ be a discontinuity of (7.1) for which there exists a structurally stable viscosity profile. Then there exists an index $1 \leq k \leq n$ such that we have*

$$\lambda_k(u_R) \leq s \leq \lambda_k(u_L). \tag{7.12}$$

Naturally, Lax's condition (7.12) is necessary but not sufficient for a viscosity profile to be stable. However, the strict inequalities $\lambda_k(u_R) < s < \lambda_k(u_L)$ constitute a sufficient condition when u_R is close enough to u_L. In fact, in this case, λ_k cannot be linearly degenerate (as u_R is on the kth Hugoniot curve which originates at u_L, (see Theorem 4.2.1), and we have $\lambda_k(u_R) \neq \lambda_k(u_L)$) so it must be simple. Therefore, we have $\lambda_{k-1}(u_L) < s < \lambda_{k+1}(u_R)$. Thus, dim $W^s(u_R) = k$ and dim $W^i(u_L) = n - k + 1$, $W^s(u_R)$ being tangent to the invariant sub-space Y_R^- of $dg(u_R)$ associated with eigenvalues of strictly negative real part and $W^i(u_L)$ being tangent to the invariant sub-space Y_L^+ of $dg(u_L)$ associated with eigenvalues of strictly positive real part. As u_R is near to u_L, s is also near to $\lambda_k(u_L)$. Let us denote by X_-, X_+ and X_0 the invariant sub-spaces of $B(u_L)^{-1}(df(u_L) - \lambda_k(u_L)I_n)$ associated with eigenvalues whose real parts are positive, negative and zero respectively. Then X_0 is a straight line and we have $\mathbb{R}^n = X_- \oplus X_0 \oplus X_+$. In addition, Y_R^- is near to $X_- \oplus X_0$ and Y_L^+ is near to $X_+ \oplus X_0$. Since $X_- \oplus X_0$ and $X_+ \oplus X_0$ are transverse to each other, the same is true of Y_R^- and Y_L^+. At every point of the trajectory, the tangent spaces

$T_R^s(U(\xi))$ and $T_L^i(U(\xi))$ are close to Y_R^- and Y_L^+, therefore they too are transverse and the profile is structurally stable.

It now remains to show that such viscosity profiles exist. This we shall do in §7.2 under the non-linearity condition $(\mathrm{d}\lambda_k \cdot r_k)(u_L) \neq 0$.

7.2 Existence of the viscosity profile for a weak shock

Being given two distinct states u_L and u_R in $\mathscr{U}$ and a number s satisfying the Rankine–Hugoniot condition (we thus have $s = \sigma(u_L, u_R)$), we want to know if there exists a trajectory of the system (7.7) going from u_L to u_R. We have seen that if the tensor B is strictly dissipative for a strongly convex entropy E, of flux F, then the existence of a viscosity profile from u_L to u_R implies the entropy inequality $F(u_R) - F(u_L) < s(E(u_R) - E(u_L))$ and hence excludes the possibility of a profile going from u_R to u_L. In what follows, we shall suppose that B is invertible, that there exists a strongly convex entropy E and that B is strictly dissipative: $(\mathrm{D}^2 E\xi, B(u)\xi) \geq c(u)\|\xi\|^2$ with $c > 0$. Finally, we abandon the use of capital letters in the notation of a profile.

We begin with the study of the scalar case, which is the simplest and the most complete.

The scalar case

If $n = 1$, B is a numerical function, denoted by $u \mapsto b(u)$, which satisfies $b > 0$. The profile equation is here

$$u' = g(u) =: \frac{f(u) - f(u_L) - s(u - u_L)}{b(u)}.$$

We have $g(u_L) = g(u_R) = 0$. The trajectories of the equation are strictly monotonic and there exists one which links u_L to u_R if and only if $(u_R - u_L)g(u)$ is strictly positive for all u lying between u_L and u_R. By this means we recover Oleĭnik's condition, where the inequalities in the broad sense have been replaced by the strict inequalities:

$u_L < u_R$: there is a viscosity profile if and only if the graph of the restriction of f to (u_L, u_R) is strictly above its chord.

$u_L > u_R$: there is a viscosity profile if and only if the graph of the restriction of f to (u_R, u_L) is strictly below its chord.

We note that the existence of a viscosity profile does not depend on the choice of the viscosity b. On the other hand, if f is not affine on any non-trivial interval, every shock $(u_L, u_R; s)$ satisfying Oleĭnik's condition can be seen as the juxtaposition of shocks for which there exist viscosity profiles.

The case of weak shocks with $B = b(u)I_n$

Let us return to the general case of a system of the form (7.1), which we assume to be strictly hyperbolic and to be provided with an eigenvalue λ_k genuinely non-linear at u_L. The Hugoniot locus $H(u_L)$ is locally the union of sub-manifolds which are tangent at u_L to the eigenspaces of $\mathrm{d}f(u_L)$. Since λ_k is genuinely non-linear at u_L, it is a simple eigenvalue. We denote by Γ_k the Hugoniot curve tangent to the eigenvector $r_k(u_L)$ at u_L, which is normalised by $\mathrm{d}\lambda_k \cdot r_k \equiv 1$. We shall denote by l_k the differential eigenform normalised by $l_k \cdot r_k \equiv 1$.

Let us recall that u_L locally separates $\Gamma_k(u_L)$ into two connected components on which we have

$$\text{if } u \in \Gamma_k^+, \; \lambda_k(u_L) < \sigma(u_L, u) < \lambda_k(u),$$
$$\text{if } u \in \Gamma_k^-, \; \lambda_k(u) < \sigma(u_L, u) < \lambda_k(u_L).$$

For a scalar diffusion, we have the following result, which is due to Foy [23]. The proof which follows is that of Goodman [36].

Theorem 7.2.1 *We suppose that λ_k is genuinely non-linear at u_L and that $B(u) = b(u)I_n$ with $b(u) > 0$. For every neighbourhood $\mathcal{V}$ of u_L, there exists a neighbourhood $\mathcal{W}$ of u_L such that if $u_R \in \mathcal{W} \cap \Gamma_k(u_L)$, the discontinuity $(u_L, u_R, \sigma(u_L, u_R))$ admits a viscosity profile with values in $\mathcal{V}$ if and only if this is a shock (that is to say if and only if $u_R \in \Gamma_k^-(u_L)$).*

Proof Even if it entails a change of the parametrisation of the profile equation, we can suppose that $b \equiv 1$. The existence of a profile for this discontinuity is equivalent to the existence of a heteroclinic orbit joining (u_L, σ) to (u_R, σ) for the augmented system

$$\frac{\mathrm{d}}{\mathrm{d}\xi}\begin{pmatrix} v \\ s \end{pmatrix} = \begin{pmatrix} g(v, s) \\ 0 \end{pmatrix} =: G(v, s),$$
$$g(v, s) = f(v) - f(u_L) - s(v - u_L). \tag{7.13}$$

At the stationary point $P = (u_L, \lambda_k(u_L))$, 0 is an eigenvalue of multiplicity 2 for

$$\mathrm{d}G(P) = \begin{pmatrix} \mathrm{d}f(u_L) - \lambda_k(u_L)I_n & 0 \\ 0 & 0 \end{pmatrix}$$

and $\ker \mathrm{d}G(P)$ is of dimension 2, spanned by $(r_k(u_L), 0)$ and $(0, 1)$. In addition, there is no other eigenvalue with real part zero, since they are all real. The theorem of the centre manifold [108] ensures the existence of a sub-manifold M locally invariant for (7.13), of dimension 2 and which contains all the orbits which remain in a sufficiently small neighbourhood $\mathscr{V}_1 \subset \mathscr{V}$ of u_L. This manifold is tangent at u_L to the kernel of $dG(P)$. Since $(v, s) \mapsto (s, x =: l_k(u_L) \cdot (v - u_L))$ is a system of affine coordinates on $dG(P)$, we can choose the same coordinates on M in $\mathscr{V}_1$. For these the flow of (7.13), restricted to M, is vertical ($s =$ const.). See Fig. 7.1.

Let us pass in review certain trajectories that M is bound to contain. First of all, we must have the zeros of G in $\mathscr{V}_1$. There are two kinds, which form two smooth curves according to Theorem 4.2.1; firstly those of the form (u_L, s), $s \in \mathbb{R}$, then those of the form $(u, \sigma(u_L, u))$ for $u \in \Gamma_k(u_L)$. For the latter, we have the formula $\sigma(u_L, u) - \lambda_k(u_L) \sim \frac{1}{2} l_k(u_L) \cdot (u - u_L)$, that is, $s - \lambda_k(u_L) \sim \frac{1}{2}x$. This curve is therefore transverse on the one hand to the preceding curve ($\{u_L\} \times \mathbb{R}$) and on the other hand to the flow. It follows that, in a neighbourhood of P, each vertical line $s = \sigma \neq \lambda_k(u_L)$ contains exactly two critical points of the flow of (7.13), let us say (u_L, σ) and (u_R, σ) with $\sigma = \sigma(u_L, u_R)$. The segment whose extremities are these two points, invariant by the flow, is therefore a heteroclinic trajectory from one to the other, whose direction of motion remains to be determined. We note that between these two points, the only heteroclinic trajectories which remain in $\mathscr{V}_1$ are obtained from the preceding by a shift of the parameter.

To know the direction of the motion of this trajectory, it is enough to know if the critical point $(s = \sigma, x = 0)$ is attractive or repulsive for the flow restricted to the vertical $s = \sigma(u_L, u_R)$ of M. The trajectory goes from (u_L, σ) to (u_R, σ) (and hence corresponds to the profile sought) if and only if $(s, 0)$ is repulsive. The flow on this line can be described by the differential equation $dx/d\xi = h(\sigma, x)$ where $h(s, x) =: l_k(u_L) \cdot (f(u) - f(u_L) - s(u - u_L))$. We have $u - u_L = x r_k(u_L) + O(x^2)$ since M contains the straight line $\{u_L\} \times \mathbb{R}$ and is tangent to $(r_k(u_L), 0)$. Thus $h(s, x) = (\lambda_k(u_L) - s)x + O(x^2)$ and $(dh/dx)(s, 0) = \lambda_k(u_L) - s$. Hence the point $(s, 0)$ is repulsive if and only if $\lambda_k(u_L) - \sigma(u_L, u_R) > 0$, that is to say if and only if $u_R \in \Gamma_k^-$. ■

Extensions of Theorem 7.2.1 ($B \neq bI_n$)

Since u_R is near to u_L, the viscosity profile which we have constructed in the case where $B = b(u)I_n$ is structurally stable (see §7.1). We deduce that if the viscosity tensor at u_L is nearly a strictly positive scalar matrix, then a weak shock $(u_L, u_R; \sigma(u_L, u_R))$ again admits a viscosity profile, still structurally stable.

When B is not near to a scalar matrix, we use the dissipative hypothesis of $B(u)$ relatively to the strongly convex entropy E. Recall the proof of Theorem 7.2.1 with

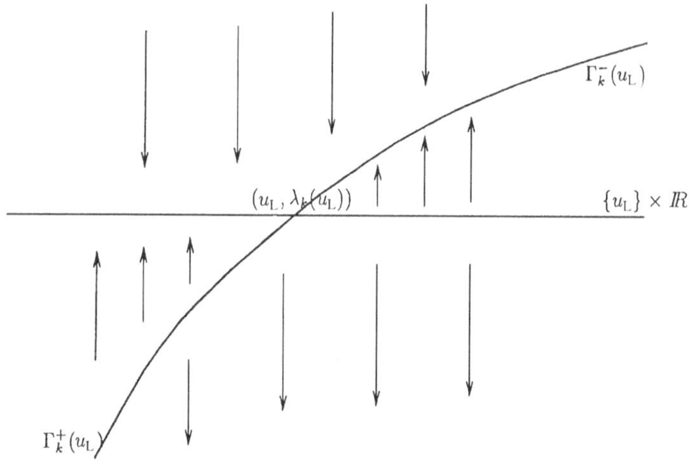

Fig. 7.1: The flow of G on the centre manifold at $(u_L, \lambda_k(u_L))$.

now

$$G_B(v, s) =: \begin{pmatrix} B(v)^{-1} g(v, s) \\ 0 \end{pmatrix}.$$

The critical points of G_B are the same as those of G. We have

$$\mathrm{d}G_B(P) = \begin{pmatrix} B^{-1}C & 0 \\ 0 & 0 \end{pmatrix},$$

with $C = \mathrm{d}f(u_L) - \lambda_k(u_L)I_n$. The matrix $B^{-1}C$ has only one eigenvalue whose real part is zero, $\lambda = 0$, which is simple (see Lemma 7.1.2 and the subsequent comments). Hence there is again a centre manifold M_B of dimension 2, which has the same tangent space as M at u_L. Therefore, the flow is again transverse to the two curves from critical points and there is always a trajectory which links (u_L, σ) and (u_R, σ), when u_R is close to u_L and $\sigma = \sigma(u_L, u_R)$. As

$$[F(u(\xi)) - \sigma E(u(\xi))]_{\xi=-\infty}^{\xi=+\infty} < 0,$$

the trajectory goes from (u_L, σ) to (u_R, σ) if and only if (u_L, u_R, σ) satisfies Lax's entropy condition. Finally:

Theorem 7.2.2 *We suppose that the system (7.1) is provided with a strongly convex entropy E for which the diffusion B is strictly dissipative. Let* $u \mapsto \lambda_k(u)$ *be a simple eigenvalue of* $\mathrm{d}f$*, genuinely non-linear at* u_L*:* $\mathrm{d}\lambda_k(u_L) \cdot r_k(u_L) \neq 0$*. Then for every neighbourhood* $\mathcal{V}$ *of* u_L*, there exists a neighbourhood* $\mathcal{W}$ *of* u_L *such that if* $u_R \in \Gamma_k(u_L) \cap \mathcal{W}$*, the discontinuity* $(u_L, u_R, \sigma(u_L, u_R))$ *admits a viscosity profile with* *values in $\mathcal{V}$ if and only if it satisfies Lax's entropy condition: $F(u_R) - \sigma E(u_R) < F(u_L) - \sigma E(u_L)$. In addition, this profile is unique to within a translation of the parametrisation.*

7.3 Profiles for gas dynamics

The reader may verify by himself (Exercise 7.6) that the existence of a viscosity profile does not depend on the choice of variables (eulerian or lagrangian), provided that the diffusion is not carried by the equation of the conservation of mass. Hence, we treat only the case of the equations in lagrangian coordinates.

Isentropic fluid with viscosity

For an isentropic fluid, the natural perturbation is the newtonian viscosity:

$$\left.\begin{aligned} v_t &= z_x, \\ z_t + p(v)_x &= \varepsilon(b(v)z_x)_x. \end{aligned}\right\}$$

For a discontinuity (u_L, u_R, s) of the system without viscosity, hence, which satisfies the Rankine–Hugoniot condition

$$[z] + s[v] = 0, \qquad [p(v)] = s[z],$$

the profile equation is

$$\left.\begin{aligned} b(v)z' &= p(v) - p(v_L) - s(z - z_L), \\ z - z_L &= -s(v - v_L). \end{aligned}\right\}$$

The existence of the viscosity profile is equivalent to that of a hetero-clinic orbit linking v_L to v_R for the reduced equation (in which $s \neq 0$)

$$-sb(v)v' = p(v) - p(v_L) + s^2(v - v_L). \tag{7.14}$$

The study of this equation is entirely analogous to that of the scalar case. We denote by I the open interval whose extremities are v_L and v_R.

Theorem 7.3.1 *We suppose that $b > 0$ and p are smooth functions of v. Let $(u_L, u_R; \sigma(u_L, u_R))$ be a discontinuity of isentropic gas dynamics.*

Case $\sigma(u_L, u_R)(u_L - u_R) > 0$: There exists a viscosity profile if and only if the graph of $v \mapsto p(v)$ restricted to I is situated strictly above its chord.

Case $\sigma(u_L, u_R)(u_L - u_R) < 0$: There exists a viscosity profile if and only if the graph of $v \mapsto p(v)$ restricted to I is situated strictly below its chord.

Remark In the above statement we have not made the hypothesis $p' < 0$ (hyperbolicity in the inviscid model). The system without viscosity can be elliptic or simply be able to change type according to the value of v. If p' takes positive values, the inviscid Cauchy problem is ill-posed, in the sense of Hadamard, with the result that the system is not a reasonable model for gas dynamics. The states $v > 0$ for which $p'(v) > 0$ must be excluded by a mathematical criterion which describes faithfully the physics of the problem. As the preceding calculation allows the construction of viscosity profiles between two states u_L and u_R for which $p'(u_L)$ and $p'(u_R)$ are of opposite signs, it seems that viscosity alone is unable to provide such a criterion. We shall see in Exercise 7.3 a more complete approach which gives plausible results, thanks to the introduction of a capillary force. When p has a local minimum at β and a local maximum at $\gamma > \beta$, the model represents a fluid able to occupy a liquid phase and a gaseous phase; the typical equation of state is that of Van der Waals.

7.4 Asymptotic stability

Generalities on the stability of profiles

Viscosity profiles are of importance only inasmuch as we can observe them in experiments or in numerical simulation. This is due to their stability in many different contexts, stability which is general for scalar problems and frequent for systems. For example, we have shown the structural stability of weak shocks associated with genuinely non-linear characteristic fields, it was a matter of stability relatively to a perturbation of the profile equation. In this section and the following one we shall consider another type of stability, where the viscosity is definitively fixed (hence we can take $\varepsilon = 1$) but where the viscosity profile is perturbed at the initial instant and where we make the time t tend to infinity. Then we seek if this profile is asymptotically stable when t tends to infinity. This general problem has principally been the object of two important studies, that of Sattinger [88] in the scalar case and that of Liu [70] for systems.[1] Sattinger treats the scalar case by a spectral analysis of the linearised problem and shows the exponential decay of the error (that is of the difference between the solution $u(t)$ and the profile). In the case of a uniformly convex flux f, this result had been obtained previously by Il' in and Oleĭnik [47]. In general the decay is shown in spaces with weights (these weights enable us, in the general procedure of Sattinger, to shift the spectrum of the linearised operator to a convenient half-plane). The weights chosen are of the

[1] See also the valuable recent work by R. Gardner and K. Zumbrun (*Comm. Pure Appl. Math.* **51** (1998), 797–855).

form $e^{\alpha|x|}$. With the spaces $L^2((1+x^2)^\alpha dx)$, Kawashima and Matsumura have also shown the algebraic decay of the error. However, Osher and Ralston [82] have shown the convergence in $L^1(dx)$ by making use of the contraction properties of the semi-group $S(t)$. In that which concerns the systems, Liu [70] uses the energy estimates and a precise analysis of the waves associated with each characteristic family and with the diffusion. Pego [83] has shown that a spectral analysis of the linearised operator is again possible for certain shocks. But, in all the cases, the stability of the viscosity profiles has been shown only for shocks satisfying strictly Lax's shock condition

$$\lambda_k(u_R) < \sigma < \lambda_k(u_L).$$

At the present time, it seems that there is no known stability theorem for viscosity profiles including the case $s = \lambda_k(u_L)$ or $s = \lambda_k(u_R)$. However, for a scalar equation, Ming Mei [76] obtains a decay rate for small initial perturbation, supposing that f''' does not vanish between u_L and u_R. We give below a result concerning the scalar case without a hypothesis concerning either the flux f or the perturbation. Obviously, this does not contain an estimate of the speed of convergence towards a profile since this convergence might be very slow.

In the next section, we shall give, without proof, a description of the results due to Liu for systems.

The stability problem is posed in the following way. We consider a hyperbolic system in space of one dimension,

$$u_t + f(u)_x = 0,$$

augmented by a diffusion term of the second order:

$$u_t + f(u)_x = (B(u)u_x)_x.$$

We suppose that a discontinuity (u_L, u_R, σ) of the hyperbolic part admits a viscosity profile, denoted by $\xi \mapsto U(\xi)$. We then consider a perturbation $\varphi \in L^1(\mathbb{R}) \cap L^\infty(\mathbb{R})$ of the profile, that is to say that we solve the Cauchy problem with diffusion for the initial condition

$$u_0(x) = U(x) + \varphi(x).$$

We ask if the solution (supposed to exist globally in time and to be unique) converges in a suitable space, let us say $L^p(\mathbb{R})$, to the profile when $t \to \infty$. Since the profile is not unique (every shift gives rise to another one) and since L^p is Hansdorff, the answer is obviously no (it is enough to choose $\varphi =: U(\cdot + x_0) - U$ and to note that we still have $\varphi \in L^1 \cap L^\infty$) and the stability sought is rather an *orbital* stability. Therefore, we ask if there exists a phase shift x_0 such that the solution of the Cauchy problem satisfies

$$\lim_{t\to+\infty} \|u(t) - U(\cdot - \sigma t - x_0)\| = 0$$

for a suitable norm. We shall see, and it is essential in this study, that the phase shift can be calculated explicitly as a function of $\int_{\mathbb{R}} \varphi(x)\,dx$, as a result of the conservation laws when (u_L, u_R, σ) is a Lax shock. In particular, it is for this reason that the perturbation must be integrable.

The scalar case

The scalar case is the simplest one and we lay out for it the most complete results. First of all, let us note that the theory of monotonic operators lets us construct a semi-group $(S(t))_{t\geq 0}$ which solves the Cauchy problem for the equation

$$u_t + f(u)_x = u_{xx} \tag{7.15}$$

when the initial datum u_0 is in $L^\infty(\mathbb{R})$. As the equation (7.15) satisfies the maximum principle, this semi-group enjoys the following properties.

(SG1) (smoothness) For all $a \in L^\infty(\mathbb{R})$, $S(t)\,a$ is infinitely differentiable on $\mathbb{R}$ for all $t > 0$.

(SG2) (maximum principle) If $a \leq b$ almost everywhere, then $S(t)\,a \leq S(t)\,b$, for all $t > 0$.

(SG3) (conservation of mass) If $a - b \in L^1(\mathbb{R})$, then $S(t)\,a - S(t)\,b \in L^1(\mathbb{R})$, and we have for all $t > 0$

$$\int_{\mathbb{R}} (S(t)\,a - S(t)\,b)\,dx = \int_{\mathbb{R}} (a - b)\,dx.$$

(SG4) (contraction) Under the same hypothesis as (SG3), the mapping $t \mapsto \|S(t)a - S(t)b\|$ is decreasing.

It is not necessary for the understanding of this section to prove the above assertions. They are classical.

We consider now a viscosity profile U for the equation (7.15), joining two values u_L and u_R. Denoting by c the speed of the shock wave between u_L and u_R, we are provided with a one-parameter family of progressive waves

$$(x, t) \mapsto u^h(x, t) =: U(x + h - ct).$$

We consider the stability of u^0. Let $\varphi \in L^1(\mathbb{R}) \cap L^\infty(\mathbb{R})$ be a perturbation of the initial condition which thus becomes $u_0(x) = U(x) + \varphi(x)$. Throughout this section we write $u(t) = S(t)u_0$. We ask if there exists a number h such that $\|u(t) - u^h(t)\|_1$

tends to zero when t tends to infinity. If such is the case, then the property (SG3) shows that $\int_{\mathbb{R}}(u_0 - u^h(0))\,dx = 0$, that is to say

$$\int_{\mathbb{R}} (U(x) + \varphi(x) - U(x+h))\,dx = 0.$$

However, Lebesgue's theorem shows that $h \to \int_{\mathbb{R}}(U(x+h) - U(x))\,dx$ is differentiable with derivative $\int_{\mathbb{R}} U'(x+h)\,dx = u_R - u_L$. We therefore have $\int_{\mathbb{R}}(U(x+h) - U(x))\,dx = h(u_R - u_L)$ and the phase shift, if it exists, is determined by

$$h = \frac{1}{u_R - u_L}\int_{\mathbb{R}} \varphi(x)\,dx. \tag{7.16}$$

It remains to show that for this value of h, u and u^h are asymptotically equivalent for the distance defined by $\|\cdot\|_1$ which we shall now state for a reasonable perturbation.

Theorem 7.4.1 *Let U be a viscosity profile which joins u_L to u_R. Let a perturbation φ be a measurable function on $\mathbb{R}$, such that $U + \varphi$ is contained between two translates of U (there exist $a, \beta \in \mathbb{R}$ such that $U(x+\alpha) \le U(x)+\varphi(x) \le U(x+\beta)$ almost everywhere). Then the solution $u(t) = S(t)(U + \varphi)$ of the Cauchy problem for (7.15) satisfies*

$$\lim_{t\to+\infty} \|u(t) - U(\cdot + h - ct)\|_1 = 0,$$

h being defined by the formula (7.16).

We notice that from the hypothesis of this theorem φ is integrable and bounded on $\mathbb{R}$. It is probable that this result remains true if $\varphi \in L^1 \cap L^\infty$, but no such result exists at the moment.[2] In any case, we can report on the work of H. Weinberger [111] in the case where the Riemann problem between u_L and u_R also involves rarefaction waves.

Proof First of all, even if it means making the change of variables $(t, x) \mapsto (t, x - ct)$, we can suppose that the shock is stationary: $c = 0$. Hence, the functions $(t, x) \mapsto U(x + \alpha)$ and $(t, x) \mapsto U(x + \beta)$ are stationary solutions of (7.15). Using the maximum principle and the hypothesis concerning the perturbation (which ensures that φ is bounded), we have $U(x+\alpha) \le u(t, x) \le U(x+\beta)$. Let us write $v(t) = u(t) - U$. Then $v(t)$ is included between two integrable functions which do not depend on the time so remains within a bounded set of $L^1(\mathbb{R})$. In addition, the contraction property yields the inequality $\|v(t, \cdot + r) - v(t)\|_1 = \|u(t, \cdot + r) -$

[2] H. Freistühler and the author have succeeded in proving stability in L^1 for every initial perturbation $\varphi \in L^1(\mathbb{R})$. This result has appeared in *Communications in Pure and Applied Mathematics* **51** (1998), 291–301.

$u(t)\|_1 \le \|u(0, \cdot + r) - u(0)\|_1 = \|\varphi(\cdot + r) - \varphi\|_1$ which tends to zero with r. By the compactness theorem of Fréchet and Kolmogorov, the family $(v(t))_{t\geq 0}$ is therefore relatively compact in $L^1(\mathbb{R})$. The ω-limit set $A =: U + \bigcap_{s\geq 0} B_s$ where B_s is the closure in $L^1(\mathbb{R})$ of $\{v(t) : t > s\}$ is non-empty since $A - U$ is the decreasing intersection of non-empty compact sets. This set is that of all cluster points for the distance $d(z, w) = \|z - w\|_1$ of sub-sequences $(u(t_n))_{n\in\mathbb{N}}$ where $t_n \mapsto \infty$.

The ω-limit set is invariant under the semi-group S since if $a \in A$, with $a = \lim_{n\to\infty} u(t_n)$, then $S(t)a = \lim_{n\to\infty} u(t + t_n)$. For the same reason, $S(t)\colon A \to A$ is onto as we also have $a = S(t)b$ where b is a cluster point of the sequence $(u(t_n - t))_{n\in\mathbb{N}}$. The smoothness property (SG1) therefore implies that A is included in $\mathscr{C}^\infty$.

Now, let $k \in \mathbb{R}$. The decreasing function $t \mapsto \|u(t) - U(\cdot - k)\|_1$ admits a limit denoted by $c(k)$ when $t \to \infty$. If $a \in A$, we deduce that $\|a - U(\cdot - k)\|_1 = c(k)$. However, $S(t)\,a$ again belongs to A, so that it follows that the function $t \mapsto \|S(t)\,a - U(\cdot - k)\|_1$ is constant. Let us write provisionally $w(t) = S(t)\,a$ and $z(t) = S(t)\,a - U(\cdot - k)$. We have

$$0 = \frac{\mathrm{d}}{\mathrm{d}t}\|z(t)\|_1 = \int_{\mathbb{R}} z_t \operatorname{sgn} z \,\mathrm{d}x.$$

Now $z_t + (f(w) - f(U(\cdot - k)))_x = z_{xx'}$ were we have used the profile equation for $U\colon U_{xx} = f(U)_x$. Multiplying this by sgn z, we deduce that

$$|z|_t + ((f(w) - f(U(\cdot - k))) \operatorname{sgn} z)_x = z_{xx} \operatorname{sgn} z,$$

which after an integration over $\mathbb{R}$ gives

$$\frac{\mathrm{d}}{\mathrm{d}t}\int_{\mathbb{R}} |z|\,\mathrm{d}x = \int_{\mathbb{R}} z_{xx} \operatorname{sgn} z \,\mathrm{d}x.$$

Finally,

$$0 = \int_{\mathbb{R}} z_{xx} \operatorname{sgn} z \,\mathrm{d}x. \tag{7.17}$$

However, the *a priori* estimates made at the time of the construction of the semi-group S show that w_{xx} is integrable over $\mathbb{R}$ and hence so also is z_{xx}. Therefore, using the theorem of dominated convergence, we have

$$0 = \lim_{\varepsilon\to 0}\int_{\mathbb{R}} z_{xx}\, j'_\varepsilon(z)\,\mathrm{d}x$$

where $j_\varepsilon(\tau) = \sqrt{(\varepsilon^2 + \tau^2)}$. Integrating by parts, we have

$$0 = \lim_{\varepsilon\to 0}\int_{\mathbb{R}} z_x^2\, j''_\varepsilon(z)\,\mathrm{d}x.$$

Let x_0 be a point where z vanishes and let $\gamma > |z_x(x_0)|$. For $\varepsilon > 0$, sufficiently small, we have $\|z\| < \varepsilon$ on $(x_0 - \gamma\varepsilon, x_0 + \gamma\varepsilon)$ since z is differentiable. Now $j''_\varepsilon(\tau) = \varepsilon^{-1} J(\tau/\varepsilon)$ with $J(\tau) = (1+\tau^2)^{-3/2}$. Thus

$$\int_{\mathbb{R}} z_x^2 j''_\varepsilon(z)\,\mathrm{d}x \geq \frac{1}{\varepsilon}\int_{x_0-\gamma\varepsilon}^{x_0+\gamma\varepsilon} J(1) z_x^2\,\mathrm{d}x,$$

of which the right-hand side tends to $2\gamma J(1) z_x(x_0)^2$ when ε tends to zero. We deduce that $z_x(x_0) = 0$. Finally we have proved (taking $t = 0$ in the preceding calculation) that

$$\forall a \in A, \quad \forall k \in \mathbb{R}, \qquad (a(x) = U(x-k)) \Rightarrow (a'(x) = U'(x-k)). \tag{7.18}$$

To conclude, we note first of all that a lies between $U(\cdot+\alpha)$ and $U(\cdot+\beta)$ as limit of such functions, hence a takes its values strictly between u_{L} and u_{R}. Thus the function $x \mapsto k(x) =: x - U^{-1} \circ a(x)$ is well-defined and smooth (recall that U is strictly monotonic). By construction, $a(x) = U(x - k(x))$, which on differentiation gives $a'(x) = U'(x - k(x))(1 - k'(x))$. Using (7.18) we find that

$$U'(x - k(x))k'(x) = 0$$

and hence that $k'(x) = 0$ since U' does not vanish. Finally, k is a constant and $a = U(\cdot - k)$. However, the elements of A satisfy

$$\int_{\mathbb{R}} (a - U - \phi)\,\mathrm{d}x = 0,$$

which, as we have seen, fixes the value of k: we have $k = -h$.

Hence, we have proved that the ω-limit set is reduced to a single-element $U(\cdot+h)$. Since the family $(v(t))_{t\geq 0}$ is relatively compact in $L^1(\mathbb{R})$ and as it has only a single limiting value when $t \to +\infty$, it is convergent, that is

$$\lim_{t\to+\infty} \|u(t) - U(\cdot+h)\|_1 = 0.$$

7.5 Stability of the profile for a Lax shock

We consider now a system with a parabolic conservation law

$$u_t + f(u)_x = u_{xx}, \tag{7.19}$$

for which the first order part is strictly hyperbolic. To simplify the notation, we suppose that all the eigenvalues of $\mathrm{d}f(u) = A(u)$ are simple (that is, including those of the linearly degenerate fields). We denote them by $\lambda_1(u) < \cdots < \lambda_n(u)$. The eigenvectors to the right and left are denoted respectively by $r_j(u)$ and $l_j(u)$ with

$l_j \cdot r_k = \delta_j^k$ and with the normalisation $\mathrm{d}\lambda_j \cdot r_j \equiv 1$ for all the genuinely non-linear fields. It is solely for the simplicity of the exposition, related to concentrating our attention on a single concept at a time, that we have chosen the matrix $B(u) \equiv I_n$. In the last sub-section of this section we shall consider other diffusion tensors.

Transport vs. diffusion

Let p be the index of a genuinely non-linear characteristic field. We consider a Lax p-shock $(u_{\mathrm{L}}, u_{\mathrm{R}}; c)$ which admits a viscosity profile (for example a weak shock, according to Theorem 7.2.1), denoted by U. For a perturbed initial condition $u(0, \cdot) = U + \varphi$ where φ is given in $(L^1 \cap L^\infty(\mathbb{R}))^n$ we seek again the asymptotic behaviour, when $t \to +\infty$, of the solution $u(t)$ of the Cauchy problem for (7.19). In particular, one stage of the argument consists of proving that this solution is defined for all time $t > 0$. But as the asymptotic description makes use of the *a priori* estimates of $u(t) - U$ in $L^1 \cap L^\infty$ and of the fact that the estimate in L^∞ is sufficient for showing that the solution remains defined, the two stages (existence, asymptotic behaviour) are done together.

The description cannot be as simple as in the scalar case. In fact, if there exists a phase shift $h \in \mathbb{R}$ such that $\lim_{t \to +\infty} \|u(t) - U(\cdot + h - ct)\|_1 = 0$, then the conservation of mass, expressed as

$$\int_{\mathbb{R}} (u(t, x) - U(x + h - ct))\,\mathrm{d}x = \text{const.},$$

entails that

$$\int_{\mathbb{R}} (U(x + h) - U(x))\,\mathrm{d}x = \int_{\mathbb{R}} \varphi(x)\,\mathrm{d}x,$$

that is, that $h(u_{\mathrm{R}} - u_{\mathrm{L}}) = \int_{\mathbb{R}} \varphi(x)\,\mathrm{d}x$. Now the mass $m = \int_{\mathbb{R}} \varphi\,\mathrm{d}x$ is a vector in $\mathbb{R}^n$ which has no reason to be collinear with $[u] = u_{\mathrm{R}} - u_{\mathrm{L}}$.

Therefore, it is necessary to introduce waves of another type which carry a constant and calculable mass. These waves, called *diffusion waves*, are of small amplitude, of the order of $t^{-1/2}$, with the result that we shall have, despite all, $\lim_{t \to +\infty} \|u(t) - U(\cdot + h - ct)\|_\infty = 0$, for an appropriate phase shift. They will be (asymptotically) localised in one of two sectors $x - ct \gg 1$ and $x - ct \ll 1$, in which $U(x - ct)$ is approximately constant. Let us look at the case of the sector $x - ct \gg 1$, where U and hence u has a value very close to u_{R}. The hyperbolic part of the system (7.19) allows waves of speeds $\lambda_j(u_{\mathrm{R}})$ to propagate. If $j < p$, these waves merge with the viscosity profile, where u is no longer nearer to u_{R}; these waves cannot therefore be present in an asymptotic description. It is the same for $j = p$ because of Lax's shock inequality $\lambda_p(u_{\mathrm{R}}) < c$. Finally, for $j > p$, these waves lengthen the zone occupied by the profile and do not interact with it; hence, we shall observe them for all times large enough. Of course, the diffusion dampens down these waves (which explains their amplitude being $O(t^{-1/2})$) at the same time as they spread out. But this spreading out is a very slow process, of the same nature as the brownian motion in which the location of a particle has expectation $(\lambda_j(u_R) - c)t$ which is linear in time, whereas its standard deviation is only of order $t^{1/2}$. Similarly, these waves will become asymptotically uncoupled from each other since they get farther apart with non-zero speeds $\lambda_j(u_R) - \lambda_k(u_R)$. Finally, we shall observe asymptotically n uncoupled waves, of which none is small in $L^1(\mathbb{R})$ though $n - 1$ are in $L^\infty(\mathbb{R})$:

the viscosity profile of the shock $(u_L, u_R; c)$, which has a phase shift h to be determined,
the diffusion waves of speed $\lambda_k(u_R)$ for $k > p$, which lengthen the shock zone to the right,
the diffusion waves of speed $\lambda_k(u_L)$ for $k < p$, which lengthen the shock zone to the left.

The asymptotic behaviour will then be able to be predicted if each one of these waves depends on a real parameter and if the principle of the conservation of mass leads to their calculation. We shall see in a later sub-section (Liu's theorem) under what condition, of a geometrical nature, that is possible.

Non-linear diffusion waves

We now construct the diffusion waves. As we anticipate that these will be of small amplitude and localised in a domain where the solution is nearly constant we begin by linearising (7.19), for example, around u_R (we shall see however that the linearisation is not precise enough):

$$v_t + A(u_R)v_x = v_{xx}. \tag{7.20}$$

Using the eigenvectors of the matrix $A(u_R)$, we decouple this system of n transport–diffusion equations. We express v in terms of the basis of eigenvectors, $v(t, x) = \sum_i w_i(t, x)r_i(u_R)$, and we obtain

$$\partial_t w_i + \lambda_i(u_R)\partial_x w_i = \partial_x^2 w_i, \quad 1 \le i \le n. \tag{7.21}$$

The change of variables $(t, x) \mapsto (t, x - \lambda_i(u_R)t)$ transforms this equation into the heat equation $w_t = w_{xx}$, for which we know the asymptotic behaviour in $L^1(\mathbb{R})$, governed by the total mass $m(w)$. If $m(w) = \int_{\mathbb{R}} w \, dy$ then $w(t) \sim m(w)k(t)$ where

$k(t)$ is the fundamental solution

$$k(t,y)=\frac{1}{\sqrt{t}}K\left(\frac{y}{\sqrt{t}}\right),\qquad K(\xi)=\frac{1}{2\sqrt{\pi}}\exp\left(-\frac{\xi^2}{4}\right).$$

Let us now see why the linearisation is not satisfactory. The error due to the linearisation of (7.19) is of the order of $\partial_x(D^2f(u_R)v\otimes v)$ which, in particular, includes $\partial_x(w_i^2)X_i$ where X_i is the vector $D^2f(u_R)r_i(u_R)\otimes r_i(u_R)$. Now $\partial_x(w_i^2)$ is a function of the form $t^{-3/2}F(t^{-1/2}(x-\lambda_i(u_R)t))$, of the same order as the two terms $(\partial_t+\lambda_i\partial_x)w$ and $\partial_x^2 w_i$ of (7.21). Hence, we cannot neglect the quadratic term (on the other hand, the terms of higher order are of negligible amplitude and mass owing to these). Therefore, we must consider *a priori* the quadratic system

$$v_t+A(u_R)v_x+\frac{1}{2}D^2f(u_R)(v\otimes v)_x=v_{xx}. \tag{7.22}$$

We observe that there do not exist explicit solutions of this equation because of the coupling due to the quadratic term. Indeed, we have to solve equations of the form

$$(\partial_t+\lambda_i\partial_x)w_i+\frac{1}{2}\sum_{j,k=1}^{n}c_{ijk}\partial_x(w_jw_k)=\partial_x^2 w_i,\qquad 1\le i\le n,$$

where $c_{ijk}=:l_i(u_R)\cdot D^2f(u_R)r_j(u_R)\otimes r_k(u_R)$. However, if w_j behaves, as we expect, as $t^{-1/2}F_j(t^{-1/2}(x-\lambda_jt))$, with $\lambda_j\neq\lambda_k$, where F_j decreases rapidly (as does its first derivative), then the terms $\partial_x(w_jw_k)$ are exponentially small for $j\neq k$, simultaneously in L^∞ and in L^1 when $t\to+\infty$. There remains for us to examine the rôle of square terms $\partial_x(w_j^2)$.

The rôle of the terms $\partial_x(w_j^2)$, $j\neq i$

To understand the effect of such a term in the ith equation when w_j is asymptotically equivalent to $t^{-1/2}F_j(t^{-1/2}(x-\lambda_jt))$ with $\lambda_j\neq\lambda_i$, F_j being rapidly decreasing, we are led, by $x\mapsto x-\lambda_it$ and the elimination of the terms of the equation which are not essential to our understanding, to the case of a heat equation with a source term:

$$w_t-w_{xx}=g(t,x).$$

Here, $g=t^{-1/2}\partial_x\tilde{g}$, $\tilde{g}=G(t^{-1/2}(x-\lambda t))$ with $\lambda\neq0$, G being rapidly descreasing. A rather tiresome calculation shows that the right-hand side of this equation is responsible for a contribution to w of the form $t^{-1}h(t,x)$ where h is bounded, rapidly decreasing and $\|t^{-1}h\|_1=O(t^{-1/2})$. Finally, the term $\partial_x(w_j^2)$ ought not to be taken into account for the calculation of the dominant terms of the asymptotic expansion which we seek. We are then led to a list of uncoupled equations of Burgers–Hopf type:

$$(\partial_t + \lambda_i(u_R)\partial_x)w_i + \frac{b_i}{2}\partial_x(w_i^2) = \partial_x^2 w_i, \quad 1 \le i \le n, \tag{7.23}$$

where $b_i = c_{iii} = (\mathrm{d}\lambda_i \cdot r_i)(u_R)$ takes the value 1 if the ith field is genuinely non-linear, and the value 0 if it is linearly degenerate.

Calculation of the diffusion waves

Equation (7.23), although non-linear, again possesses self-similar solutions of the form

$$w_i(t,x) = \frac{1}{\sqrt{t}} W\left(\frac{x - \lambda_i t}{\sqrt{t}}\right).$$

The profiles W are the integrable solutions of the differential equation $W'' = b_i W W' - \frac{1}{2}(\xi W' + W)$ which has a first integral

$$2W' = b_i W^2 - \xi W. \tag{7.24}$$

The constant of integration is zero, as otherwise, W with a behaviour at infinity of the order of ξ^{-1} would not be integrable. The solutions of (7.24) are written

$$W(\xi) = \frac{\mathrm{e}^{-\xi^2/4}}{C - \frac{b_i}{2}\int_{-\infty}^{\xi} \mathrm{e}^{-s^2/4}\,\mathrm{d}s} = \frac{\kappa \mathrm{e}^{-\xi^2/4}}{1 - \frac{b_i}{2}\kappa \int_{-\infty}^{\xi} \mathrm{e}^{-s^2/4}\,\mathrm{d}s} \tag{7.25}$$

where $C = \kappa^{-1}$, the constant of integration, belongs to $\mathbb{R}\backslash[0, b_i\sqrt{\pi}]$. We link C with the mass m of W by the following calculation (if $b_i \neq 0$):

$$m = \int_{\mathbb{R}} \frac{\mathrm{e}^{-\xi^2/4}\,\mathrm{d}\xi}{C - \frac{b_i}{2}\int_{-\infty}^{\xi} \mathrm{e}^{-s^2/4}\,\mathrm{d}s} = \int_0^{2\sqrt{\pi}} \frac{\mathrm{d}e}{C - \frac{b_i}{2}e} = \frac{2}{b_i}\log\left(\frac{C}{C - b_i\sqrt{\pi}}\right).$$

Finally, the mass being given, we find the value of the parameter

$$\kappa = \frac{1 - e^{-b_i m/2}}{b_i\sqrt{\pi}} \quad (b_i \neq 0) \quad \text{or} \quad \kappa = \frac{m}{2\sqrt{\pi}} \quad (b_i = 0). \tag{7.26}$$

We denote by $W_i(\cdot, \kappa)$ the profile defined in (7.25).

Remarks (1) As for the viscosity profiles, we can envisage composing a diffusion wave for a translation of the space variable (or of the time). However, such an operation does not change the asymptotic expansion since if $w_i = t^{-1/2} W_i(\cdot, \kappa)$, we have $\|w_i(\cdot + h) - w_i\|_1 = \mathrm{O}(t^{-1/2})$ and $\|w_i(\cdot + h) - w_i\|_\infty = \mathrm{O}(t^{-1/2})$.

(2) The asymptotic behaviour for the system (7.19) is extremely different from that of a hyperbolic system since the diffusion waves are of constant sign and depend on only a single parameter, their mass. Tai-Ping Liu has shown ([71] and [68]) that the solution of a hyperbolic system whose initial function has compact support and is small enough behaves for large times as a superposition of N-waves. The N-waves, thus called because of their form in the scalar case,[3] take opposite signs on opposite sides of their mean position and depend on two scalar parameters. Also, the N-waves represent only the effect of genuinely non-linear fields and the situation is even worse for linearly degenerate fields. For these, their oscillations are not damped in the absence of dissipation and their profiles are arbitrary instead of depending on a finite number of parameters.

Liu's theorem

Now, we are able to calculate the terms of the asymptotic expansion by supposing that its main terms are of the form

$$U(\cdot + h - ct) + \sum_{j<p} w_j^{\mathrm{L}}(t) r_j(u_{\mathrm{L}}) + \sum_{j>p} w_j^{\mathrm{R}}(t) r_j(u_{\mathrm{R}}),$$

with

$$w_j^{\mathrm{L}}(t,x) = \frac{1}{\sqrt{t}} W_j\left(\frac{x - \lambda_j(u_{\mathrm{L}})t}{\sqrt{t}}; \kappa_j\right), \quad w_j^{\mathrm{R}}(t,x) = \frac{1}{\sqrt{t}} W_j\left(\frac{x - \lambda_j(u_{\mathrm{R}})t}{\sqrt{t}}; \kappa_j\right).$$

The parameters to be determined are the phase shift h and the numbers κ_j linked to the masses of the terms w_j^{L} and w_j^{R}. Let us suppose that this expansion is valid in $L^1(\mathbb{R})$, that is to say that

$$\lim_{t\to+\infty} \|u(t) - U(\cdot + h - ct) - \sum_{j<p} w_j^{\mathrm{L}}(t) r_j(u_{\mathrm{L}}) - \sum_{j>p} w_j^{\mathrm{R}}(t) r_j(u_{\mathrm{R}})\|_1 = 0. \tag{7.27}$$

Then the conservation of mass, expressed by

$$\int_{\mathbb{R}} (u(t,x) - U(x - ct))\,\mathrm{d}x = \text{const.},$$

implies

$$\begin{aligned}\int_{\mathbb{R}} \phi(x)\,\mathrm{d}x &= \int_{\mathbb{R}} (u(t,x) - U(x-ct))\,\mathrm{d}x \quad (\forall t > 0)\\ &= \lim_{t\to+\infty} \int_{\mathbb{R}} (u(t,x) - U(x-ct))\,\mathrm{d}x\end{aligned}$$

[3] However, for the Burgers equation, this form is rather that of a cyrillic vowel И than an N.

$$= \int_{\mathbb{R}} \left(U(y+h) - U(y) + \sum_{j<p} W_j(y;\kappa_j) r_j(u_L) \right.$$
$$\left. + \sum_{j>p} W_j(y;\kappa_j) r_j(u_R) \right) dy$$
$$= h(u_R - u_L) + \sum_{j<p} m_j r_j(u_L) + \sum_{j>p} m_j r_j(u_R),$$

where the relation between m_j and κ_j is given by the formula (7.26).

The above equality enables us to calculate the n-tuple $(m_1, \ldots, m_{p-1}, h, m_{p+1}, \ldots, m_n)$ when the family of vectors $\mathscr{B}_p(u_L, u_R) = (r_1(u_L), \ldots, r_{p-1}(u_L), u_R - u_L, r_{p+1}(u_R), \ldots, r_n(u_R))$ is a basis of $\mathbb{R}^n$. Finally, knowing the m_j, we can deduce the κ_j by the formula (7.26).

The geometrical condition which we are about to state generalises in a certain way the hyperbolicity hypothesis. In fact, if the shock $(u_L, u_R; c)$ is weak, then the direction of $u_R - u_L$ is close to that of $r_p(u_L)$ while the vectors $r_j(u_R)$ are close to $r_j(u_L)$ respectively. Therefore, we have

$$\det(\mathscr{B}_p(u_L, u_R)) \sim l_p(u_L) \cdot (u_R - u_L) \det(r_1(u_L), \ldots, r_n(u_L)) \neq 0$$

and $\mathscr{B}_p(u_L, u_R)$ is really a basis. On the other hand, for a shock of large amplitude, this condition is not necessarily satisfied and can be used to eliminate the shocks which are not stable for generic perturbations: A. Majda has shown that this condition plays a rôle in the stability (local in time, without diffusion) of shock waves for hyperbolic systems [73, 74]. It expresses Lopatinski's condition for a mixed problem which is equivalent to the linearisation of the system which governs the evolution of the shock front (see Chapter 14).

Although the property (7.27) has not yet been proved (in fact Tai-Ping Liu asserts that it is false in general as it lacks a term of zero mass), it provides a good method of calculating the parameters, at least that which concerns the phase shift. The result, obtained by Liu [70] and completed by Szepessy and Xin [99], is

Theorem 7.5.1 (Liu) *We suppose that the eigenvalues of* $df(u)$ *are simple, and that* $d\lambda_p(u_L) \cdot r_p(u_L) \neq 0$. *There exist a neighbourhood* $\mathscr{V}$ *of* u_L *in* $\mathscr{U}$ *and a number* $c_1 > 0$ *such that if* $(u_L, u_R; c)$ *is a p-shock with* $u_R \in \mathscr{V}$, *then*

(1) *there exists a viscosity profile U of equation (7.19) linking* u_L *to* u_R,
(2) *for all* $\varphi \in L^1(\mathbb{R})^n$, *there exists one and only one n-tuple* $(m_1, \ldots, m_{p-1}, h, m_{p+1}, \ldots, m_n)$ *such that*

$$\int_{\mathbb{R}} \varphi(x)\, dx = h(u_R - u_L) + \sum_{j<p} m_j r_j(u_L) + \sum_{j>p} m_j r_j(u_R),$$

(3) *if in addition* $\varphi \in H^1(\mathbb{R})$, $\sum_{j\neq p} |m_j| \le c_1 \|u_R - u_L\|$ *and* $\int_{\mathbb{R}}(1+(x-h)^2)\|\varphi(x) + U(x) - U(x-h)\|^2\,dx \le c_1$, *then the Cauchy problem for (7.19), provided with the initial condition* $u(0) = U + \varphi$, *has one and only one smooth solution, global in time, which satisfies in addition*

$$\lim_{t\to+\infty} \|u(t) - U(\cdot + h - ct)\|_2 = 0,$$

$$\lim_{t\to+\infty} \|u(t) - U(\cdot + h - ct)\|_\infty = 0.$$

We note that no estimate of the decay of the norms of the error in L^2 or in L^∞ is known. However, Pego [83] has shown that the viscosity profile is linearly stable in that which concerns the 1-shocks or the n-shocks provided that the perturbation decreases exponentially at infinity. The limitation of this study to the shocks of the fastest characteristic fields is not troublesome in applications, such as in gas dynamics or in the elasticity of an elastic string. However, a family of shocks in magnetohydrodynamics is not covered by this study.

7.6 Influence of the diffusion tensor

We now consider the Cauchy problem for a general diffusion,

$$u_t + f(u)_x = (B(u)u_x)_x. \tag{7.28}$$

We are given a Lax shock $(u_L, u_R; c)$ which we suppose to have a viscosity profile U:

$$B(U)U' = f(U) - f(u_L) - c(U - u_L),$$
$$U(-\infty) = u_L,$$
$$U(+\infty) = u_R.$$

The perturbation φ of the initial condition ($u(0) = U + \varphi$) is given, smooth and sufficiently small, at least as in Theorem 7.5.1. However, we shall not give here a precise statement, leaving the reader to consult [70]. We again look for a description of $u(t)$ in the form

$$U(\cdot + h - ct) + \sum_{j<p} w_j^{\mathrm{L}}(t) r_j(u_L) + \sum_{j>p} w_j^{\mathrm{R}}(t) r_j(u_R)$$

with

$$w_j^{\mathrm{L,R}} = \frac{1}{\sqrt{t}} W_j \left(\frac{x - \lambda_j(u_{\mathrm{L,R}})t}{\sqrt{t}}\right),$$

W_j being a rapidly decreasing function which has to be determined.

Owing to the conservation of mass, the phase shift and the masses of the diffusion waves are again determined by the formula

$$\int_{\mathbb{R}} \varphi(x)\,dx = h(u_R - u_L) + \sum_{j<p} m_j r_j(u_L) + \sum_{j>p} m_j r_j(u_R).$$

Hence, they do not depend on the diffusion chosen. However, to justify the construction, it is necessary to be able to calculate the terms (!), here the diffusion waves W_j. For $j > p$, they are essentially supported by the zone $x \gg ct$, located to the right of the profile, where the value of u is very close to u_R. The same argument as that of the §7.5 leads to the retention, on the left-hand side of (7.28), only of the terms $(\partial_t + \lambda_j(u_R)\partial_x)w_j$ and $\frac{1}{2}b_j\partial_x(w_j^2)$. As for the right-hand side, it develops in the following way:

$$\begin{aligned} l_j(u_R)\cdot\partial_x(B(u_R)\partial_x u) &= l_j(u_R)\cdot B(u_R)\partial_x^2 u + O(t^{-2}) \\ &= (l_j B)(u_R)\partial_x^2 U + \sum_{k\neq p} l_j(u_R)B(u_R)r_k(u_R)\partial_x^2 w_k + O(t^{-2}). \end{aligned}$$

The remaining terms are $O(t^{-2})$ in uniform norm and in L^1-norm. They have a negligible influence on the correction to the asymptotic expansion; for example, they are negligible owing to the term $(l_j Br_j)\partial_x^2 w_j$. Concerning the terms of the sum for which $k \neq j, p$, they are of the form $t^{-3/2}W_k(t^{-1/2}(x - \lambda_k t))$ with $\lambda_k \neq \lambda_j$. Hence they have an influence of the same order as those of the terms $\partial_x(w_j w_k)$ which we have already neglected in §7.2. It is the same for the term $(l_i B)\partial_x^2 U$ since it is transported with speed $c \neq \lambda_j$. In the two cases ($k \neq j, p$ and $k = p$), it is essential that the terms considered should have a zero mean with respect to x; this is the case since these are derivatives of functions vanishing at infinity.

Hence, there only remains the term corresponding to $k = j$ and we again have a system of uncoupled Burgers–Hopf equations:

$$(\partial_t + \lambda_j(u_L)\partial_x)w_j + \frac{1}{2}b_j\partial_x\left(w_j^2\right) = \alpha_j\partial_x^2 w_j, \qquad \alpha_j = (l_j Br_j)(u_L), \quad j < p,$$

$$(\partial_t + \lambda_j(u_R)\partial_x)w_j + \frac{1}{2}b_j\partial_x\left(w_j^2\right) = \alpha_j\partial_x^2 w_j, \qquad \alpha_j = (l_j Br_j)(u_R), \quad j > p.$$

This equation has integrable diffusion waves if and only $\alpha_j > 0$, that is if and only if these equations are well-posed for increasing time. Hence, we require that the diffusion satisfy the general condition

$$(l_j Br_j)(u) > 0, \qquad \forall u \in \mathcal{U}, \qquad 1 \leq j \leq n. \tag{7.29}$$

We notice that this condition is satisfied when the diffusion is strictly dissipative for a strongly convex E, or simply when it is dissipative and, in addition, satisfies

$Br_j \neq 0$ (this allows us to apply the theory to the case of gas dynamics [50]). In fact, the eigenvector basis $(r_j)_{1\le j\le n}$ is orthogonal for the quadratic form D^2E, that is, there exist numbers $e_j(u)$ such that $D^2Er_j = e_jl_j$. These numbers are strictly positive since $D^2E(r_j, r_j) > 0$. In addition, if $D^2E(B\eta, \eta) \geq c(u)\|B\eta\|^2$ (dissipation hypothesis) we have

$$0 < c(u)\|Br_j\|^2 \leq e_jl_jBr_j.$$

Finally, $l_jBr_j > 0$.

Example: gas dynamics

Let us consider the system of gas dynamics in lagrangian coordinates, with viscosity and heat conduction (we shall also envisage the case where the viscosity is negligible):

$$\begin{aligned} v_t &= z_x, \\ z_t + p(v, e)_x &= (b(v)z_x)_x, \\ \left(e + \frac{1}{2}z^2\right)_t + (pz)_x &= (bzz_x + k(v, e)T_x)_x. \end{aligned}$$

Denoting by $c = \sqrt{(pp_e - p_v)}$ the speed of sound, the eigenvectors of $\mathrm{d}f$ are

$$r_1 = \begin{pmatrix} 1 \\ c \\ zc - p \end{pmatrix}, \quad r_2 = \begin{pmatrix} p_e \\ 0 \\ -p_v \end{pmatrix}, \quad r_3 = \begin{pmatrix} 1 \\ -c \\ -zc - p \end{pmatrix},$$

while the eigenforms are

$$\begin{aligned} l_1 &= \frac{1}{2c^2}(-p_v, c + zp_e, -p_e), \\ l_2 &= \frac{1}{c^2}(p, -z, 1), \\ l_3 &= \frac{1}{2c^2}(-p_v, -c + zp_e, -p_e). \end{aligned}$$

Finally,

$$Br_2 = \begin{pmatrix} 0 \\ 0 \\ -kT_ep_v \end{pmatrix}, \quad Br_{1,3} = \begin{pmatrix} 0 \\ \pm bc \\ k(T_v - pT_e) \pm czb \end{pmatrix}.$$

We have seen that the hyperbolicity of the system of gas dynamics is equivalent to the convexity of the entropy (opposite to the physical entropy), which from now on we shall suppose realised, and also that the tensor of viscous and thermal diffusion is dissipative for this case. We deduce that the condition $l_j Br_j > 0$ holds if and only if $Br_j \neq 0$. For $j = 2$, this is satisfied by all real gases as $T_e > 0$ and $p_v < 0$. For $j = 1$ or $j = 3$, this is again true when the viscosity is present ($b \neq 0$). If it is not ($b \equiv 0$), it will be enough that $pT_e - T_v$ is not zero. In this case, we shall have $pT_e - T_v > 0$ since $\alpha_1 = \alpha_3 = \frac{1}{2}c^{-2}kp_e(pT_e - T_v)$ and $p_e > 0$ for all real gases. This inequality is generally satisfied. For example, for a polytropic gas, $T_v = 0$, and $T_e > 0$. Again, other ways of writing this inequality are

$$\left.\frac{\partial S}{\partial v}\right|_{T=\text{const.}} > 0, \quad \text{or} \quad \left.\frac{\partial T}{\partial p}\right|_{S=\text{const.}} > 0.$$

7.7 Case of over-compressive shocks

Definition 7.7.1 We call an *over-compressive shock* of a strictly hyperbolic system $u_t + f(u)_x = 0$ every triplet $(u_\mathrm{L}, u_\mathrm{R}; c)$ which satisfies the Rankine–Hugoniot condition and is such that $\lambda_{j-1}(u_\mathrm{L}) < c < \lambda_j(u_\mathrm{L})$, $\lambda_k(u_\mathrm{R}) < c < \lambda_{k+1}(u_\mathrm{R})$ with $j < k$ (we recall that if $j = k$, $(u_\mathrm{L}, u_\mathrm{R}; c)$ is a Lax shock).

Let us consider an over-compressive shock $(u_\mathrm{L}, u_\mathrm{R}; c)$ with the fixed perturbation u_{xx} and profile equation $U' = f(U) - f(u_\mathrm{L}) - c(U - u_\mathrm{L})$. The stable manifold $W^s(u_\mathrm{R})$ and the unstable manifold $W^i(u_\mathrm{L})$ have dimensions k and $n - j + 1$ respectively. Their intersection $\mathscr{V}$ is thus in general a sub-manifold of dimension $k - j + 1 \geq 2$, for example in the structurally stable case where these manifolds are transverse to each other. The set of viscosity profiles for this shock is identified with $\mathscr{V}$ by the mapping $U \mapsto b(U) := U(0)$. Concerning the diffusion waves which lengthen the zone of the shock (and hence which are able to be superposed on a profile in an asymptotic expansion) they have speeds $\lambda_i(u_\mathrm{L})$ for $i \leq j - 1$ or $\lambda_i(u_\mathrm{R})$ for $i \geq k + 1$. Hence there are $n + j - k - 1$ families of diffusion waves, each being parametrised by m_i, $m_i r_i(u_{\mathrm{L,R}})$ being the mass of the wave. The asymptotic expansions which we use to express the behaviour of a solution $u(t)$ when $t \to +\infty$, when the initial condition is a perturbation $U(\cdot; b_0) + \varphi$ of a viscosity profile, are therefore of the form

$$U(\cdot; b) + \sum_{i \leq j-1} \frac{1}{\sqrt{t}} W_i\left(\frac{x - \lambda_i(u_\mathrm{L})t}{\sqrt{t}}; \kappa_i\right) r_i(u_\mathrm{L})$$
$$+ \sum_{i \geq k+1} \frac{1}{\sqrt{t}} W_i\left(\frac{x - \lambda_i(u_\mathrm{R})t}{\sqrt{t}}; \kappa_i\right) r_i(u_\mathrm{R}) + \mathrm{o}(1).$$

In this formula, the remainder is small in $L^1 \cap L^\infty$ and there are n parameters to be determined by making use of the conservation of mass. If we denote by $\delta(b; b_0) = \int_{\mathbb{R}}(U(\xi; b) - U(\xi, b_0))\,\mathrm{d}\xi$ the difference of mass between two profiles, the equation we have to solve to determine b and the κ_i (or equivalently the masses m_i) is

$$\delta(b; b_0) + \sum_{i \leq j-1} m_i r_i(u_\mathrm{L}) + \sum_{i \geq k+1} m_i r_i(u_\mathrm{R}) = \int_{\mathbb{R}} \varphi(x)\,\mathrm{d}x. \tag{7.30}$$

The mapping $b \mapsto \delta(b; b_0)$ is differentiable at b_0 and the range $\pi(b_0)$ of its differential contains the straight line generated by $u_\mathrm{R} - u_\mathrm{L}$ since $\int_{\mathbb{R}}(U(\xi + h; b_0) - U(\xi; b_0))\mathrm{d}\xi = h(u_\mathrm{R} - u_\mathrm{L})$. When the spaces $\pi(b_0)$, $\oplus_{i \leq j-1}\mathbb{R}r_i(u_\mathrm{L})$, and $\oplus_{i \geq k+1}$ $\mathbb{R}r_i(u_\mathrm{R})$, the sum of whose dimensions equals n, are in direct sum, equation (7.30) has a solution

$$(m_1, \ldots, m_{j-1}, b, m_{k+1}, \ldots, m_n)$$

when the mass $m = \int_{\mathbb{R}} \varphi(x)\,\mathrm{d}x$ is small enough. More generally, this equation has a solution for all m belonging to a cylinder $C = \omega + X$, where $X = \mathbb{R}(u_\mathrm{R} - u_\mathrm{L}) + \oplus_{i \leq j-1}\mathbb{R}r_i(u_\mathrm{L}) + \oplus_{i \geq k+1}\mathbb{R}r_i(u_\mathrm{R})$. When $\pi(b_0) + X = \mathbb{R}^n$ (which held for a Lax shock with weak amplitude because $X = \mathbb{R}^n$), C is a neighbourhood of the origin.

When the equation (7.30) has a solution, we can hope that $\lim_{t \to +\infty} \|u(t) - U(\cdot - ct; b)\|_\infty = 0$. Such a result is obviously as difficult to prove as for a Lax shock and must necessitate supplementary hypotheses concerning the perturbation φ.

The essential difference with the case of a Lax shock is that the cylinder C is not, in general, the whole of $\mathbb{R}^n$. Hence, there are perturbations φ of the initial condition for which the equation (7.30) does not have a solution. For these, the asymptotic behaviour of $u(t)$ cannot be described by a viscosity profile. We shall illustrate this by an example.

Example 7.7.2

The simplest system which possesses super-compressive shocks is that which has been popularised by Keyfitz and Kranzer (which has $n = 2$):

$$u_t + (r^2 u)_x = 0, \qquad r = \|u\|. \tag{7.31}$$

The study which follows draws its inspiration from the work of T.-P. Liu and H. Freistühler [25, 72].

Over-compressive shocks

The characteristic speeds of the system (7.31) are $\lambda(u) = r^2$ and $\mu(u) = 3r^2$. The first corresponds to a linearly degenerate field. Being given a state $u_L \neq 0$, $u_L = r_L e_L$, the shocks (here, we exclude contact discontinuities) $(u_L, u_R; c)$ are of two kinds:

Regular shocks: $u_R = r_R e_L$ and $c = r_L^2 + r_L r_R + r_R^2$,
Irregular shocks: $u_R = -r_R e_L$ and $c = r_L^2 - r_L r_R + r_R^2$.

Also, we require that the shocks have a viscosity profile for the parabolic perturbation u_{xx}.

Now, let us fix $r_L > 0$ and the unit vector e_L and choose a shock speed $c \in (\frac{3}{4}r_L^2, r_L^2)$. For such a choice, there is no shock such that $u_R = r_R e_L$. On the other hand, there are two states u_1 and u_2 of the form $-r_R e_L$ with $0 < r_1 < r_2$. We have $r_2 < r_L$. The eigenvalues of the system, calculated with the states u_1, u_2, u_L, have the following properties:

$$\lambda(u_1), \lambda(u_2), \mu(u_1) < c < \mu(u_2), \lambda(u_L), \mu(u_L).$$

Thus, $(u_L, u_2; c)$ is a Lax shock and $(u_L, u_1; c)$ is an over-compressive shock.

The profile equation

The profile equation is written

$$u' = G(u) =: (r^2 - c)u - \left(r_L^2 - c\right) r_L e_L. \tag{7.32}$$

Its critical points are u_1, u_2 and u_L, respectively attractive focus, saddle point and repulsive focus. None among these is degenerate.

As the vector field G is emerging from the disk $D(0; r_L)$, every trajectory which passes into this disk at an instant ξ_0 remains there for the times $\xi < \xi_0$. As this disk is compact, the Poincaré–Bendixson theorem ensures that trajectories tend to a stationary point or are wound round a limit cycle (which can contain critical points) when $t \to -\infty$. We are going to show that this latter eventuality is impossible.

Let Δ be the axis $\mathbb{R}e_L$. Let us consider a limit cycle γ, which is a piece-wise $\mathscr{C}^1$ curve, invariant for the flow of G. Since the axis Δ is invariant for the flow, γ is contained in one of the half-planes delimited by Δ. Since γ encircles a simply connected invariant compact set, this must contain a stationary point of the flow, that is u_1, u_2 or u_L. As these points are on Δ, we deduce that in fact γ passes through one of these points. Since γ is oriented by the flow, this stationary point cannot be either attractive or repulsive, this is therefore the saddle point u_2. But then γ contains one of the two trajectories of $W^{i}(u_2)$, which must be bounded. Now of these two trajectories, contained in Δ, one goes off to infinity while the other ends in u_1. Finally, γ contains u_1, which is absurd since u_1 is attractive.

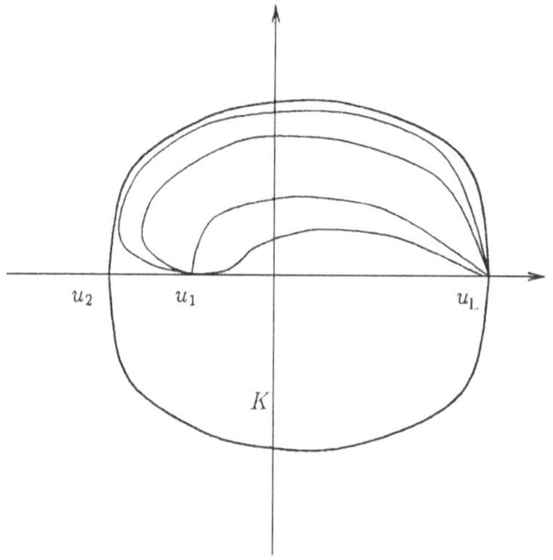

Fig. 7.2: The set, open but bounded, of the profiles between u_L and u_1, over-compressive shock.

Every trajectory which passes into $D(0; r_L)$ has therefore issued from one of the stationary points of G. There are only four cases, of which the last is generic:

The trajectory $u \equiv u_1$.
The trajectory $u \equiv u_2$.
The two trajectories contained in Δ and issuing from u_2.
The other trajectories, which have issued from u_L.

An important application of this result is that the two trajectories which end in u_2 (they are symmetric with respect to Δ and their union forms $W^s(u_2)$) have issued from u_L. They encircle a compact set K (see Fig. 7.2) with non-empty interior, invariant by the flow. In the interior of K, all the trajectories leave u_L and end up in u_1, with the exception of the trajectories $u \equiv u_i$ and of that which goes from u_2 to u_1 following the axis. Since K is a neighbourhood of u_1, invariant by the flow, there is no other trajectory going from u_L to u_1. Finally, the image V of the viscosity profiles for the shock $(u_L, u_1; c)$ is the interior of K, without the segment $[u_1, u_2]$.

Here, the cylinder C has $\mathbb{R}e_L$ for direction. Let us take, for example, b_0 on the segment $[u_L, u_1]$. The profile that we are going to disturb is thus that which follows the axis Δ. Without loss of generality, we can put $e_L = (1, 0)$. The cylinder C is defined by a relation $y_2 \in J$, y_2 being the coordinate along to the vector $(0, 1)$. The interval J is the set of values taken by

$$\delta_2(b; b_0) = \int_{\mathbb{R}} (U_2(\xi; b) - U_2(\xi; b_0))\, d\xi = \int_{\mathbb{R}} U_2(\xi; b)\, d\xi.$$

Let us write the profile equation in polar coordinates ($U = r \exp i\theta$):

$$r' = (r^2 - c)r - \left(r_L^2 - c\right)r_L \cos\theta,$$
$$r\theta' = \left(r_L^2 - c\right)r_L \sin\theta.$$

As each profile lies wholly in a half-plane $U_2 > 0$ or $U_2 < 0$, we have $J = -J$ and we can treat only the case $U_2 > 0$. Hence, the range of the angle θ is $[0, \pi]$. Thus $\theta' > 0$, and we can parametrise the trajectory by θ. It turns out that

$$U_2 \, d\xi = r \sin\theta \, d\xi = \frac{r^2\theta' d\xi}{\left(r_L^2 - c\right)r_L} = \frac{r^2}{\left(r_L^2 - c\right)r_L} \, d\theta.$$

Thus,

$$\delta_2(b; b_0) = \frac{1}{\left(r_L^2 - c\right)r_L} \int_0^{\pi} r^2 \, d\theta.$$

This expression does not depend on b but only on the trajectory on which b occurs. We can parametrise the trajectories by $\theta \mapsto r(\theta, \beta)$ where $(0, \beta)$ is a point on the vertical axis. By the Cauchy–Lipschitz theorem, the mapping $\beta \mapsto r(\theta, \beta)$ is continuous and strictly increasing. It follows that $J = [-Y, Y]$ where Y is the integral of U_2 when U is the viscosity profile which links u_L to u_2 (and not to u_1) situated in the upper half-plane (it is the upper boundary of K). As $K \subset D(0; r_L)$, we have $r < r_L$ in the integral and so $Y < \pi r_L/(r_L^2 - c)$. Similarly the vector field G is entering into the disk $D(0, r_2)$ (verify that $r_2^2 < c$). Hence, the trajectory is wholly outside of this disk and we again have $Y > \pi r_2/(c - r_2^2)$. Finally, we have shown the following result.

Theorem 7.7.3 *Being given an over-compressive shock $(u_L, u_1; c)$ of the system (7.31) (see the above construction) and an initial condition such that $\int_{-\infty}(a(x) - u_L)\,dx$ and $\int^{\infty}(a(x) - u_1)\,dx$ converge, the following conditions are equivalent:*

(1) *There exists a viscosity profile U between u_L and u_1 such that $\int_{\mathbb{R}}(U(x) - a(x))\,dx = 0$;*
(2) $\left|\int_{\mathbb{R}} a_2(x)\,dx\right| < Y$.

The number Y satisfies

$$\frac{\pi r_2}{c - r_2^2} < Y < \frac{\pi r_L}{r_L^2 - c}.$$

The asymptotic behaviour of the solution of the parabolic Cauchy problem for the initial condition a can therefore be described by a viscosity profile when the inequality (2) is satisfied (a necessary but not a sufficient condition).

Instability of the over-compressive shock

As T.-P. Liu notes, the useful notion of stability is that of uniform stability with respect to the coefficient $\varepsilon > 0$ of the parabolic perturbation of a hyperbolic system of a conservation law. In other words, being given an initial condition $a \in \mathscr{C}_{\mathrm{b}}(\mathbb{R})$, such that $\int_{-\infty}(a(x) - u_{\mathrm{L}})\,\mathrm{d}x \int^{+\infty}(a(x) - u_{\mathrm{R}})\,\mathrm{d}x$ converge, we ask if the solution u^{ε} of the Cauchy problem for the parabolic system

$$\begin{aligned} u_t + f(u)_x &= \varepsilon(B(u)u_x)_x, \\ u(0, x) &= a(x) \end{aligned}$$

has the under-noted properties:

(1) u^{ε} is defined on $\mathbb{R}^+ \times \mathbb{R}$,
(2) $u^{\varepsilon}(t)$ is asymptotic, in $L^1(\mathbb{R}) \cap L^{\infty}(\mathbb{R})$, to a progressive wave $U(\varepsilon^{-1}[x - ct - x_0(\varepsilon)]; \varepsilon)$ when $t \to +\infty$,
(3) the profile $y \mapsto U(y - x_0(\varepsilon); \varepsilon))$ possesses a uniform limit when $\varepsilon \to 0$.

In the example presented above, Theorem 7.7.3 shows that the condition (2) does not hold when $\int_{\mathbb{R}} a_2(x)\,\mathrm{d}x \neq 0$. Indeed, the condition for the existence of a profile U between u_{L} and u_{R} having the desired mass defect is that $|\int_{\mathbb{R}} a_2(x)\,\mathrm{d}x| < \varepsilon Y$, this is obtained by making the change of variables $(t, x) \mapsto (\varepsilon t, \varepsilon x)$ to lead to the case where $\varepsilon = 1$. It is this instability of the shock, when the velocity is sufficiently small, which leads T.-P. Liu to reject this. He notes finally that by excluding the over-compressive shocks, we can solve the Riemann problem for the system (7.31) uniquely.

7.8 Exercises

7.1 We consider again the proof of Theorem 7.2.1 in the case where $(\mathrm{d}\lambda_k \cdot r_k)(u_{\mathrm{L}}) = 0$. But as contact discontinuities are not able to admit profiles, we exclude the case where $\mathrm{d}\lambda_k \cdot r_k$ is constant on an integral curve of r_k. More precisely, we suppose that $\gamma_k := (\mathrm{d}(\mathrm{d}\lambda_k \cdot r_k) \cdot r_k)(u_{\mathrm{L}}) \neq 0$.

(1) Show that the curve of the critical points of G of the form $(u_{\mathrm{R}}, \sigma(u_{\mathrm{L}}, u_{\mathrm{R}}))$ for $u_{\mathrm{R}} \in \Gamma_k(u_{\mathrm{L}})$, $u_{\mathrm{R}} \neq u_{\mathrm{L}}$, is locally situated on one side of the straight line $s = \lambda_k(u_{\mathrm{L}})$, this side depending on the sign of γ_k.
(2) Depending on this sign, deduce that there exist viscosity profiles for all the triplets $(u_{\mathrm{L}}, u_{\mathrm{R}}, \sigma(u_{\mathrm{L}}, u_{\mathrm{R}}))$ or none, as we stay in a neighbourhood of $(u_{\mathrm{L}}, u_{\mathrm{L}}, \lambda_k(u_{\mathrm{L}}))$.

7.2 We consider the p-system

$$\left.\begin{aligned} v_t &= z_x, \\ z_t + p(v)_x &= 0, \end{aligned}\right\}$$

with $p \in \mathscr{C}^3(\mathbb{R})$, $\sup_{v\in\mathbb{R}} p'(v) < 0$. We suppose that the points where p'' is zero satisfy $p''' \neq 0$. We adopt the admissibility condition of shock waves such as is stated in Theorem 7.3.1.

(1) A simple 2-wave is by definition a self-similar solution

$$(x, t) \mapsto (v(x/t), z(x/t)),$$

piecewise smooth, where we have connected two constant states u_- and u_+ by 2-rarefaction-waves ($x/t = \sqrt{-p'(v)}$) and 2-admissible-shock-waves ($s > 0$).

Let $u_L = (v_L, z_L) \in \mathbb{R}^2$. Show that the states $u_R = (v_R, z_R)$ to which u_L can be linked by a simple 2-wave form a curve parametrised by v and defined in the following way. Denoting by I the interval with extremities u_L and u_R, we denote by p_{*I} the envelope of the restriction of p to I, lower convex (if $u_R < u_L$) or upper concave (if $u_R > u_L$). Then

$$z_R = z_L - \int_{v_L}^{v_R} \sqrt{-\frac{dp_{*I}}{dv}}\, dv =: \varphi(v_R; v_L).$$

Show that $v \mapsto \varphi(v; v_L)$ is continuous.

(2) Similarly, study the curves of the simple 1-wave and show that the Riemann problem admits one and only one solution made up of one 1-wave and one 2-wave separated by a constant state.

7.3 We consider the isentropic gas dynamics with viscosity *and* capillarity. This latter is expressed by a perturbation of the law of conservation of momentum (a is a positive constant):

$$\left.\begin{aligned} v_t &= z_x, \\ z_t + p(v)_x &= \varepsilon(b(v)z_x)_x - a\varepsilon^2 v_{xxx}. \end{aligned}\right\}$$

After the elimination of the speed, the profile equation becomes

$$av'' + sb(v)v' + p(v) - p(v_L) + s^2(v - v_L) = 0. \tag{7.33}$$

So far as the question (5) inclusive, we assume that $b > 0$.

(1) We suppose that $(u_L, u_R, \sigma(u_L, u_R))$ admits a visco-capillary profile (we must properly call it thus!). Show that the expression

$$\int_{v_L}^{v_R} (p(v) - p(v_L) + \sigma^2(v - v_L))\, dv$$

is of opposite sign to that of σ and that it is identically zero if and only if $\sigma = 0$. Deduce that if $\sigma \neq 0$, one at least of the discontinuities $(u_L, u_R; \sigma)$ and $(u_R, u_L; \sigma)$ does not admit a visco-capillary profile.

(2) Show that this inequality is always satisfied be an entropy shock when p'' is of constant sign (we can take Lax's entropy condition or Lax's shock condition since, here, they are equivalent).

(3) We suppose that p' is of the sign of $(\beta - v)(v - \alpha)$ where $-\infty < \alpha < \beta < +\infty$. We consider the stationary discontinuities $((v_L, 0), (v_R, 0); 0)$, that is the couples (non-ordered) (v_L, v_R) for which $p(v_L) = p(v_R)$. Show that there exists one and only one which satisfies the condition

$$\int_{v_L}^{v_R} (p(v) - p(v_L))\, dv = 0.$$

(4) For this couple, show that $(u_L, u_R; 0)$ and $(u_R, u_L; 0)$ each admit a visco-capillary profile.

(5) We denote this couple by (v_-, v_+) with $v_- < v_+$. Show that $v_- < \alpha < \beta < v_+$.

(6) We consider the case without viscosity ($b \equiv 0$). Show that for all v_L in the neighbourhood of v_-, there exists a single couple (v_R, σ) in the neighbourhood of $(v_+, 0)$ satisfying

$$\left.\begin{aligned} p(v_R) - p(v_L) + \sigma^2(v_R - v_L) &= 0, \\ \int_{v_L}^{v_R} (p(v) - p(v_L) + \sigma^2(v - v_L))\, dv &= 0. \end{aligned}\right\}$$

Show that for all $z_L \in \mathbb{R}$, there exists a capillary profile in each direction between (v_L, z_L) and $(v_R, z_L + \delta z)$ where $\delta z = \sigma(v_L - v_R)$.

7.4 We consider the dynamics of a perfect gas with state equation $pv = (\gamma - 1)e$ and for sole perturbation a thermal diffusion obeying Fourier's law ($k = k(v, e) > 0$). As here the temperature can be taken equal to e, the equations are equivalent to

$$\left.\begin{aligned} v_t &= z_x, \\ z_t + p_x &= 0, \\ \left(e + \frac{1}{2}z^2\right)_t + (pz)_x &= \varepsilon(k(v, e)e_x)_x. \end{aligned}\right\}$$

(1) Show that the profile equation leads to a single differential equation of the form $f(v)' = g(v)$ where $g(v_{\mathrm{l,r}}) = 0$ and $f'(v) < 0$ for $v > v^*$ and $f'(v) > 0$ for $v < v^*$, v^* being defined by

$$v^* = \frac{1}{2\sigma^2}(p_{\mathrm{L}} + \sigma^2 v_{\mathrm{L}}) = \frac{1}{2\sigma^2}(p_{\mathrm{R}} + \sigma^2 v_{\mathrm{R}}).$$

(2) Verify that g is quadratic, hence of constant sign between v_{L} and v_{R}. Deduce that if v_{L} and v_{R} are on opposite sides of v^*, then those singular points of the reduced equation $f(v)' = g(v)$ are of the same nature (attractive or repulsive) and hence that there is not a thermal profile for the discontinuity $(u_{\mathrm{L}}, u_{\mathrm{R}}; \sigma(u_{\mathrm{L}}, u_{\mathrm{R}}))$ in this case.

(3) On the other hand, show that if v_{L} and v_{R} are situated on the same side with respect to v^*, there is a thermal profile from one state towards the other, in the direction which respects Lax's entropy condition.

7.5 We consider once again gas dynamics in lagrangian variables with the perfect gas equation of state $pv = (\gamma - 1)e$ with $\gamma > 1$, but with viscosity ($b = b(v, e)$) and without thermal conduction. Hence the equations are

$$\left.\begin{aligned} v_t &= z_x, \\ z_t + p_x &= \varepsilon(b(v, e)z_x)_x, \\ \left(e + \frac{1}{2}z^2\right)_t + (pz)_x &= \varepsilon(b(v, e)zz_x)_x. \end{aligned}\right\}$$

(1) Show that for this perturbation, the contact discontinuities ($z_{\mathrm{R}} = z_{\mathrm{L}}$, $p_{\mathrm{R}} = p_{\mathrm{L}}$, $s = 0$) have viscosity profiles.

(2) Let $(u_{\mathrm{L}}, u_{\mathrm{R}}; \sigma(u_{\mathrm{L}}, u_{\mathrm{R}}))$ be a discontinuity for which $\sigma \neq 0$. Show that the existence of a viscosity profile is equivalent to the existence of a heteroclinic trajectory from z_{L} to z_{R} for a differential equation

$$vb(v, e)z' = q(z)$$

where q is a quadratic polynomial which depends on σ, is zero at z_{L} and at z_{R} and where v, e are expressed as functions of z.

(3) Deduce that there exists a viscosity profile between u_{L} and u_{R} but only in a single sense, that which respects Lax's entropy condition.

7.6 Let us consider non-isentropic gas dynamics with diffusion, whose general form in eulerian variables is

$$u_t + f(u)_x = \varepsilon(B(u)u_x)_x.$$

The first of these equations is, on supposing that there is no diffusion for the mass,

$$\rho_t + (\rho z)_x = 0, \tag{7.34}$$

where we have denoted the density by ρ and the velocity by z. We recall that the lagrangian coordinates (t, y) are defined by the formula $\mathrm{d}y = -\rho z\,\mathrm{d}t + \rho\,\mathrm{d}x$, justified by the equation (7.34).

(1) Show that outside of the vacuum, the equations of motion are written, in lagrangian coordinates, in the form

$$\left(\frac{u}{\rho}\right)_t + (f(u) - zu)_y = \varepsilon(\rho B(u) u_y)_y,$$

which contains a trivial equation $1_t + 0_y = 0$, which we replace by the equation $v_t = z_y$ which itself comes from the trivial equation $1_t + 0_x = 0$ (we denote by $v = \rho^{-1}$ the specific volume).

(2) Let $(u_{\mathrm{L}}, u_{\mathrm{R}}; s)$ be a discontinuity satisfying the Rankine–Hugoniot condition for the eulerian system:

$$[f] = s[u].$$

We suppose that $z_{\mathrm{R}} \neq z_{\mathrm{L}}$ (we thus exclude contact discontinuities). Show that there corresponds a discontinuity $((v_{\mathrm{L}}, z_{\mathrm{L}}), (v_{\mathrm{R}}, z_{\mathrm{R}}); \sigma)$ of the eulerian system, whose speed of propagation σ is given by the formula

$$\sigma = \frac{s - z_{\mathrm{L}}}{v_{\mathrm{L}}} = \frac{s - z_{\mathrm{R}}}{v_{\mathrm{R}}}.$$

(3) Write the profile equations for the lagrangian system and for the eulerian system. Show that the existence of an eulerian viscosity profile for $(u_{\mathrm{L}}, u_{\mathrm{R}}; s)$ is equivalent to that of a lagrangian viscosity profile for $(v_{\mathrm{L}}, u_{\mathrm{L}}, v_{\mathrm{R}}, u_{\mathrm{R}}; \sigma)$ and that we pass from one to the other by a change of parameter.

Bibliography

[1] Bergh, J. and Löfström, J. *Interpolation spaces*, Springer Verlag, Berlin, 1976.

[2] Boillat, G. *La Propagation des ondes*, Gauthier-Villars, Paris, 1965.

[3] Boillat, G. 'Chocs caractéristiques' *C.R. Acad. Sci. Paris, Série I* **274** (1972), 1018–21.

[4] Brenner, P. 'The Cauchy problem for the symmetric hyperbolic systems in L_p.' *Math. Scand* **19** (1966), 27–37.

[5] Carasso, C., Rascle, M., and Serre, D. 'Etude d'un modèle hyperbolique en dynamique des câbles' *Mod. Math. Anal. Numer.* **19** (1985), 573–99.

[6] Chen, Gui-Qiang. 'Convergence of the Lax–Friedrichs scheme for isentropic gas dynamics (i)' *Acta Math. Scientia* **6** (1986), 75–120.

[7] Chen, Gui-Qiang, Lu, Y.-G. 'Convergence of the approximate solutions to isentropic gas dynamics' *Acta Math. Scientia* **10** (1990), 39–45.

[8] Ciarlet, P. *Mathematical elasticity: vol. I, three-dimensional elasticity*, North Holland, Amsterdam, 1988.

[9] Coleman, B. D. and Dill, E. H. 'Thermodynamic restrictions on the constitutive equations of electromagnetic theory' *Z. Angew. Math. Phys.* **22** (1971), 691–702.

[10] Collela, P. 'Glimm's method for gas dynamics' *SIAM J. Sci. Statist. Comp.*, **3** (1982), 76–109.

[11] Courant, R. and Friedrichs, K. O. *Supersonic flow and shock waves*, John Wiley & Sons – Interscience, New York, 1948.

[12] Courant, R. and Hilbert, D. *Methods of mathematical physics, vol. II*, John Wiley & Sons – Interscience, New York, 1962.

[13] Crandall, M. and Ligett, T. M. 'Generation of semi-groups of nonlinear transformations on general Banach spaces' *Amer. J. Math.* **93** (1971), 265–98.

[14] Crandall, M. 'The semi-group approach to first order quasilinear equations in several space variables' *Israel J. Math.* **12** (1972), 108–32.

[15] Dafermos, C. 'Characteristics in hyperbolic conservation laws, a study of the structure and asymptotic behaviour of solutions' In R. J. Knops (editor), *Nonlinear analysis and mechanics*, vol. **39**, Research Notes in Mathematics, Pitman, London, 1977, pp. 1–58.

[16] Dafermos, C. and Ling Hsiao 'Development of singularities in solutions of the equations of nonlinear thermoelasticity' *Quart. Appl. Math.* **44** (1986), 463–74.

[17] DiPerna, R. J. 'Convergence of approximate solutions to conservation laws' *Arch. Rat. Mech. Anal.* **82** (1983), 27–70.

[18] DiPerna, R. J. 'Convergence of the viscosity method for isentropic gas dynamics' *Comm. Math. Phys.* **91** (1983), 1–30.
[19] Ekeland, I. and Temam, R. *Analyse convexe et problèmes variationnels*, Dunod, Paris, 1974.
[20] Evans, L. C. *Weak convergence methods for nonlinear partial differential equations.* vol. **74,** CBMS Regional Conference Series in Math., Amer. Math. Soc., Providence, R.I., 1990.
[21] Feynman, R. P., Leighton, R. B., and Sands, M. *Feynman lectures on physics*, Addison-Wesley, Reading, Mass., 1963.
[22] Fife, P. 'Mathematical aspects of electrophoresis' in R. J. Knops (editor), *Reaction–diffusion equations*, Oxford University Press, 1988.
[23] Foy, L. 'Steady state solutions of hyperbolic systems of conservation laws with viscosity terms' *Comm. Pure Appl. Math.* **17** (1964), 177–88.
[24] Freistühler, H. 'Rotational degeneracy of hyperbolic systems of conservation laws' *Arch. Rat. Mech. Anal.* **113** (1991), 39–64.
[25] Freistühler, H. and Liu, Tai-Ping. 'Nonlinear stability of overcompressive shock waves in a rotationally invariant system of viscous conservation laws' *Comm. Math. Phys.* **153** (1993), 147–58.
[26] Friedrichs, K. O. and Lax, P. D. 'Systems of conservation laws with a convex extension' *Proc. Nat. Acad. Sci. U.S.A.* **68** (1971), 1686–8.
[27] Garcke, H. 'Travelling wave solutions as dynamic phase transitions in shape memory alloys' *J. Diff. Eqns.* **121** (1995), 203–31.
[28] Gårding, L. 'Problèmes de Cauchy pour les systèmes quasi-linéaires d'ordre un strictement hyperboliques' *Les EdPs. Colloques Internationaux du CNRS* **117** (1963), 33–40.
[29] Gilquin, H. 'Une famille de schémas numériques TVD pour les lois de conservation hyperboliques' *Math. Mod. Numer. Anal.* **20** (1986), 429–60.
[30] Gisclon, M. 'Etude des conditions aux limites pour un système strictement hyperbolique, *via* l'approximation parabolique' *J. Math Pures Appl.* **75** (1996), 485–508.
[31] Gisclon, M. and Serre, D. 'Etude des conditions aux limites pour un système strictement hyperbolique *via* l'approximation parabolique' *C.R. Acad. Sci. Paris, Série I* **319** (1994), 377–82.
[32] Glimm, J. 'Solutions in the large for nonlinear hyperbolic systems of equations' *Comm. Pure Appl. Math.* **18** (1965), 697–715.
[33] Glimm, J. and Lax, P. D. *Decay of solutions of systems of nonlinear hyperbolic conservation laws*, Memoir **101**, A.M.S., Providence, R.I., 1970.
[34] Godlewski, E. and Raviart, P. -A. *Hyperbolic systems of conservation laws*, Ellipses, Paris, 1991.
[35] Godunov, S. K. 'Lois de conservation et intégrales d'énergie des équations hyperboliques' In C. Carasso, P.-A. Raviart, and D. Serre (editors), *Nonlinear hyperbolic problems, St Etienne*, vol. **1270**, Springer Lecture Notes in Mathematics, Berlin, 1987, pp. 135–49.
[36] Goodman, J. 'Nonlinear asymptotic stability of viscous shock profiles for conservation laws' *Arch. Rat. Mech. Anal.* **95** (1986), 325–44.
[37] Hagan, R. and Slemrod, M. 'The viscosity–capillarity condition for shocks and phase transitions' *Arch. Rat. Mech. Anal.* **83** (1983), 333–61.
[38] Hammersley, J. M. and Handscomb, D. C. *Monte Carlo methods*, Methuen, London, 1965.
[39] Heibig, A. 'Régularité des solutions du problème de Riemann' *Comm. Partial Diff. Eqns.* **15** (1990), 693–709.

[40] Heibig, A. 'Unicité des solutions du problème de Riemann' *C.R. Acad. Sci. Paris, Série I* **312** (1991), 793–7.
[41] Heibig, A. and Serre, D. 'Une approche algébrique du problème de Riemann' *C.R. Acad. Sci. Paris, Série I* **309** (1989), 157–62.
[42] Hoff, D. 'Invariant regions for systems of conservation laws' *Trans. Amer. Math. Soc.* **289** (1985), 591–610.
[43] Hoff, D. and Smoller, J. 'Solutions in the large for certain nonlinear parabolic systems' *Ann. Inst. Henri Poincaré* **3** (1985), 213–35.
[44] Hoff, D. and Smoller, J. 'Global existence for systems of parabolic conservation laws in several space variables' *J. Diff. Eqns.* **68** (1987), 210–20.
[45] Hörmander, L. *Lectures on nonlinear hyperbolic differential equations*, Springer-Verlag, Berlin, 1997.
[46] Hsiao, Ling and Tong Zhang. *The Riemann Problem and interaction of waves in gas dynamics*, Longman, London, 1989.
[47] Il'in, A. M. and Oleĭnik, O. A. 'Behaviour of the solutions of the Cauchy problem for certain quasilinear equations for unbounded increase of time' *AMS Translations* **42** (1964), 19–23.
[48] Ivrii, V. Ya. *Linear hyperbolic equations*, vol. **33**, *Encyclopedia of mathematical sciences*, Springer-Verlag, Heidelberg, 1993.
[49] Joly, J.-L., Métivier, G., and Rauch, J. 'A nonlinear instability for 3×3 systems of conservation laws' *Comm. Math. Phys.* **162** (1994), 47–59.
[50] Kawashima, S. and Matsumura, A. 'Asymptotic stability of travelling wave solutions of systems for one-dimensional gas motion' *Comm. Math. Phys.* **101** (1985), 97–127.
[51] Kazhikhov, A. V. 'Cauchy problem for viscous gas equations' *Sibirsk. Mat. Zh.* **23** (1982), 60–4.
[52] Keyfitz, B. L. and Kranzer, H. C. 'A system of non-strictly hyperbolic conservation laws arising in elasticity theory' *Arch. Rat Mech. Anal.* **72** (1980), 219–41.
[53] Klainerman, S. 'Global existence for nonlinear wave equations' *Comm. Pure Appl. Math.* **33** (1980), 43–101.
[54] Klainerman, S. 'Uniform decay estimates and the Lorentz invariance of the classical wave equation.' *Comm. Pure Appl. Math.* **38** (1995), 321–32.
[55] Kreiss, H.-O. and Lorenz, J. *Initial–boundary value problems and the Navier–Stokes equations*, Academic Press, San Diego, 1989.
[56] Kružkov, S. N. 'Generalized solutions of the Cauchy problem in the large for nonlinear equations of first order' *Dokl. Akad. Nauk SSSR* **187** (1969), 29–32.
[57] Kunik, M. 'A solution formula for a non-convex scalar hyperbolic conservation law with monotone initial data' *Math. Methods Appl. Sci.* **16** (1993), 895–902.
[58] Kuznetsov, N. N. 'Accuracy of some approximate methods for computing the weak solution of a first order quasi-linear equation' *USSR Comp. Math. Math. Phys.* **16** (1976), 105–19.
[59] Lax, P. D. 'Hyperbolic systems of conservation laws II' *Comm. Pure Appl. Math.* **10** (1957), 537–66.
[60] Lax, P. D. *Hyperbolic systems of conservation laws and the mathematical theory of shock waves* vol. **11**, CBMS-NSF Regional Conference Series in Applied Mathematics, SIAM, Philadelphia, 1973.
[61] Lax, P. D. and Levermore, D. 'The small dispersion limit for the KdeV equation' *Comm. Pure Appl. Math.* **36** (1983), I 253–90, II 571–93, III 809–30.
[62] Leveque, R. *Numerical methods for conservation laws*, Lectures in Mathematics, Birkhäuser, Basle, 1990.

[63] Levermore, C. D. 'The hyperbolic nature of the zero dispersion KdeV limit' *Comm. Partial Diff. Eqns.* **13** (1988), 495–514.

[64] Li Ta-Tsien. 'Solutions régulières globales des systèmes hyperboliques quasi-linéaires d'ordre 1' *J. Math. Pures Appl.* **4** (1982), 401–9.

[65] Li Ta-Tsien and Yu Wen-Ci. *Boundary value problems for quasilinear hyperbolic systems*, vol. **V**, Math. Series, Duke University, Durham, N.C., 1985.

[66] Liu, Tai-Ping. 'The Riemann problem for general 2×2 conservation laws' *Trans. Amer. Math. Soc.* **199** (1974), 89–112.

[67] Liu, Tai-Ping. 'The Riemann problem for general systems of conservation laws' *J. Diff. Eqns.* **18** (1975), 218–34.

[68] Liu, Tai-Ping. 'Decay to N-waves of solutions of general systems of nonlinear hyperbolic conservation laws' *Comm. Pure Appl. Math.* **30** (1977), 585–610.

[69] Liu, Tai-Ping. 'The deterministic version of Glimm's scheme' *Comm. Math. Phys.* **57** (1977), 135–48.

[70] Liu, Tai-Ping. *Nonlinear stability of shock waves for viscous conservation laws, Memoir* **328**, Amer. Math. Soc., Providence, R.I., 1985.

[71] Liu, Tai-Ping. 'Pointwise convergence to N-waves for solutions of hyperbolic conservation laws' *Bull. Inst. Math. Acad. Sinica* **15** (1987), 1–17.

[72] Liu, Tai-Ping. 'On the viscosity criterion for hyperbolic conservation laws' In M. Shearer (editor), *Viscous profiles and numerical methods for shock waves*, SIAM, Philadelphia, 1991, pp. 105–14.

[73] Majda, A. *The existence of multi-dimensional shock fronts, Memoir* **281**, Amer. Math. Soc., Providence, R.I., 1983.

[74] Majda, A. *The stability of multi-dimensional shock fronts, Memoir* **275**, Amer. Math. Soc., Providence, R.I., 1983.

[75] Majda, A. *Compressible fluid flow and systems of conservation laws in several space variables*, Springer-Verlag, Berlin, 1984.

[76] Mei, Ming. 'Stability of shock profiles for nonconvex scalar viscous conservation laws' *Math. Mod. Methods Appl. Sci.* **5** (1995), 279–96.

[77] Mihăilescu, M. and Suliciu, I. 'Riemann and Goursat step data problems for extensible strings with non-convex stress–strain relation' *Rev. Roum. Math. Pures et Appl.* **20** (1975), 551–9.

[78] Mihăilescu, M. and Suliciu, I. 'Riemann and Goursat step data problems for extensible strings' *J. Math. Anal. Appl.* **52** (1975), 10–24.

[79] Morawetz, C. S. *Notes on time decay and scattering for some hyperbolic problems*, vol. **19**, Regional Conference Series in Applied Mathematics, SIAM, Providence, R.I., 1975.

[80] Nishida, T. 'Global solutions for an initial boundary value problem of a quasilinear hyperbolic system' *Proc. Jap. Acad.* **44** (1968), 642–6.

[81] Oleĭnik, O. A. 'Uniqueness and stability of the generalized solution of the Cauchy problem for a quasi-linear equation' *Uspekhi Mat. Nauk.* **14** (1959), 165–70.

[82] Osher, S. and Ralston, J. 'L^1 stability of travelling waves with application to convective porous media flow' *Comm. Pure Appl. Math.* **35** (1982), 737–51.

[83] Pego, R. L. 'Linearized stability of extreme shock profiles in systems of conservation laws with viscosity' *Trans. Amer. Math. Soc.* **280** (1983), 431–61.

[84] Poupaud, F., Rascle, M. and Vila, J.-P. 'Global solution to the isothermal Euler–Poisson system with arbitrarily large data' *J. Diff. Eqns.* **123** (1995), 93–121.

[85] Rauch, J. 'BV estimates fail for most quasilinear hyperbolic systems in dimension greater than one' *Comm. Math. Phys.* **106** (1986), 481–4.

[86] Richtmyer, R. D. and Morton, K. *Difference methods for initial value problems*, Wiley–Interscience, New York, 1967.

[87] Sakamoto, R. *Hyperbolic boundary value problems*, Cambridge University Press, 1982.

[88] Sattinger, D. H. 'On the stability of waves of nonlinear parabolic systems' *Advances in Math.* **22** (1976), 312–55.

[89] Schatzman, M. 'Fonctionnelle de Glimm continue' *C.R. Acad. Sci. Paris, Série I* **291** (1980), 519–22.

[90] Schatzman, M. 'Continuous Glimm functionals and uniqueness of the solution of the Riemann problem' *Indiana Univ. Math. J.* **34** (1985), 533–89.

[91] Schecter, S. and Shearer, M. 'Transversality for undercompressive shocks in Riemann problems' In M. Shearer (editor), *Viscous profiles and numerical methods for shock waves*, SIAM, Philadelphia, 1991, pp. 142–54.

[92] Serre, D. 'Temple's fields and integrability of hyperbolic systems of conservation laws' In Guangchang Dong and Fanghua Lin (editors), *International conference on nonlinear PDEs*, International Academic Publishers, New York, 1993, pp. 233–51.

[93] Sévennec, B. *Géométrie des systèmes de lois de conservation*, vol. **56**, *Mémoires*, Soc. Math. de France, Marseille, 1994.

[94] Shearer, M. 'The Riemann problem for the planar motion of an elastic string' *J. Diff. Eqns.* **61** (1986), 149–63.

[95] Slemrod, M. 'Global existence, uniqueness and asymptotic stability of classical smooth solutions in one-dimensional, non-linear thermoelasticity' *Arch. Rat. Mech. Anal.* **76** (1981), 97–133.

[96] Slemrod, M. 'Dynamic phase transitions in a Van der Waals fluid.' *J. Diff. Eqns.* **52** (1984), 1–23.

[97] Smoller, J. *Shock waves and reaction diffusion equations*, Springer-Verlag, Berlin, 1983.

[98] Sod, G. A. *Numerical methods in fluid dynamics*, Cambridge University Press, 1985.

[99] Szepessy, A. and Zhouping Xin. 'Nonlinear stability of viscous shock waves' *Arch. Rat. Mech. Anal.* **122** (1993), 53–103.

[100] Tadmor, E. 'Local error estimates for discontinuous solutions of nonlinear hyperbolic equations' *SIAM J. Numer. Anal.* **28** (1991), 891–906.

[101] Tartar, L. 'Compensated compactness and applications to partial differential equations' In R. J. Knops (editor), *Nonlinear analysis and mechanics*, vol. **39**, Research Notes in Mathematics, Pitman, London, 1979, pp. 136–92.

[102] Taub, A. H. 'Relativistic Rankine–Hugoniot equations' *Phys. Rev.* **74** (1948) 328–34.

[103] Temple, B. 'Systems of conservation laws with invariant manifolds' *Trans. Amer. Math. Soc.* **280** (1983), 781–95.

[104] Temple, B. 'No L^1-contractive metrics for systems of conservation laws' *Trans. Amer. Math. Soc.* **288** (1985), 471–80.

[105] Temple, B. and Young, R. 'The large time stability of sound waves' *Comm. Math. Phys.* **179** (1996), 417–66.

[106] Tsarev, S. P. 'On Poisson brackets and one-dimensional systems of hydrodynamic type' *Dokl. Akad. Nauk SSSR* **282** (1985), 534–37.

[107] Tsarev, S. P. 'The geometry of hamiltonian systems of hydrodynamic type. The generalized hodograph method.' *Izv. Akad. Nauk SSSR* **54** (1990), 1048–68.

[108] Vanderbauwhede, A. 'Centre manifolds, normal forms and elementary bifurcations.' *Dynamics Reported* **2** (1989), 89–169.

[109] Vichnevetsky, R. and Bowles, J. B. *Fourier analysis of numerical approximations of hyperbolic equations*, SIAM, Philadelphia, 1982.

[110] Wagner, D. H. 'Equivalence of the Euler and Lagrangian equations of gas dynamics for weak solutions' *J. Diff. Eqns.* **68** (1987), 118–36.
[111] Weinberger, H. F. 'Long-time behaviour for a regularized scalar conservation law in the absence of general nonlinearity' *Ann. Inst. Henri Poincaré* **7** (1990), 407–25.
[112] Whitham, J. *Linear and nonlinear waves*, John Wiley & Sons, New York, 1974.

Index